# Space Science and Technologies

**Series Editor**

Peijian Ye, China Academy of Space Technology, Beijing, China

Space Science and Technologies publishes a host of recent advances and achievements in the field – quickly and informally. It covers a wide range of disciplines and specialties, with a focus on three main aspects: key theories, basic implementation methods, and practical engineering applications. It includes, but is not limited to, theoretical and applied overall system design, subsystem design, major space-vehicle supporting technologies, and the management of related engineering implementations.

Within the scopes of the series are monographs, professional books or graduate textbooks, edited volumes, and reference works purposely devoted to support education in related areas at the graduate and post-graduate levels.

More information about this series at http://www.springer.com/series/16385

Jun Xie · Haihong Wang · Peng Li ·
Yansong Meng

# Satellite Navigation Systems and Technologies

北京理工大学出版社
BEIJING INSTITUTE OF TECHNOLOGY PRESS

Springer

Jun Xie
Institute of Spacecraft System Engineering
CAST
Beijing, China

Haihong Wang
Institute of Spacecraft System Engineering
CAST
Beijing, China

Peng Li
Institute of Spacecraft System Engineering
CAST
Beijing, China

Yansong Meng
CAST-Xi'an
Xi'an, Shanxi, China

ISSN 2730-6410 ISSN 2730-6429 (electronic)
Space Science and Technologies
ISBN 978-981-15-4862-8 ISBN 978-981-15-4863-5 (eBook)
https://doi.org/10.1007/978-981-15-4863-5

Jointly published with Beijing Institute of Technology Press
The print edition is not for sale in China (Mainland). Customers from China (Mainland) please order the print book from: Beijing Institute of Technology Press.



This Springer imprint is published by the registered company Springer Nature Singapore Pte Ltd.
The registered company address is: 152 Beach Road, #21-01/04 Gateway East, Singapore 189721, Singapore

# Series Editor's Preface

China's space technology and science research have earned a place in the world, but have not been compiled into a series of systematic publications yet. In 2018, the series Space Science and Technology edited mainly by me and co-authored by the leading figures in China's space industry was published in China, when China Academy of Space Technology was celebrating the 50th anniversary of its founding. This collection contains 23 volumes in Chinese, only 10 of which have been selected, re-created and translated into English. In addition, each English volume has been re-created at the suggestion of the Springer, by deleting the contents similar to Springer's existing publications and adding the contents that are internationally advanced and even leading, and bear both Chinese characteristics and worldwide universality. This series fully reflects the knowledge and engineering experience recently accumulated by Chinese scientists and engineers in space technology and science research.

As the Editor-in-Chief of this series, I always insist that this collection must be of high-quality, either in Chinese version or English version. First, the contents of this series must be condensed and sublimated based on the combination of theory and practice, so as to provide both a theoretical value and engineering guidance. Second, the relationships between past knowledge and state-of-the-art and between other people's work and our own new findings should be properly balanced in the book contents to ensure the knowledge systematicness and continuity and to highlight new achievements and insights. Each volume intends to introduce the readers something new. Third, the English version should be customized for international exposure and play a solid supporting role for China to contribute to the world's space field.

This collection consists of 10 volumes, including *Spacecraft Thermal Control Technologies, Spacecraft Power System Technologies, Spacecraft Electromagnetic Compatibility Technologies, Technologies for Spacecraft Antennas Engineering Design, Satellite Navigation Systems and Technologies, Satellite Remote Sensing Technologies, Spacecraft Autonomous Navigation Technologies Based on Multi-source Information Fusion, Technologies for Deep Space Exploration, Space Robotics, Manned Spacecraft Technologies.*

*Satellite Navigation Systems and Technologies* focuses on the space segment of satellite navigation system, studies and discusses the navigation techniques, mainly including data uploading link, navigation and broadcasting link and inter-satellite link techniques, and highlights the technical characteristics of this system such as high-precision measurement, high-continuity information processing, high-stability channel transmission, high-power signal emission, high availability and good integrity.

In addition, the key technologies of navigation satellite system are studied and discussed here in depth to provide a reference for the design of navigation satellite system. Those technologies include time-frequency reference system, navigation signal evaluation, autonomous satellite operation and other technologies.

The publication of this series adds a new member to the international family of space technology and science publications, and intends to play an important role in promoting academic exchanges and space business cooperation. It provides comprehensive, authentic and rich information for international space scientists and engineers, enterprises and institutions as well as government sectors to have a deeper understanding of China's space industry. Of course, I believe that this series will also be of great reference value to the researchers, engineers, graduate students and university students in the related fields.

Peijian Ye
Academician
Chinese Academy of Sciences
Beijing, China

# Preface

Satellite navigation technology refers to the technology of providing the users with three-dimensional position information service and time information service by using the radio navigation signal broadcast by the satellites.

As the infrastructure of high-precision spatial reference and time reference, satellite navigation system has the technical characteristics such as high-precision measurement, high-continuity information processing, high-stability time transmission, high-power signal transmission, high availability and integrity. It embodies the advanced achievements in science and engineering technology fields such as spacecraft design, communication system, electromagnetics and microwave engineering, precision measurement, signal and information processing, and quantum physics. Satellite navigation system can provide all-time and all-weather positioning, navigation and timing services to various users on the earth's surface, at sea and in the near-earth space. It is an absolutely necessary space information infrastructure for national economy, social development and national defense construction nowadays.

Based on the design theory and development experience of Beidou navigation satellite system (BDS), this book highlights the space segment and the related satellite technologies as well as satellite-ground integration design from the perspective of engineering. According to the role of satellite system in the GNSS and the process of signal information flow, through the study on the developments of several Global Navigation Satellite Systems (GNSS), the satellite navigation technology in this book is divided into uplink and reception technology, broadcasting link technology, inter-satellite link technology, time-frequency system technology, navigation signal generation and assessment technology, navigation information management technology, autonomous operation technology of navigation satellite. Each chapter not only studies on the theoretical function and main technologies, but also focuses on engineering development. In the last chapter, we have a research on the technological development status and trend of BDS and other GNSS, and propose the technologies of future development, such as new navigation signal structure, spaceborne cold atomic clock technology, MEO/HEO and deep-space satellite navigation technology, Long-term Operation Technology of Autonomous Constellation, advanced inter-satellite link technology, X-ray pulsar navigation application, LEO satellite navigation augmentation technology, multi-source information fusion navigation technology, and autonomous health management of navigation satellite.

This book consists of 10 chapters, which was written by Jun Xie, Haihong Wang, Peng Li and Yansong Meng from the China Academy of Space Technology (CAST). The preparation of this book has received care and support from the experts and scholars from the CAST, Institute of Spacecraft System Engneering (ISSE) and the CAST-Xi'an. Bian Lang, Zhang Zhongying, Li Yang and Wang Guoyong et al. participated in the preparation of Chap. 3. Tao Yan, Ying Wang, Lang Bian, Haiqing Dong et al. participated in the preparation of Chap. 4. Chengbin Kang, Xiangjun Wu and Yanan Gu participated in the preparation of Chap. 5. Lijun Du, Guoyong Wang, Dong He, Feng Xu et al. participated in the preparation of Chap. 6. Xiaozhun Cui and Ying Wang participated in the preparation of Chap. 7. Jia Guo et al. participated in the preparation of Chap. 8. Lang Bian, Jia Guo, Gong Zhang et al. participated in the preparation of Chap. 9. Xiangjun Wu, Qingjun Liu et al. participated in the preparation of Chap. 10. The authors acknowledge the contributions and achievements made by these individuals in the process of writing this book. In particular, the guidance and help of academician Peijian Ye has enabled the authors to constantly revise and improve this book. The authors who have benefited a lot would like to express heartfelt thanks to him.

This book is the research achievements made by the CAST navigation satellite technology team over the years. It can be used as a technical manual for the engineers and technicians engaged in the design of satellite navigation system, and as a textbook for senior undergraduate students and graduate students.

With the rapid development of satellite navigation technology, the application of satellite navigation has become more extensive and will be involved in more technical contents. We will keep on research and look forward to sharing our research achievements, engineering experience and ideas with readers. The correction by readers and experts is kindly requested.

Beijing, China Jun Xie
Beijing, China Haihong Wang
Beijing, China Peng Li
Xi'an, China Yansong Meng

# Contents

# About the Authors

**Jun Xie**, research fellow, graduated from the University of Defense Science and Technology, an expert in the field of space electronic information in China, and one of the leading figures in satellite navigation technology. He is currently a member of the standing committee of Science and Technology Committee of CAST. He has been engaged in the research of navigation satellite system, navigation payload, space microwave technology and so on for a long time. It has won 1 special prize of National Science and Technology Progress Award, 3 first prize and 2 third prize of Scientific and Technological Progress Award in the army and ministries, and authorized 11 invention patents.

**Haihong Wang**, research fellow, graduated from Nanjing University. He is the deputy director designer of Beidou satellite autonomous navigation subsystem. He has been engaged in research on orbital mechanics of spacecraft and autonomous operation of navigation satellite systems for over ten years. He was awarded the second prize of National Science and Technology Progress Award of the ministry of education, the second prize of Science and Technology Progress of the Army, the first prize of national innovation and authorized 5 invention patents.

**Peng Li**, senior engineer, graduated from Tsinghua University, is currently the general manager of the aerospace technology application and international business department of ISSE. He is the deputy director designer of the Beidou satellite payload. He has been engaged in the research of navigation system demonstration, application, navigation system ground validation, and navigation satellite payload design for over ten years. He was awarded the second prize of Science and Technology Progress of the Army, and authorized 5 invention patents.

**Yansong Meng**, researcher follow, graduated from CAST-Xi'an. He is currently a member of the Science and Technology Committee and director of the Navigation CAST-Xi'an. He is the assistant chief designer of Beidou satellite and technical director of the Inter-satellite link system. He has been engaged in the research of satellite navigation and inter-satellite link technology for over ten years. He has published more than 10 papers indexed by SCI and EI and authorized 13 invention patents.

# Chapter 1
# Introduction

The Chinese phrase "Yu" and "Zhou" (the Universe) first appeared in the ancient Chinese book *Zhuangzi*. "Yu" stands for the three-dimensional position in the space, and "Zhou" represents information about time. They all can be put into the category of navigation.

Navigation is the science of studying the determination, maintenance, delivery, and acquisition of space position and time information. It is about how to direct (plan, record, and control) various objects (e.g., planes, ships, and vehicles) or people from one location to another under the real-world scenarios. The widely used Global Navigation Satellite System (GNSS) provides position information worldwide, making travel much more convenient for people, and making it possible for people to share bicycles, to make deliveries using drones, and to achieve autonomous driving. An ecosystem of the sharing economy is taking shape thanks to the GNSS. In addition, the GNSS can also provide remote synchronization. This helps to realize the synchronization of remote and smart power transmission and distribution, the synchronization between the base-stations of wireless communication networks on the ground, and the synchronization between multiple radar stations.

However, it's a long and challenging process for us to realize high-availability, high-continuity, high-precision, and high-integrity navigation. Navigation technologies have already gone through a long process of evolution with a lot of twists and turns. Satellite navigation systems are among the most complex systems in the history of human beings. Scientists have developed them by overcoming numerous difficulties, and by comparing various measurement technologies, positioning methods, and multiple-access modes. They can provide seamless, all-weather, real-time, and high-precision coverage across the globe.

In this chapter, we look back at the development history of navigation technologies in general and satellite navigation technologies in particular,



J. Xie et al., *Satellite Navigation Systems and Technologies*, Space Science and Technologies, https://doi.org/10.1007/978-981-15-4863-5_1

summarize the functions and services of satellite navigation systems, and introduce the mechanisms of satellite navigation including measuring, positioning as well as multiple-access modes. We also talk about the concepts of time system and spatial coordinate system, the composition of a satellite navigation system, as well as the standards for system performance.

## 1.1 Evolution of Navigation Technology

It is of great value to acquire the information about one's own location and have the ability to move from one place to another in the development history of human society. The problems people often encounter in daily life when leaving or returning to a familiar environment and expanding their living range are: Where am I now? How can I get to my destination? These problems belong to the sphere of navigation study.

In the history of civilizations, only those who mastered the navigation technology could conquer vast space, engage in trading for economic construction, deploy military forces for national defense, and develop from tribes and city-states to large countries covering a vast geographical area. According to the story of the Sinbad navigation, the Arabs discovered and mastered the laws of the monsoon, relied on Kamal to measure the ship latitude for navigation, and thus ruled the central and southern parts of Eurasia. The fleet led by Zheng He, a Chinese navigator in Ming Dynasty (1368–1644), adopted the integrated navigation technique. As recorded in "Ping Zhou Ke Tan" a Chinese book written in Northern Song Dynasty (960–1127) by "watching stars at night, observing the sun in the daytime, and utilizing the Sinan (the prototype of compass) on the cloudy days", Zheng He's fleet accomplished seven voyages to the Western Pacific and the Northern Indian Ocean. Legend has it that in about 2600 BC, the tribe led by Yellow Emperor fought a significant war against Chi You in Zhuolu. By using the Southward Pointing Chariot to identify the direction in a heavy storm, the Yellow Emperor won the war and later founded the Chinese civilization.

According to the degree of autonomy, navigation technologies can be divided into autonomous navigation and non-autonomous navigation. Autonomous navigation refers to the navigation in which only the equipped instruments, rather than external input, can be used. The autonomous navigation techniques in common use include astronomical navigation, inertial navigation, gravity field navigation, and geomagnetic navigation. Non-autonomous navigation is a type of technique that requires the combination of the user's equipment with other external instruments or systems to achieve navigation functions, including ground radio navigation, satellite navigation, etc.

With the development of human civilization, the navigation technology has been continuously improved. In ancient times, navigation accuracy was generally low due to the lack of understanding of science and technology. As the understanding of natural science increases, the navigation technology and services have been improved with higher navigation accuracy. Generally speaking, the evolution of navigation technology can be divided into four stages.

**The first stage** was using natural phenomena for navigation before the fourteenth century. The people at that time lacked the knowledge of natural sciences and an effective way of location description. The navigation sources were mainly natural landmarks and geomagnetic physics including topographical features and matching, Southward Pointing Chariot, Sinan, and so on, which mainly served the fields such as human migration, hunting, and farming. In prehistoric times, people marked their travel routes on barks or animal skins so that the earliest maps of human beings came into being. As recorded in "Kuei Ku-tzu" a Chinese book authored in the Warring States Period (475–221 B.C.), the ancient Chinese looked for jades with Sinan. In 120 BC, Zhang Heng, an ancient Chinese astronomer, created a coordinate grid on the map to indicate the position of the objects. In about the same period, the ancient Greek astronomer Hipparchus proposed a method of using geographical longitude and latitude to represent the position of a certain point on the Earth, thereby establishing the basis for celestial navigation.

In ancient times, people established the concept of time from the alternating of day and night. From the Shang Dynasty (1600–1046 B.C), various clocks, including sundial, water clock, sand clock, etc., had emerged. In the Northern Song Dynasty, Su Song organized the construction of a large water-based astronomical clock tower.

In general, the locating and timing accuracy at this stage was relatively low.

**The second stage** was using mechanical devices for navigation in the fifteenth–nineteenth century. The natural sciences such as geometry, astronomy, and mechanics developed in full swing after the Renaissance. The new navigation tools at that time mainly included sextant, marine chronometer, compass, and star ephemeris. Navigation mainly relied on sailing time, the ephemeris of celestial bodies, and star elevation angle. Their application gradually became widespread, especially in the marine navigation and colonial development, which have benefited from the advancement of navigation technology in this period.

In 1464, Johannes Müller in Germany wrote *De triangulis omnimodis*, the first book on trigonometry published in 1533, where he elaborated on the geometric relationship between celestial bodies and ships and the way to calculate the ship position according to the celestial almanac and elevation observation, thus laying the theoretical foundation for the use of mechanical devices in celestial navigation. Since 1470, nautical exploration activities have begun to increase in Europe. Especially, the Portuguese have mastered the method of determining the latitude by observing the sun and moon. In 1569, Gerhardus Mercator in Belgium published a world map made of cylindrical projections. In the map, there were straight longitude and latitude lines that could help the navigator determine the direction. In 1731, the sextant was invented in the UK. In 1761, the British watchmaker John Harrison invented the first marine chronometer. With this, the problem of astronomical measurement in marine navigation had been solved. Meanwhile, the original dead-reckoning instrument appeared as the prototype of inertial navigation technology and entered its initial application.

This period also witnessed the emergence of mechanical clocks in Europe to unveil an independent scientific time. In 1673, Christiaan HuyGens invented the pendulum clock, pushing the timing accuracy up to a few second level. In 1833, the British established the falling-ball time tower on the Thames River in London. Later the countries around the world followed suit.

**The third stage** was using radio electronic instruments and precision instruments for navigation from the late nineteenth century to the 1960s. This period witnessed the overflow of electromagnetic knowledge (including Coulomb's law, Ampere's law and Faraday's Law, Maxwell's electromagnetic field theory) and mathematic theories (including calculus, Hamiltonian quaternions method, the analytic theory of fixed-point rotating rigid bodies, Gibbs vector analysis, and numerical calculation theory), and the establishment of high-precision navigation data calculation methods. At the same time, the development of mechanical manufacturing and radio engineering provided favorable conditions for the design and manufacturing of navigation signal sources and equipments. With the emergence of high-precision quartz clock technology and computer technology, the scope of navigation has been broadened and the navigation accuracy improved. The navigation techniques used by people at this stage include gyrocompass, radio beacon, radar, inertial navigation, etc. The use of artificial radio resource contributed to the emergence of high-precision regional navigation technique. The wide application of navigation technology in land long-distance transportation, scientific exploration, shipping, and air transportation pushed forward the progress of human civilization.

In 1852, Jean-Bernard-Léon Foucault produced a gyroscope consisting of a high-speed rotor and a gimbal, which was the ancestor of inertial instrumentation. In 1903, Hermann Anschütz in Germany invented the gyrocompass. Compared with the traditional compass, the gyrocompass was independent of the interference of the magnetic field, thus improving the ship's heading accuracy. During the First World War, a wireless telecommunication platform emitting 375 kHz radio waves was deployed on the coast. In 1919, the British Robert Watson Watt invented a device tracking and locating aircrafts with short-wave signals, which was the predecessor of modern radar. In 1942, the United States began to design and construct the Long Range Navigation (LORAN) system, an important maritime navigation device used in the 1940s and 1960s. In 1953, Charles S. Draper in the United States successfully developed an independent inertial navigation system that could work without external signal input. In 1960, the American Seymour Cray invented the transistor computer, making fast navigation and positioning solutions possible.

In the early twentieth century, the technique of radio timekeeping appeared and ushered the timing service into a new era. In the 1930s, the quartz clock appeared. In 1942, the Greenwich Observatory replaced its astronomical pendulum clock with a quartz clock to provide standard time. In 1958, the International Astronomical Society defined an ideal time standard based on the Earth revolution and Newcomb solar calendar.

**The fourth stage** was using astronautic technologies and microelectronics for navigation from the 1960s to the present. This period witnessed the booming of a variety of advanced technologies, including the design of microelectronic systems, the establishment and maintenance of high-precision spatial reference frame, the establishment and maintenance of high-precision time-frequency reference, the orbital mechanics and precision orbit determination of artificial earth satellites, the transmission and reception of electromagnetic signals, the technique of spread spectrum communication, signal and information processing technology, the technologies of computer and microelectronics, the technique of spaceborne atomic clock, and the astronautic technologies. As the satellite navigation technique appeared, the scope of navigation service has been gradually expanded to the whole globe.

In 1957, the Soviet Union launched the first artificial earth satellite. When monitoring the satellite movement, American scientists found the relationship between the Doppler shift curve of the radio signal and the satellite orbit, and hit upon the idea of moving the radio navigation signal source from the ground to the satellite. In December 1958, the United States officially began to develop the navigation satellite system "Transit" (Navy Navigation Satellite System, NNSS). This system was put into official use in January 1964, and retired in 1996. The application of Transit demonstrated the great superiority of satellite navigation. In 1967–1968, the Soviet Union established a similar system named "CICADA". In 1973, the United States began to develop the NAVSTAR system, also known as Global Positioning System (GPS), which officially started services from 1994. This system has higher global positioning accuracy than Transit and can continuously provide accurate information on 3D position/velocity and time for real-time navigation and positioning. In 1978, the Soviet Union began to develop a satellite navigation system similar to GPS, namely GLONASS (Global Navigation Satellite System), which officially started services from 1995. At the end of 2003, China BDS-1 double-satellites positioning system was officially put into operation. This is the world's first satellite navigation system that provides the Radio Determination Satellite Service (RDSS). At the end of 2012, China built BDS-2, the regional satellite navigation system, which provided RDSS and RNSS (Radio Navigation Satellite Service).

At the International Metrology Conference in 1967, the zero magnetic field transition of cesium atoms was used to give a standard definition of the second length called International Atomic Time (TAI), and the starting point of TAI was defined as 0:00 on January 1, 1958. In 1955, the United States successfully developed the first practical cesium atomic beam frequency standard and began to put it into commercial use. The frequency accuracy of this standard was up to $1 \times 10^{-12}$, and its long-term frequency stability up to $10^{-14}$. In the 1960s, the use of storage bubble technique invented by N.Ramsey et al. contributed to the successful development of hydrogen atomic frequency standard. The one second stability of this standard was up to $1 \times 10^{-13}$, and its one day stability up to $10^{-16}$. In 1978, the GPS atomic clock of GPS Block I satellite could achieve the one day stability of $10^{-14}$. In 2009, the NASA JPL studied the miniaturized $Hg^{+}$ ion microwave clock with one second stability of $1 \times 10^{-13} - 2 \times 10^{-13}$, and one day stability better than $1 \times 10^{-15}$.

With the further development of astronautic and microelectronic navigation techniques and the emergence of new technologies such as quantum technology and information network, the future development of navigation technology will inevitably involve the seamless integration of land, sea, aviation, and aerospace. The development of a variety of navigation sources and techniques has gradually evolved into a new phase of navigation for the construction of a comprehensive positioning, navigation, and timing (PNT) system.

## 1.2 Function and Application of Satellite Navigation

The basic function of the satellite navigation system is to establish, maintain, and transmit a spatial position reference and a time reference, and provide users with PNT services. The key techniques of satellite navigation are high-precision measurement technique, high-continuity information processing technique, high-stable time transmission technique, high-power signal transmission technique, and high-availability/high-integrety technique.

The establishment and maintenance of spatial position/time references is the basic premise for the construction of satellite navigation systems and services. By defining unified spatial and temporal references, the state of navigation satellites and users (i.e. navigation receivers), including location coordinates and time status parameters, can be accurately described.

The Earth-centered Earth-fixed (ECEF) coordinate system is usually selected as a spatial reference, while the satellite navigation time system traced to Coordinate Universal Time (UTC) selected as a time reference.

The transmission of spatial and temporal references is a process in which the satellite navigation system provides services to the user for positioning, navigation and timing, including: ① the navigation satellite continuously acquires and broadcasts the information on its coordinates in the spatial reference frame and on the deviation of navigation satellite time from reference time; ② the user acquires the distance information from navigation satellites and the navigation messages, and calculates its position and local time based on spatial and temporal references, either continuously or intermittently as needed.

Positioning is the process or technique for determining the user's position by measurements. The positioning service of the satellite navigation system is that the user receives the radio navigation signal from the satellite, and calculates the spatial position of the satellite at the time of signal transmission based on the ephemeris information (including time information) contained in the navigation message and the distance between satellite and user receiver, including pseudorange and carrier phase. The basic positioning principles of the satellite navigation system include the three-ball intersection adopted by GNSS and the hyperbolic intersection adopted by Transit.

Navigation is the process of guiding (planning, recording, and controlling) various motion platforms (aircrafts, ships, vehicles, etc.) and individuals to move from one location to another, or is the aggregation of all the sciences and technologies associated with this process. The system provides the navigation services, including a series of positioning processes, by combining the distance measurement with the target position to guide the user to move from the current position to the target position.

Timing is the process or technique of transmitting standard time by means of broadcasting. Timing by the satellite navigation system is the process based on weekly counting, weekly seconds, pseudo-range delay, satellite clock information, and the user's clock difference in positioning solution when passing the GNSS standard time, which can be converted to UTC time. The timing principle of the satellite navigation system is equivalent to the principle of pseudorange time difference, that is, to accurately measure the pseudorange from satellite to user, minus the geometric distance between satellite and receiver, and the delay of the device, to obtain the deviation.

In addition, the satellite navigation system provides extended services such as search and rescue, nuclear explosion detection, and short message sending and location reporting through the corresponding equipment for measurement, detection, transmission, and processing.

The PNT functions provided by satellite navigation system make it possible to establish and maintain temporal and spatial references for users on land and sea and in space. Satellite navigation technique is characterized by wide coverage, high real-time performance, high -precision, and all-weather operation. It is widely used in geographic data acquisition, surveying and mapping, traffic monitoring and dispatching, aviation navigation, national defense construction, precision strike, and communication, and even found in the fields such as power, finance, mass consumption, and leisure.

## 1.3 Fundamentals of Satellite Navigation

In the satellite navigation system, the satellite broadcasts the navigation signal as a location-known spatial point based on unified time/space references. From the signal, the user obtains the information on the distance between satellite and user or the Doppler frequency shift, through which the services such as positioning, speed measurement, navigation, and timing are provided. The process of satellite navigation involves the selection of measurement method, positioning method, and multiple-access method. Technological progress has been made in the determination of satellite navigation mechanism, while considering the risk mitigation and marketing.

### 1.3.1 Measurement Method

The acquisition of satellite navigation measurements is the first step to calculate the user's velocity and position. The measurement method determines the measuring techniques between satellites, ground control stations, monitoring stations, and users. The primary measurement data that can be acquired from the radio signal include the Doppler shift, code phase, carrier phase. In the satellite navigation system, the selection of measurement data is closely related to system design, receiver technique, clock technique, and product engineering. The measurement methods currently used in the satellite navigation systems are Doppler velocity measurement and spread spectrum ranging measurement (including the code and carrier phase pseudorange measurement).

"Transit", the first satellite navigation system, adopted the Doppler shift measurement. The Transit receiver with an altimeter received satellite signals, measured the Doppler frequency shift, accumulated the number of Doppler cycles in a period of time, calculated the slope difference between user and satellite, and determined the user's two-dimensional position. Owing to the use of dual-frequency (400 and 150 MHz) ionospheric correction and orbit determination techniques, Transit achieved the optimal positioning accuracy of 30 meters and the timing accuracy of 3 $\mu$s. This was a great technological achievement for the satellites that were equipped with only a high-stability crystal oscillator.

TOA measurement or ranging is realized by measuring the propagation time of radio signals intuitively. However, it's not easy to achieve high-precision ranging. To do this, a stable and synchronous clock technique and an appropriate measurement technique (such as pseudo-random code measurement) are needed. For example, TACAN (tactical air navigation), a ground-based radio navigation system, achieved ranging by measuring the pulsed signal, which produced an error of several hundred meters or even kilometers. In the late 1960s and early 1970s, these two preconditions became ready. The US Navy launched the Timation program to achieve the on-orbit application of spaceborne atomic clock (cesium clock). At the same time, the US Air Force initiated the test of a 621B satellite navigation system, in which the pseudo-random code ranging was proposed and the test was demonstrated. The Timation system and the 621B system jointly completed the preparation for the two key techniques, thereby laying the foundation for the birth of GPS.

The user receiver can achieve ranging actively or passively. The ranging process where the user needs to transmit signal is called active ranging or RDSS (Radio Determination Satellite Service). On the contrary, the ranging process where the user does not transmit any signal is called passive ranging or RNSS (Radio Navigation Satellite Service). In the active ranging, the user receives a signal from the satellite and then sends it back to realize two-way ranging without counting the user's clock bias. Whereas in the passive ranging, positioning is based on one-way ranging and three-ball intersection, in which the user's clock bias is considered as a variable to resolve.

The active ranging, which required fewer transmitting sources or beacons, was adopted by nearly all the early radio positioning systems. However, the passive ranging has many advantages over the active ranging. On the one hand, the passive receiver wouldn't be exposed easily as it doesn't transmit signals, so that it is beneficial to military applications. On the other hand, the receiver is relatively small and lightweight with low power consumption, as it has no transmission device. In addition, the passive ranging system has an infinite number of users, which is important for the massive navigation application.

Chinese BDS-1 is an active navigation satellite system that positions Chinese users through two GEO satellites and has extended functions such as short message sending and location reporting. Chinese BDS-2 and BDS-3 provide both passive and active navigation services in the Asia Pacific region and even around the globe.

### *1.3.2 Positioning Principle*

The positioning principle is the process of calculating the position with the measurement data including Doppler shift, code, and carrier phase. It is also related to the demands for two-dimensional or three-dimensional positioning and for the user's clock bias quantity.

The Transit system received the signals transmitted by each satellite, and measured the Doppler shift according to the difference of the carrier phase, as shown in Eq. (1.1).

$$f_d(t) = \frac{\phi(t+T) - \phi(t)}{T} \tag{1.1}$$

Where$\varphi$ *(t* + *T)* and $\varphi$ *(t)* are the two carrier phase measurements at the interval *T*.

The Doppler shift is integrated to obtain the integral Doppler $\Delta\varphi_{1i}$, which is the increment of the distance between user and satellite in a period of time, as shown in the Eq. (1.2).

$$\Delta\phi_{l_i} = -\int_{t_n}^{l_i} f_d(t)\mathrm{d}t = \phi(t_i) - \phi(t_1) \tag{1.2}$$

Satellite positioning is to receive the information transmitted by each satellite, accurately calculate the satellite position as a function of time, and measure the distance between the satellite and the observation point based on the calculated position (integral Doppler). Given that the user's velocity is known, the user's position can be calculated. The three-dimensional (longitude, latitude, and altitude) data for one point can be obtained by processing the measurement data from multiple satellites.The satellite position in the ECEF system at the $n$th time point $t_n$, denoted as ($X_{tn}$, $Y_{tn}$, $Z_{tn}$), can be derived from the satellite message, while the user position, denoted as (x, y, z), can be calculated with integral Doppler by the following equation.

$$\begin{cases} \Delta\phi_{12} = \sqrt{(X_{t_2} - x)^2 + (Y_{t_2} - y)^2 + (Z_{t_2} - z)^2} \\ -\sqrt{(X_{t_1} - x)^2 + (Y_{t_1} - y)^2 + (Z_{t_1} - z)^2} + \varepsilon_1 \\ \Delta\phi_{13} = \sqrt{(X_{t_3} - x)^2 + (Y_{t_3} - y)^2 + (Z_{t_3} - z)^2} \\ -\sqrt{(X_{t_1} - x)^2 + (Y_{t_1} - y)^2 + (Z_{t_1} - z)^2} + \varepsilon_2 \\ \Delta\phi_{14} = \sqrt{(X_{t_4} - x)^2 + (Y_{t_4} - y)^2 + (Z_{t_4} - z)^2} \\ -\sqrt{(X_{t_1} - x)^2 + (Y_{t_1} - y)^2 + (Z_{t_1} - z)^2} + \varepsilon_3 \end{cases} \tag{1.3}$$

Where $\varepsilon_1$, $\varepsilon_2$, $\varepsilon_3$ are measurement noises. Transit could achieve the positioning accuracy of 30 m and the timing accuracy of 3 μs.The positioning method of GNSS is a geometric three-ball intersection, where three navigation satellites are used as the centers of three balls and the distance between user receiver and navigation satellite is ball radius. The three balls meet at one intersection point, which is just the position of the user receiver. As shown in Fig. 1.1, the positions and clock biases of navigation satellites are $(x_i, y_i, z_i, \delta t_{S,i})$, $i = 1, 2, \ldots, 4$, which are known. Whereas the position and clock bias of user receiver are $(x, y, z, \delta t_u)$, which are unknown. The distances between the user receiver and the four satellites are recorded as $d_i$, $i = 1, 2, \ldots, 4$.The pseudorange $\rho$ is defined as the difference between the signal-receiving time $t_{re}$ and

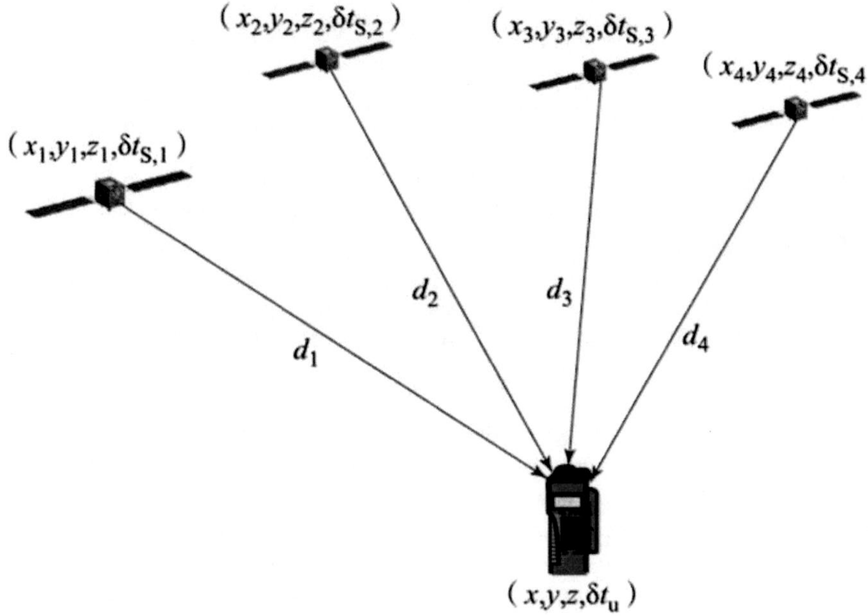


**Fig. 1.1** GNSS positioning method

the signal-transmitting time $t_{tr}$ multiplied by the speed of light c, i.e.

$$\rho = c(t_{re} - t_{tr}) \tag{1.4}$$

The positions and clock biases of the four navigation satellites, denoted as $(x_i, y_i, z_i, \delta t_{S,i})$, can be derived from the satellite message. The user's position and clock bias, denoted as $(x, y, z, \delta t_u)$, can be calculated with pseudorange by the following equation.

$$\begin{cases} \sqrt{(x_1 - x)^2 + (y_1 - y)^2 + (z_1 - z)^2} = \rho_1 - c(\delta t_u - \delta t_{S,1}) \\ \sqrt{(x_2 - x)^2 + (y_2 - y)^2 + (z_2 - z)^2} = \rho_2 - c(\delta t_u - \delta t_{S,2}) \\ \sqrt{(x_3 - x)^2 + (y_3 - y)^2 + (z_3 - z)^2} = \rho_3 - c(\delta t_u - \delta t_{S,3}) \\ \sqrt{(x_4 - x)^2 + (y_4 - y)^2 + (z_4 - z)^2} = \rho_4 - c(\delta t_u - \delta t_{S,4}) \end{cases} \tag{1.5}$$

### *1.3.3 Multiple-Access Technology*

In the GNSS positioning, the user needs to receive more than four signals from different satellites at the same time, and the signals should be distinguished by multiple access methods. The satellite navigation system essentially shares the same multiple-access method with a mobile communication system, that is, the channels can be separated in the time dimension (TDMA), frequency dimension (FDMA), code dimension (CDMA) or spatial dimension (SDMA). The four multiple-access methods are compared in the Fig. 1.2, and their advantages and disadvantages are shown in Table 1.1.

At present, CDMA is widely used in the GNSS systems including GPS, BDS and Galileo, while FDMA used in the GLONASS system. TDMA is generally applied in the pseudolites in order to mitigate the near-far effect. SDMA is usually applied in the communication system such as MIMO (Multiple Input and Multiple Output), rather than in GNSS.

CDMA is based on spread spectrum technology, in which the message data is modulated by the pseudorandom code (PRN) with a bandwidth much wider than signal bandwidth, so that the spectrum of the original signal is spread. Each satellite

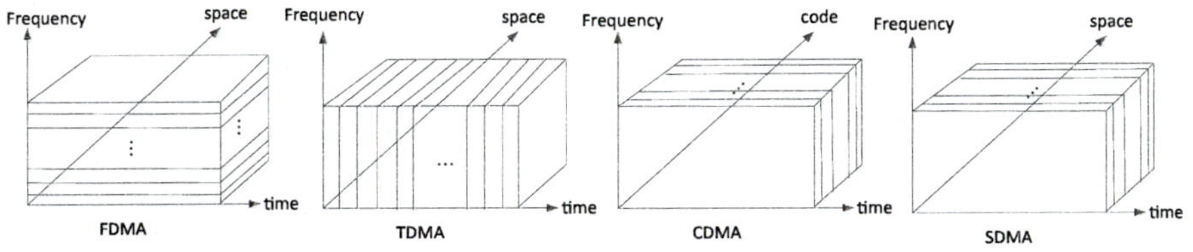


**Fig. 1.2** Comparison of four multiple-access method

**Table 1.1** Advantages and disadvantages of four multiple-access method

| Multiple-access method | Advantages | Disadvantages |
|---|---|---|
| CDMA | • Continous<br>• Anti-interference<br>• High-precision ranging<br>• Highly extentable | • Complex receiver structure<br>• Near-far problem in pseudolite application |
| FDMA | • Continous<br>• Low internal interference<br>• Simple receiver | • Low-spectrum utilization<br>• Low interoperability |
| TDMA | • No near-far problem in pseudolite application | • Pulsed signal |
| SDMA | • High-spectrum utilization | • Complex receiver structure |

has an allocated code sequence. The code sequences are orthogonal to each other to achieve multiple accesses. The advantages of CDMA are that the navigation signals are continuously transmitted, and that the code sequences can be used conveniently for ranging with high measurement accuracy. Since the channels of CDMA signals can be separated in the code dimension, the frequency, time, and space can all be overlapped. Therefore, the spectrum utilization efficiency of CDMA is higher, and the system is easier to expand with only one new code sequence assigned to the new satellite. CDMA has been used in GPS as well as Galileo and BDS.

The FDMA divides the entire signal spectrum into a series of equally spaced sub-channels that are allocated to different satellites. Each satellite has a different carrier frequency. The interference between satellites is relatively low, while the anti-narrowband interference is high. This is one of the reasons that FDMA is selected for GLONASS. In addition, a simple first-order equalizer can be used to compensate for the channel transmission distortion in the receiver. However, the satellite transmission band-pass filter is required to have a steep frequency, and the receiver clock is required to have high frequency accuracy and stability. Furthermore, the utilization and interoperability of the FDMA spectrum are relatively low, and the receiver compatible with multiple constellations has relatively high power consumption. As a result, the next generation GLONASS will introduce the CDMA navigation signal.

The TDMA separates the signals in the time domain, divides the signal period into time slots, and allocates different time slots to different signal-transmitting stations or pseudolites that are synchronized with each other. It mitigates the multiple-access interference, and reduces the near-far effect. However, the pseudolites fail to continuously transmit signals, so the receiver cannot receive the signals at the same time. As a result, the accuracy and dynamic performance of TDMA are relatively poor. Furthermore, as the number of signals increases, the slots will become smaller, so that the tracking performance of the signal will be further deteriorated.

## 1.4 Time-Frequency Reference

As one of the seven basic physical units, time is provided with two concepts: "time epoch" and "time interval". The former represents the moment when a certain phenomenon occurs, while the latter represents the duration of the phenomenon. Frequency is the number of times that a certain event occurs in a unit of time. It is closely related to time. The time-frequency system used for satellite navigation is of great importance. Without a high-precision time system, it is impossible to establish a high-precision satellite navigation system.

The main independent satellite navigation systems have their own time-frequency systems by setting the atomic clock groups (GPST for GPS, GLONASST for GLONASS, BDT for BDS, and GST for Galileo). These time-frequency systems are associated with each other through Coordinated Universal Time (UTC), a time system based on International Atomic Time (TAI).

### *1.4.1 International Atomic Time*

Atomic frequency standard, called atomic time (AT), was obtained from atomic transition frequency. The new atomic time second [Adopted by the 13th General Conference of Weights & Measures (CGPM in French), in October 1967] is defined as: the duration of 9,192,631,770 periods of the $Cs^{133}$ radiation transition between two hyperfine energy levels at sea level in the zero magnetic field. The time is denoted as SI. The AT is a time reference more accurate and stable than UTC, and easier to be measured without the need for long-term astronomical observation [16].

The International Atomic Time (TAI) was established by the International Bureau of Weights and Measures (BIPM in French) through the collaboration of more than 400 independent atomic clocks (in October 2017) distributed in more than 50 countries around the world. The atomic clock laboratories sent the comparison data of UTC ($k$)-Clock ($k$, $i$) (where $k$ is lab code and $i$ is clock code) to BIPM monthly. The BIPM aggregated these data, derived the TAI with high-stability and accuracy by using a specific algorithm, and defined 00:00 January 1, 1958, as the start of TAI.

### *1.4.2 Coordinated Universal Time*

The atomic time can provide a very stable and accurate time reference. However, TAI has no practical meaning for the fields in relation to the Earth's rotation, such as geodesy, astronomy, and navigation. The universal time (UT) based on the Earth's rotation reflects the Earth's position and is closely related to the people's daily life. However, the Earth's rotational speed is not constant, so the UT is about 1 s slower than the TAI per year. In order to balance the difference between TAI and UT, the

International Astronomical Union (IAU) and the International Radio Consultative Committee (CCIR) decided to adopt a new time reference with the second defined as that of TAI and with the practical meaning as UT. The new time reference is the result of coordination, so it is called Coordinated Universal Time (UTC).

In the UTC using the TAI second, the epoch is required to stay within the deviation of $\pm 0.9$ s from UT. In 1971, the CGPM passed a resolution to adjust the UTC with the leap second if the difference between UTC and UT exceeds 0.9 s. According to the resolution, 1 s plus is called a positive leap second, and 1 s minus is called a negative leap second. The leap second is executed at the last one second of June 30th or December 31st in the current year, upon the notification by BIPM to each time service center two months in advance. The initial epoch of the TAI was at 00:00 on January 1st, 1958, the first UTC adjustment was on June 30th, 1972, and the latest UTC adjustment was on January 1st, 2017. By October 2017, a total of 27 leap seconds had been executed.

### *1.4.3 Time Reference of Navigation Satellite System*

The timing service standard of the International Telecommunication Union (ITU) requires that the synchronization with UTC should be within 100 ns. As a result, the time of the satellite navigation system must be traced to UTC time, with the difference of less than 100 ns. The satellite navigation user can learn the deviation between system time and UTC from navigation message, so that the time of each satellite navigation system can be traced to UTC. This further indirectly establishes the time relationships among major satellite navigation systems.

1. **GPS time**

The GPS system uses an independent time reference, called GPS time (GPST), for navigational and positioning calculation. The GPS time belongs to the atomic time reference, that is, its second is defined as that of TAI. The starting epoch of GPST is at 00:00 on January 6th, 1980, coincided with the UTC. The GPST time accumulates by atomic second length, but without leap second adjustment. Therefore, the gap between GPST and UTC will gradually increase as an integer multiple of second. The relationship between GPST and TAI is relatively simple, with a fixed difference of 19 s between the two at any epoch [16].

2. **GLONASS time**

Similar to GPS time, the GLONASS time also has an independent time reference, called GLOGNASST. The GLONASST belongs to atomic time reference so that its second is also defined as that of TAI and it is traced to the UTC (SU) in the Moscow region, as shown below:

$$\text{GLONASST} = \text{UTC (SU)} + 03\ \text{h}\ 00\ \text{min} \tag{1.6}$$

Unlike GPS, the GLONASST executes the leap second adjustment to keep itself synchronous with UTC (SU), so the difference between GLOGNASST and UTC is not a fixed integer multiple of second [18].

3. **Galileo time**

The Galileo time is derived from the calculation of several types of UTC time in European, and is traced to UTC by Precise Timing Facility (PTF). The PTF generates and maintains a stable and accurate Galileo Time (GST) traced to UTC, and provides a continuous UTC reference to Galileo system. It also monitors the GST and compares it with other time references such as TAI, UTC, and GPST through the cooperation with Time Service Provider (TSP), which is outside the Galileo system.

4. **BDS time**

Like GPST, the BDS time (BDT) uses the second defined in the International System of Units (SI) as the basic unit for continuous accumulation. Starting at UTC 00:00 on January 1st, 2006, the BDT counts by week and second, without leap second adjustment. The BDT is traced to UTC (NTSC: China National Time Service Center). The difference between BDT and UTC remains within 100 ns, as shown in the navigation messages including BDT-UTC time offset, BDT-GPST time offset, BDT-GST time offset, and BDT-GLONASST time offset [1].

## 1.5 Spatial Coordinate System

The spatial coordinate system for satellite navigation is critically important, because precise spatial coordinate is the basis for accurately describing the positions of navigation satellites and user receiver, and for building the mathematical model of satellite navigation.

### *1.5.1 Overview of Coordinate System*

Motion is ubiquitous—from the motion of stars in the Milky Way, the rotational and orbital motion of planets and satellites in the solar system, to the rotation and pole shift of the Earth, the motion of spacecrafts near the Earth or in the solar system, and the motion of terrestrial plates. The position and motion of these celestial bodies or points on the Earth must be described relative to a reference, namely coordinate system. A coordinate system usually consists of: ① a set of theories and data processing methods; ② a set of models and constants; and ③ a corresponding coordinate reference frame. The selection of an appropriate coordinate system can make the research process clear and the dynamic model simple. On the contrary, a coordinate system improperly selected will complicate the study.

A coordinate system can be defined by a mathematical model and dynamical description. A frame consisting of a set of physical points that have been scaled to coordinate values is called a coordinate frame. The coordinate frame is the physical implementation of a mathematically defined coordinate system implemented through a set of coordinates calibrated by physical points.

### 1.5.2 Geocentric Celestial Reference System (GCRS)

In order to determine the orbit of the navigation satellite, it is convenient to use the Earth-centered celestial coordinate system, the origin of which is at the Earth's center of mass, and the coordinate axes of which point to the fixed stars.

Current observation data, such as solar system ephemeris, are described in the International Celestial Reference System (ICRS), whose coordinate representation is called ICRF. The ICRS defined by International Earth Rotation and Reference System Service (IERS) includes: ① IERS celestial coordinate frame, called International Celestial Reference Frame (ICRF); ② a set of models used to determine the coordinate frame, i.e., IERS specifications; ③ a set of theories and data processing methods for determining the coordinate values of the reference point (out-of-Milky-Way radio source or optical counterpart). Its origin is at the center of the solar system. The orientation of the coordinate axes is determined by 212 precisely measured coordinates of out-of-Milky-Way radio source. The $x$-$y$ plane of ICRF is close to the J2000.0 mean equator, and its $x$-axis roughly points to the J2000.0 mean equinox.

Before the implementation of ICRS and the introduction of out-of-Milky-Way radio source to the reference frame, the basic astronomical reference system is Fifth Fundamental Catalog (FK5), which is strictly defined by dynamics and considers the kinematics correction of fixed stars. FK5 is based on the observations of bright fixed stars and the IAU1976 astronomical constants. In the FK5, the basic plane is J2000.0 mean equator, and the direction of the $x$-axis is J2000.0 mean equinox. Obviously, FK5 is a kind of kinetic reference system related to epoch. It is usually referred to as J2000.0 mean equator reference system. J2000.0 is a type of GCRS commonly used in the determination of spacecraft orbits (especially the orbits of various artificial Earth satellites including navigation satellites). In the geocentric celestial coordinate system, the navigation satellites obey the Newton's law of motion and the law of gravity.

Considering the frame continuity, the axes of ICRS should be kept as close to those of FK5 as possible in the J2000.0 epoch. The basic plane of ICRS is determined by the observation of very long baseline interference (VLBI), and the difference between ICRS pole and FK5 pole is about 20 milliarcseconds. In order to establish the connection between ICRS and FK5, the average right ascension origin of 23 out-of-Milky-Way radio sources has been selected as the origin of ICRS. The relationship of ICRS and FK5 is determined by three parameters, including the deviations of celestial poles $\xi_0$, $\eta_0$, and the longitude difference at zero point $d\alpha_0$:

$$\begin{cases} \xi_0 = -0''.016617 \pm 0''.000\,010 \\ \eta_0 = -0''.006819 \pm 0''.000\,010 \\ d\alpha_0 = -0''.0146 \pm 0''.0005 \end{cases} \tag{1.7}$$

The relationship between ICRS and J2000.0 can be written as:

$$\begin{cases} \vec{r}_{J2000.0} = B\vec{r}_{\text{ICRS}} \\ B = R_x(-\eta_0)R_y(\xi_0)R_z(d\alpha_0) \end{cases} \tag{1.8}$$

where: $\vec{r}_{J2000.0}$ and $\vec{r}_{\text{ICRS}}$ are the representations of the same vector in different reference frames; the constant matrix $B$, called reference frame deviation matrix, consists of three small angular rotations. Given no special requirement, the reference frame deviation will not be discussed.

### *1.5.3 Earth-Fixed Coordinate System*

The Earth-fixed coordinate system is the Terrestrial Reference System (TRS), a spatial reference system that rotates with the Earth. In this system, the position of the observing station fixed to the Earth's surface is almost constant over time, except occasionally small changes due to geophysical effects such as tectonic movement or tidal deformation. TRS is implemented by the Terrestrial Reference Frame (TRF), which is a group of ground points precisely defined in a specified coordinate frame attached to the TRS. It is convenient to use the Earth-fixed coordinate system to calculate the position of the navigation receiver, whose latitude, longitude, and altitude can be calculated more easily in this system. In addition, the navigation satellite broadcasts the ephemeris in the form of navigation message to the user, and describes the information on orbit and velocity in the TRF.

In a satellite navigation system, both the navigation satellite and the user must reside in a unified coordinate system, such as the WGS84 by GPS, the PZ90 by GLONASS, the ITRF96 by the Galileo, or the BDCS by the BDS. If two or more GNSS systems are used together for navigation, there they should be transformed to each other.

1. **WGS84 coordinate system**

The WGS84 coordinate system was initially established in 1987. It is the modification of the US Navy's navigation coordinate system NSWC9Z-2 by changing the origin and scale, and rotating the zero meridian plane to the plane defined by the International Time Bureau (BIH in French). In the WGS84, the coordinate origin is the Earth's Center of Mass $M$. The $z$-axis points to the Conventional Terrestrial Pole (CTP) defined by BIH1984.0, the $x$-axis points to the intersection of the zero meridian plane defined by BIH1984.0 and the CTP equator, and the $y$-axis is perpendicular to the $x$–$z$ plane to constitute a right-hand orthogonal coordinate system (Fig. 1.3).

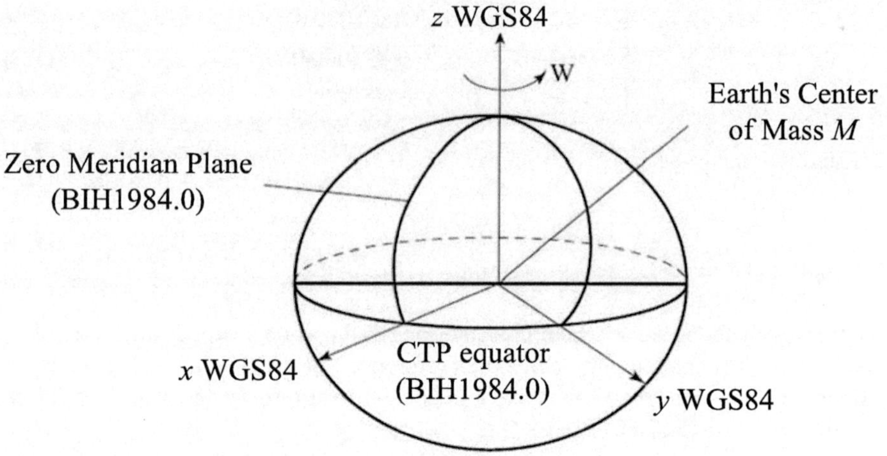


**Fig. 1.3** WGS84 coordinate system

The ellipsoid model used in the WGS84 coordinate system is called the WGS84 ellipsoid model, in which the constants were recommended by the 17th General Assembly of the International Union of Geodesy and Geophysics (IUGG). Its four basic parameters are:

Long semi-major axis: $a = (6{,}378{,}137 \pm 2)$m

Geocentric (including atmospheric) gravitational constant: $\text{GM} = (3{,}986{,}005 \times 10^8 \pm 0.6 \times 10^8)\text{m}^3/\text{s}^2$

Normalized second-order harmonic coefficient: $\bar{C}_{2.0} = -484.16685 \times 10^{-6} \pm 1.3 \times 10^{-9}$

Rotational angular velocity of the Earth: $\omega = (7{,}292{,}115 \times 10^{-11} \pm 0.1500 \times 10^{-10})\text{rad/s}$

Generally, the normalized second-order harmonic coefficient $\overline{C}_{2,0}$ ($\overline{C}_{2,0} = -J_2/\sqrt{5}$) is used instead of $J_2$. Based on the above four parameters, the flattening of the WGS84 ellipsoid can be calculated as $f = 1/298.257223563$. In 1994, these parameters were improved by the National Imagery and Mapping Agency (NIMA) and defined as four constants $a, f$, GM, and $\omega$ (see Table 1.2).

**Table 1.2** WGS84 constants

| Constant | Value |
|---|---|
| Long semi-major axis | $a = 6{,}378{,}137 \pm 2$ m |
| Geocentric (including atmospheric) gravitational constant | $\text{GM} = (3{,}986{,}004.418 \times 10^8 \pm 0.008 \times 10^8)\ \text{m}^3/\text{s}^2$ |
| Normalized second-order harmonic coefficient | $f = 1/298.257223563$ |
| Rotational angular velocity of the Earth | $\omega = (7292115 \times 10^{-11} \pm 0.1500 \times 10^{-10})$ rad/s |

**Table 1.3** CGCS2000 constants

| Constant | Value |
|---|---|
| Long semi-major axis | $a = 6378137.0$ m |
| Geocentric (including atmospheric) gravitational constant | $\mathrm{GM} = (3.986004418 \times 10^{14})$ $\mathrm{m}^3/\mathrm{s}^2$ |
| Normalized second-order harmonic coefficient | $f = 1/298.257222101$ |
| Rotational angular velocity of the Earth | $\omega = (7.292115 \times 10^{-5})$ rad/s |

2. **BDCS2000 coordinate system**

The BDCS has the same definition as CGCS2000. The origin of the CGCS2000 is located at the mass center of the Earth including the ocean and the atmosphere. The *Z*-Axis is the direction of the IERS Reference Pole (IRP). The *X*-Axis is at the intersection of the IERS Reference Meridian (IRM) and the plane passing through the origin and normal to the *Z*-Axis. The *Y*-Axis, together with *Z*-Axis and *X*-Axis, constitutes a right-handed orthogonal coordinate system. The ellipsoid parameters used in the CGCS2000 are shown in Table 1.3.

It can be seen from Table 1.3 that three parameters of the CGCS2000 ellipsoid are the same as those of the WGS ellipsoid, and only the normalized second-order harmonic coefficient is different. Because of this, the geodetic coordinates (or Gaussian plane coordinates) of the same spatial point in the CGCS2000 are different from those in the WGS84. To be specific, the latitude difference between them is 0–0.105 mm, and the altitude difference is 0–0.105 mm, but there is no difference in longitude. At the current coordinate measurement accuracy level of 1 mm, the coordinate difference for the same point between CGCS2000 and WGS84 is negligible.

### *1.5.4 Transformation Between Coordinate Systems*

Celestial axis is defined as the extension line of the Earth's rotation axis, which is intersected with celestial sphere at celestial poles. Due to the Earth's precession, the orientation of the Earth's rotation axis in the celestial reference system is an instantaneous variable that can change over time with celestial poles and equator. In order to further clarify the relationship between the celestial reference system and Earth-fixed reference system, the instantaneous celestial pole and celestial equator are called Celestial Intermediate Pole (CIP) and Intermediate Equator (IE), respectively, in the IAU2003 specification.

To take measurements in the celestial reference frame, a point in IE that does not rotate in the celestial reference frame is chosen as Celestial Intermediate Origin (CIO). Similarly, in order to take measurements in the Earth-fixed reference system,

Fig. 1.4 Intermediate equator

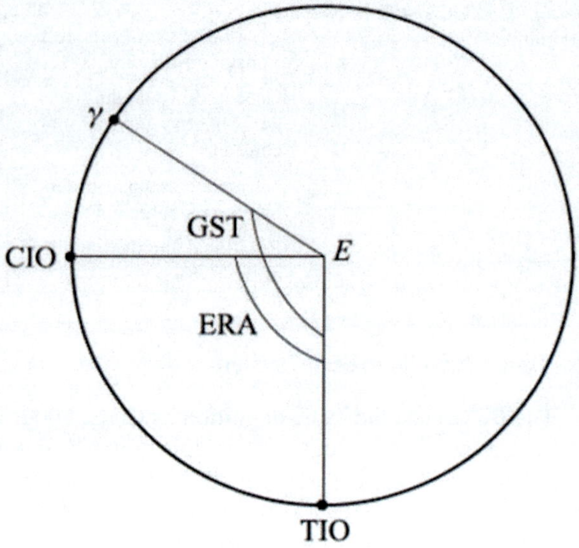


the point in IE not rotating in the Earth-fixed reference frame is chosen as Terrestrial Intermediate Origin (TIO). The selection of CIO is based on a group of quasars called celestial reference frame, which is close to the right ascension origin (vernal equinox) of ICRS. Whereas the selection of TIO is based on a group of ground stations called Earth-fixed reference frame, which is close to the zero longitude direction of ITRF or the direction of the prime meridian (the Greenwich direction). The circle in Fig. 1.4, shows the Intermediate Equator in the geocentric celestial reference system, where $E$ is the Earth's mass center, $\gamma$ is the vernal equinox, and GST is the Greenwich Sidereal Time.

When being observed in the celestial reference system, the intermediate equator is fixed with the CIO [thus called Celestial Intermediate Equator (CIE)] and the TIO moves counterclockwise along the equator in a cycle of 1 sidereal day. Conversely, when being observed in the Earth-fixed reference system, the intermediate equator is fixed with the TIO [thus called Terrestrial Intermediate Equator (TIE)] and the CIO moves clockwise along the equator in the same cycle. The observations in both cases reflect the movement of the Earth around the axis. The angle between CIO and TIO is the measure of the Earth's rotation, called Earth Rotation Angle (ERA).

The transformation between ITRF ($O$-$XYZ$) and ICRF ($O$-$xyz$) is defined in the IAU1980 specification, where the position vectors of an object in the ICRF and ITRF are denoted by $\vec{r}$ and $\vec{R}$ respectively. The positional relationship between the position vectors of the detector in the two coordinate frames is

$$\vec{R} = (\mathrm{HG})\vec{r} \tag{1.9}$$

where: the transformation matrix HG includes four rotation matrices,

$$(\text{HG}) = (\text{EP})(\text{ER})(\text{NR})(\text{PR}) \tag{1.10}$$

where: PR is the precession matrix, NR is the nutation matrix, ER is the Earth rotation matrix, and EP is the Earth pole shift matrix. They are expressed by the following equations:

$$(\text{EP}) = R_y\left(-x_p\right)R_x\left(-y_p\right) \tag{1.11}$$

$$(\text{ER}) = R_z(S_G) \tag{1.12}$$

$$(\text{NR}) = R_x(-(\varepsilon + \Delta\varepsilon))R_z(-\Delta\psi)R_x(\varepsilon) \tag{1.13}$$

$$(\text{PR}) = R_z(-z_A)R_y(\theta_A)R_z(-\zeta_A) \tag{1.14}$$

where: $x_p$, $y_p$ are the polar shift components; $S_G$ is the Greenwich sidereal angle; $\varepsilon$ is the mean obliquity; $\Delta\psi$ is the celestial longitude nutation; $\Delta\varepsilon$ is the obliquity nutation; $\zeta_A$, $z_A$ and $\theta_A$ are the precession constants. The calculation formulas of $R_x(\theta)$, $R_y(\theta)$, $R_z(\theta)$ are shown in the Eqs. (1.15)–(1.17). Note that the rotation matrices of $R_x(\theta)$, $R_y(\theta)$, $R_z(\theta)$ are orthogonal matrices, that is, $R_x^T(\theta) = R_x^{-1}(\theta) = R_x(-\theta)$, $R_y^T(\theta) = R_y^{-1}(\theta) = R_y(-\theta)$ and $R_z^T(\theta) = R_z^{-1}(\theta) = R_z(-\theta)$.

$$R_x(\theta) = \begin{pmatrix} 1 & 0 & 0 \\ 0 & \cos\theta & \sin\theta \\ 0 & -\sin\theta & \cos\theta \end{pmatrix} \tag{1.15}$$

$$R_y(\theta) = \begin{pmatrix} \cos\theta & 0 & -\sin\theta \\ 0 & 1 & 0 \\ \sin\theta & 0 & \cos\theta \end{pmatrix} \tag{1.16}$$

$$R_z(\theta) = \begin{pmatrix} \cos\theta & \sin\theta & 0 \\ -\sin\theta & \cos\theta & 0 \\ 0 & 0 & 1 \end{pmatrix} \tag{1.17}$$

## 1.6 Time Reference and Coordinates of Navigation Satellite

The navigation satellite uses the broadcast navigation message to describe the time offset between satellite time and system time, as well as the position in the Earth-fixed coordinate system. The satellite time, which is measured by satellite-ground time synchronization method, varies in different systems and is broadcast as a clock correction parameter. The satellite position information is acquired through precise orbit determination, high-precision orbit prediction, and parameter fitting, and is broadcast as an ephemeris parameter.

### 1.6.1 Offset Between Satellite Time and System Time

The clock correction parameters include $t_{oc}$ (clock reference time), $a_0$ (clock bias), $a_1$ (first-order clock correction coefficient) and $a_2$ (second-order clock correction coefficient). The user can calculate the transmission time according to the following formulas:

$$t_{\mathrm{tr}}^{\mathrm{GNSS}} = t_{\mathrm{tr}} - \delta t \tag{1.18}$$

$$\delta t = a_0 + a_1(t - t_{oc}) + a_2(t - t_{oc})^2 + \Delta t_{\mathrm{rel}} \tag{1.19}$$

where: $t$ can be replaced by $t_{tr}$. $\Delta t_{\mathrm{rel}}$ is the correction term to relativistic effect:

$$\Delta t_{\mathrm{rel}} = F \cdot e \cdot \sqrt{A} \cdot \sin E_k \tag{1.20}$$

where: $e$ is the orbit eccentricity, $\sqrt{A}$ is the square root of semi-major axis of the satellite orbit, and $E_k$ is the eccentric anomaly of the satellite orbit. The three parameters are obtained from the satellite broadcasting ephemeris. $F = -2\mu^{1/2}/C^2$, where $\mu = 3.986004418 \times 10^{14}\ \mathrm{m}^3/\mathrm{s}^2$ and $C = 2.99792458 \times 10^8\ \mathrm{m/s}$ (the speed of light).

### 1.6.2 Ephemeris Parameters

The ephemeris parameters describe the coordinates of the navigation satellite in the Earth-fixed coordinate system and differ in GPS and GLONASS. The GPS ephemeris parameters include a reference time, 6 Kepler elements, and a set of correction coefficients used to calculate the orbit by the formula. The GLONASS ephemeris parameters include a reference time, the position and velocity of satellites, and solar and lunar perturbation acceleration that should be calculated with the 4th-order Runge-Kutta iterative integration method. As GPS ephemeris parameters are more computationally efficient in the orbit calculation, similar ephemeris parameters are also found in the BDS and Galileo systems.

Generally, the ephemeris fitting accuracy can be improved by increasing and optimizing the correction parameters. In the classic 16-parameter ephemeris of GPS/BDS, 3 long-term correction factors correct the long-term changes along the orbital and normal directions on the orbital plane, while 6 short-term correction factors correct the short-term changes in three directions with a period of $T/2$. The

User Ranging Error (URE) generally is not more than 0.35 m (1σ) (better than 0.1 m as reported) fitted with 4-hour ephemeris data, and less than 1.5 m (1σ) fitted with 6-hour ephemeris data. It has been reported that the URE can be improved to $\leq 0.1$ m (1σ) by using the improved GPS 18-parameter ephemeris, and that the URE can be further improved to the centimeter level by using the 24-parameter ephemeris, an ephemeris potentially to be adopted in the further BDS.

The motion of satellite is a matter of two perturbed bodies involved mathematically in a complex nonlinear dynamic system. The mathematical models of satellites are time-variant and need to be corrected in the form of ephemeris parameters through a series of observations. This process is called orbit determination. In fact, the orbit determination is composed of two phases: ① initial orbit calculation, or the determination of the initial orbit in the shorter arc; ② precise orbit determination, or the improvement of the orbit in the longer arc. Initial orbit calculation is an issue of two unperturbed bodies; while precise orbit determination, or called orbit improvement traditionally, is an issue of two perturbed bodies. At the latter stage, the measurement is to determine some geometric and physical parameters related to orbit based on a large amount of observation data. The process of orbit determination is an iterative process, where the calculation will be repeated until the accuracy requirement is met.

### *1.6.3 GNSS Precise Orbit Determination and Time Synchronization*

As introduced above, the optimized efficient techniques for precision orbit determination and time synchronization are crucial to ensure the accuracy of navigation satellites at the two stages. They are also the basis for users to obtain high-precision PNT services. The current GNSS systems, including the US GPS, Russian GLONASS, European Galileo, and Chinese BDS, have attached great importance to precision orbit determination and time synchronization.

1. **GPS**

The GPS relies on the processing method based on pseudorange and carrier phase to realize the integral calculation of satellite orbit and clock offset. The satellites are equipped with spaceborne atomic clocks. The monitoring stations and the Main Control Stations (MCS) are also equipped with high-performance atomic clocks calibrated by the US Naval Observatory (USNO) in real time. The synchronization between GPS time and satellite system time is realized by a closed-loop system with self-calibrating function and through one-way ranging and orbit determination. While meeting the accuracy requirement of the system, the one-way ranging method can simplify the spaceborne equipments.

In the early application of GPS, there were only five ground-based observatories. At the end of 2005, six monitoring stations of the National Geospatial Intelligence Agency (NGA) were introduced to the GPS satellite-ground monitoring network, significantly improving the precision and robustness of clock synchronization and orbit determination. By 2006, five more NGA monitoring stations had been added to reach a total of 16 GPS ground monitoring stations, so that any GPS satellite could track at least 3 stations at any time. As a result, the observation continuity and the effectiveness of orbit determination and integrity monitoring have been greatly enhanced. The URE has also been improved from 4.6 m in 1990 to 1.1 m now. Generally, the improvement of clock synchronization and orbit determination by GPS is related to three aspects: ① the expansion of global monitor network; ② the increase of the uploading frequency of ground stations; ③ the improvement of an algorithm.

2. **GLONASS**

The GLONASS system relies on an independent observation method to ensure the accuracy of clock synchronization and orbit determination. Although GLONASS ground stations are not globally distributed, they are also widely distributed due to the vastness of Russian territory. Furthermore, the satellites are equipped with laser reflectors that can be tracked by high-precision Satellite Laser Reflectors (SLRs), so that the ephemeris accuracy reaches 10–25 m. The GLONASS ground control system realizes the time synchronization through data synthesis. In the data synthesis, the time offset is calculated from the satellite-ground ranging data obtained through SLR ranging and radar ranging and sent by monitoring stations provided that the precise satellite orbit is known. The calculation of satellite clock correction parameters is performed by linear time offset fitting algorithm, in which the observation data is updated once every 30–60 min.

The ground segment of GLONASS consists of a control center, 5 TT&C stations (including laser tracking stations), and 7 monitoring stations. After the disintegration of the Soviet Union, 2 TT&C stations (in Ukraine and Kazakhstan, respectively) no longer participated in the maintenance of GLONASS. Since 2010, Russia has established GLONASS monitoring stations in Australia, Venezuela, Cuba, and Brazil, thereby significantly improving the accuracy of navigation and positioning. Generally, the improvement of GLONASS clock synchronization and orbit determination is attributed to two factors: ① the completion of the first phase of modernization, which has contributed to the improvement of synchronization subsystems and monitoring stations; ② the upgrading of coordinate system to PZ-90.02.

3. **Galileo**

The Galileo relies on the same processing method as GPS to realize the integral calculation of satellite orbit/clock offsets. Its monitoring stations are equipped with cesium atomic clocks precisely synchronized with the system time. The ground segment of Galileo consists of 40 monitoring stations, 2 control centers, 9 upstream injection stations, and 2 sets of orbital and synchronous processing equipments. The Galileo updates the satellite atomic clock once every 30 min to confine the error of clock and orbit to 0.65 m so as to reduce the impact of satellite clock slip.

As a late navigation system, Galileo has an advanced design concept based on long-term extensive analysis and testing. According to the result of GIOVE-A, the

Galileo's orbit error along the trace is less than 50 cm, and its clock error is less than 0.15 ns. The performance of Galileo is built on higher satellite clock performance and higher updating frequency of navigation message.

4. **BDS**

Different from the integral calculation by GPS/Galileo and the data synthesis by GLONASS SLR, the BDS determines the orbit/time offsets independently. The BDS mainly relies on a two-way radio method to realize satellite-ground time synchronization, with the aid of laser ranging and pseudo-code ranging. In addition, it relies on the two-way time transmission of C-band to realize inter-station time synchronization, with the aid of the satellite common-view method. The precision orbit determination of BDS is realized dynamically through data and time synchronization, by using a strategy of combining single-satellite determination with multi-satellites determination.

The ground segment of BDS in China consists of a main control station, type-I monitoring stations, type-II monitoring stations, and time synchronization/upload stations. The main control station, together with time synchronization/upload stations, is responsible for time synchronization. According to the test result of BDS, the fitting accuracy is better than 0.8 ns, the 2-hour forecast better than 1 ns, the 6-hour forecast better than 2 ns, and the 12-hour forecast better than 5 ns. The type-I monitoring stations are responsible for precise orbit determination. According to the testing data in china, the Root Mean Square Error (RMSE) of orbital residuals is 0.4 m, 0.3 m or 0.2 m for GEO, IGSO or MEO, respectively. The consistency between orbit determination data and SLR data is 0.6 m, 0.3 m or 0.2 m for GEO, IGSO or MEO, respectively.

## 1.7 Functional Composition of Satellite Navigation System

In the operation of the satellite navigation system, the ground monitoring station collects the navigation signal and the satellite measurement data, the data processing center generates the broadcast message including ephemeris and clock offset, and the upload station injects the information into the satellite. The navigation satellite receives information from the ground, saves and arranges it into a navigation message, and broadcasts the message to the user. The user obtains the information on time and coordinates from the message, as well as the pseudo-range and carrier phase from measurement, and then calculates its three-dimensional position and velocity in the system coordinate frame and its time in the time reference.

It can be seen that the above process involves the functions and equipments of ground, satellite, and user, namely ground segment, space segment, and user terminal. The three segments are closely related and coordinated. Satellite, ground control system, and user terminal are in the same time reference and coordinate reference frame.The upload from the ground to satellite realizes the information transmission and time synchronization. The broadcast link realizes the precision ranging among satellite, ground control system, and user terminal, and links the three segments together in the same reference frame. The functions for each segment and subsystem in the BDS are introduced as follows.

### 1.7.1 Ground Segment

The main functions of the ground segment are: to establish, deliver and maintain the temporal and spatial references for navigation system by maintaining the ground-based atomic clock group, monitoring the navigation signals, and generating and transmitting the message to satellite for navigation, augmentation and communication; and to complete the constellation maintenance and management, and the inter-satellite link operation and management through TT&C, orbit determination, and emergency treatment.

The ground segment is generally composed of four kinds of stations: main control station, monitoring station (*s*), time synchronization/ upload station (*s*), and TT&C center (*s*) or station (*s*). Their main functions are addressed below.

1. **Main Control Station**
   a. Establish and maintain the time reference for the navigation system, and achieve the time comparison and synchronization between ground stations.
   b. Establish and maintain the spatial coordinate frame for the navigation system, and keep the coordinates of ground stations unchanged.
   c. Collect satellite observation data and meteorological data from monitoring stations, time synchronization/upload stations, SLRs, etc.
   d. Carry out precise orbit determination and ephemeris prediction, system time synchronization and satellite clock error prediction, ionospheric correction forecast, wide-area differential correction, and system integrity monitoring.

2. **Monitoring Station**
   a. Measure the pseudorange, carrier phase, and Doppler observations of satellite signals.
   b. Receive the navigation message.
   c. Collect meteorological data.
   d. Carry out the time synchronization between stations.

3. **Time Synchronization/Upload Station**
   a. Upload and inject the navigation information to the satellite.
   b. Carry out the satellite-ground time synchronization and measurement.
   c. Carry out the laser ranging.

4. **TT&C Center and TT&C Station**
   a. Carry out the satellite TT&C in both the early orbital phase and long-term operation phase.

b. Carry out the TT&C management and emergency treatment for navigation satellites.
c. Maintain and manage the navigation constellation and, inter-satellite link.

### *1.7.2 Space Segment*

The main functions of space segment are: to receive the upload-time synchronization signal from the ground segment, and keep the satellite time-frequency reference unchanged; to receive the information for navigation, augmentation, and communication from the ground segment, and generate the navigation signals; to receive, execute and transmit various measurements, remote control commands, and management information from/to ground segment, and also conduct inter-satellite transmission; and to achieve independent integrity monitoring, independent health management, and autonomous navigation.

The space segment is composed of a certain number of navigation satellites with independent functions. The navigation satellites constitute the configured constellation according to the specified design, and cover at least four navigation signals in the service area (global or regional). A navigation satellite is mainly composed of the subsystems dealing with spaceborne time frequency, upload signal reception and measurement, navigation and broadcasting, satellite information management, inter-satellite measurement and transmission, self-management, TT&C, attitude and orbit maintenance, power supply, and thermal control, etc. The main functions of each subsystem are as follows:

1. **Spaceborne Time-Frequency Subsystem** (Atomic Clock)
a. Generate and distribute a continuous and stable time-frequency reference network inside the satellite.
b. Receive the time calibration command from the ground segment to complete the unification between satellite time and system time.

2. **Upload Signal Reception and Measurement Subsystem**
a. Receive the uploaded signal from the ground and complete the satellite-ground time comparison based on the satellite time.
b. Receive the uploaded navigation message from the ground and the remote control message in an anti-interference manner.

3. **Navigation and Broadcasting Subsystem**
a. Save and process the uploaded navigation message to generate the spaceborne navigation message.
b. Monitor (with alarm) the navigation signals and autonomous integrity information, and integrate such information into navigation message.
c. Generate the navigation signals through modulation, and amplify and broadcast them via the antenna.

4. **Satellite Information Management Subsystem**
a. Manage the satellite navigation service information, remote control information, and monitoring information.
b. Manage the multi-source, multi-frequency, and multi-channel information that is inter-satellite or satellite-ground.

5. **Inter-Satellite Measurement and Transmission Subsystem**
a. Establish the inter-satellite constellation network.
b. Carry out the high-precision inter-satellite ranging.
c. Carry out the inter-satellite information transmission.
d. Carry out the satellite-ground ranging and information transmission to obtain a backup.

6. **Self-management Subsystem**
a. Carry out the autonomous health management of satellite.
b. Carry out the autonomous integrity monitoring of satellite.
c. Carry out the autonomous navigation of satellite.

7. **TT&C Subsystem**
a. Receive and execute the remote control commands from the ground.
b. Generate the telemetry information such as satellite status, and transmit it to the ground.

8. **Attitude and Orbit Maintenance Subsystem**
a. Control the three-axis attitude of navigation satellite and the solar wing direction, and orient the navigation antenna to the ground.
b. Complete the orbit maintenance during the lifetime of the navigation satellite.

9. **Power Supply Subsystem**
a. Regulate the battery charging and discharging to achieve full energy supply and self-management of the satellite.

10. Thermal Control Subsystem
a. Monitor and control the temperature of all the spaceborne instruments, equipments, and components.

### *1.7.3 User Terminal*

User terminals are various receivers for military and civilian uses. Currently, hundreds of satellite navigation receivers are available on the market. According to their purpose, they can basically be classified into the following three categories:

a. Navigation receiver: standard type or precise type.
b. Precise survey receiver: to measure the carrier phase, in addition to pseudorange, with the centimeter-level accuracy.

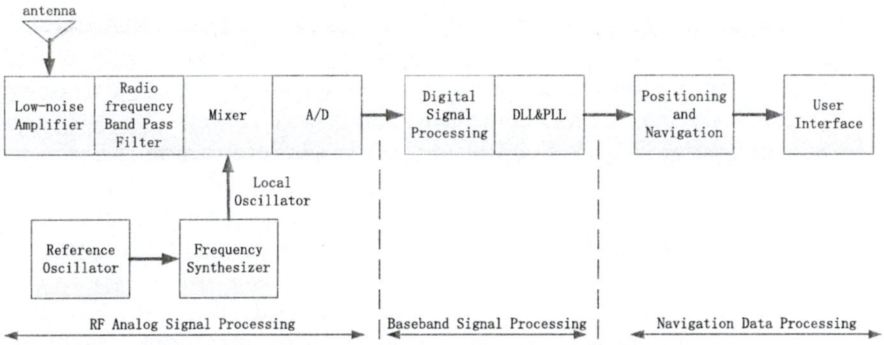


**Fig. 1.5** Typical functional diagram of receiver

c. Timing receiver: to mainly provide users with high-precision time information, with the usual timing accuracy better than 100 ns, even up to 10 ns.

The receivers have different designs and constructions, but the similar principles of basic hardware and software function blocks. The main functions of a satellite navigation receiver are:

a. Receive the satellite navigation signals, and acquire and track the pseudorandom codes and carriers of each signal.
b. Demodulate the navigation messages such as satellite ephemeris, clock offset, wide-area correction, ionospheric grid and integrity parameters, etc.
c. Determine the pseudorange, Doppler shift, and carrier phase observations.
d. Calculate the local time in the time reference of the navigation system, as well as the three-dimensional position and velocity of the user receiver, in the spatial reference frame of the navigation system.
e. Display the navigation and positioning results, and output and save the results.

A navigation receiver is generally the device used after receiving a signal from the antenna and before outputting the result to the display. It is usually divided into three functional modules: RF signal processing, baseband digital signal processing, and navigation data processing, as shown in Fig. 1.5.

## 1.8 System Service Performance Standard

The performance standard of the satellite navigation system is the combination of all the system performance parameters with system capability and performance. As the basic performance indicator of the system, service accuracy is composed of the measurement accuracy of position, time, and velocity. Service reliability, mainly including availability, continuity, and integrity, is the system's ability to provide continuous, stable, and reliable services. It depends on not only the constellation performance, satellite reliability, and control segment performance, but also the user environment.

Navigation signal performance is a major concern for industrial users, as it decides the way and mode of use of a satellite navigation system. However, for navigation user, the main concern is the PNT performance of a specific user receiver, which is not only related to navigation signal performance, but also to the user environment and receiver. As a result, the service performance standard of the system usually consists of: the standard of Signal in Space (SIS) performance, including the definition, characteristics, and criteria of SIS unrelated to user receiver; and the standard of service or position/time domain performance, including the definition, characteristics and criteria of positioning and timing service under the assumed receiver condition.

Since the GPS was put into service in 1994, the GPS-oriented guiding documents including *GPS Positioning Service Performance Standard*, *GPS Precision Positioning Service Performance Standard,* and *GPS Wide-area Enhanced System Performance Standard* have been publicly released. Similar documents have also been published for GLONASS and Galileo systems. In December 2012, the BDS-2 regional satellite navigation system was put into service. In December 2013, the *Open-service Performance Standard for Beidou Navigation Satellite System* based on the test results of system service performance was publicly released.

### *1.8.1 Definitions of Standard Indicators*

1. **Availability**

The system availability includes SIS availability and service availability. The SIS availability refers to the percentage of the time when a constellation satellite in the specified orbital slot provides a trackable and healthy SIS.

The service availability refers to the percentage of the time when the navigation and positioning service provided by the system meets the specified service performance requirements. The satellite geometric dilution and observation error vary with user position and observation times. The availability of a certain point at a certain time is called instantaneous availability. The availability of the same position at different times is called single point availability. The availability of different single point availability in a service area is called service area availability or PDOP (Position Dilution of Precision) availability. The PDOP availability can be determined by dividing the service area into small grids, and then statistically calculating the single point availability in the constellation operation period.

2. **Continuity**

The system continuity includes SIS continuity and service continuity. The SIS continuity refers to the percentage of the time when a trackable and healthy SIS can continuously work without any unscheduled outages within a specified period.

The service continuity refers to the probability of continuously satisfying the criteria of NPT accuracy and integrity under the condition that these criteria are met at the initial time. Correspondingly, the situation that these criteria are not met is called continuity risk.

3. **Integrity**

The service integrity refers to the degree of trust placed in the correctness of the information provided by the system. It is expressed as the system's ability to alert the user in time when the degradation of the provided positioning service exceeds the limit. The service integrity is usually described by three parameters: Alarm Limit, Time to Alarm, and Integrity Risk.

a. Alarm Limit: the value of positioning error that triggers the system alarm if exceeded. An alarm limit is further divided into a horizontal alarm threshold (HAL) and a vertical alarm threshold (VAL).
b. Time to Alarm: the delay between the time when the user's positioning error exceeds the alarm limit and the time when the user receives the alarm information.
c. Integrity Risk: the probability that the system doesn't send a signal to the user to warn against the positioning error already above the alarm limit, and thus results in the user's loss. This risk happens when the over-limit error is not detected by the system or detected but not sent at the specified time.

4. **PNT Service Accuracy**

For satellite navigation users, the greatest concern is the accuracy of the positioning and timing service obtained through a specific user receiver, which depends on not only the SIS performance and service performance, but also the space environment and user receiver. In general, the accuracy of the positioning and timing service obtained by satellite navigation user can be calculated by the following formula:

$$(\text{Positioning/Timing Accuracy}) = (\text{DOP}) \times (\text{UERE}) \tag{1.21}$$

where: DOP is precision dilution, and UERE is user equivalent range error, namely pseudorange error. The UERE can be expressed by the following formula, if all the error sources are independent of each other.

$$\text{UERE} = \sqrt{(\text{SISRE})^2 + (\text{STRE})^2 + (\text{UEE})^2} \tag{1.22}$$

Among them, SISRE, the abbreviation of SIS Ranging Error, is related to the errors associated with satellite and ground segment. The SISRE is mainly determined by the distribution and position error of the ground tracking station, the observation error of satellite, the orbit determination error, and the stability of the satellite clock (expressed by the ephemeris forecast error, clock error, and non-ideality of navigation signals).

STRE, the acronym of Signal Transmit Ranging Error, is mainly determined by ionosphere delay error, tropospheric delay error, and multipath error. The first two errors are caused by the inaccuracy of the navigation signal propagation model, and change with region and season. The third error is related to the user environment and the elevation angle of the observation-receiving antenna.

UEE, the acronym of User Equipment Error, is related to the design and quality of the user receiver. The UEE is mainly determined by system errors and random errors. The system errors are the residual of the fixed delay of user equipment channel after correction, such as antenna phase center error. The random errors include the track error caused by thermal noise, the measurement quantization error, and other unpredictable errors.

### *1.8.2 Relationships Among Service Reliability Indicators*

The system service reliability can be expressed by service availability, continuity, and integrity on the condition that the positioning accuracy meets the requirement. The relationships among the accuracy, availability, continuity, and integrity of a satellite navigation system are shown in Fig. 1.6.

Accuracy is the foundation of the standard indicators. It is directly related to the DOP of each satellite in the space constellation and to the measurement error URE caused by low SIS quality. Integrity is the ability of SIS to send an alarm to the receiver when spatial signals are not available or the service performance is unqualified. The integrity and continuity of SIS are related to each other. The system maintains the integrity and sends an alarm against the satellite unavailability, which will lead to the loss of continuity. However, if the satellite does not give an alarm in due time, the loss of integrity will be caused.

Both the continuity and availability of SIS are related to satellite interrupts, either scheduled or unscheduled. The scheduled interrupt refers to the interrupt of the signal

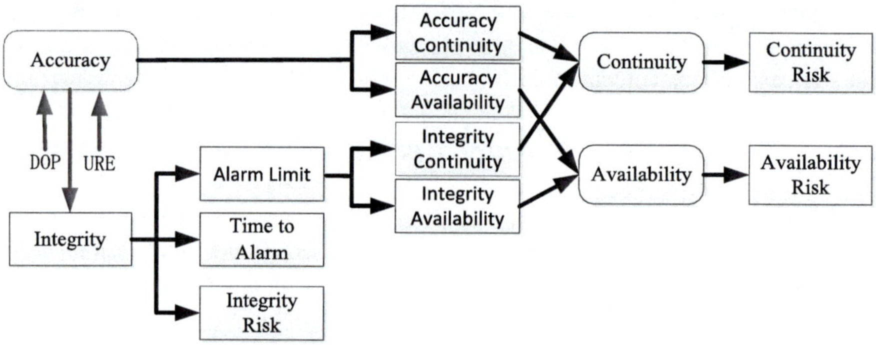


**Fig. 1.6** Relationships among accuracy, integrity, availability and continuity

sent to the user in advance within the specified time, such as satellite maneuver, on-orbit adjustment or testing. Such an interrupt can cause satellite unavailability, but will not cause continuity risk. However, the unscheduled interrupt (including unscheduled failure) will affect availability and continuity. Integrity availability refers to the availability to decide the Alarm Limit, which is similar to the accuracy threshold.

### *1.8.3 BDS Service Standard*

In the *Open-service Performance Standard for Beidou Navigation Satellite System*, the SIS performance indicators and service performance indicators of BDS OS have been defined.

1. **BDS OS SIS Performance Standard**

The BDS SIS performance criteria consist of coverage, accuracy, continuity, and availability. The SIS coverage is expressed by per-satellite coverage, namely the portion of region visible from the satellite orbit. This near-Earth region extends 1000 km on the Earth's surface.

The SIS accuracy is the statistical description of errors, for example, the error statistics (95% confidence) of any healthy satellite under normal operating conditions, including SIS Ranging Error (SISRE), SIS Ranging Rate Error (SISRRE), SIS Ranging Acceleration Error (SISRAE) and Coordinated Universal Time Offset Error (UTDOE).

The SIS continuity refers to the probability that a healthy OS signal can continue to work without unscheduled interrupts within a specified period. It is seriously affected by unscheduled interrupts, rather than those scheduled with a notice in advance.

The SIS availability is expressed by per-satellite availability, the probability that a BDS satellite in the specified orbit provides a healthy SIS signal.

2. **BDS OS Service Performance Standard**

The BDS service performance criteria consist of service accuracy, service availability, and interrupt-information dissemination time.

The service accuracy includes positioning accuracy and timing accuracy. The positioning accuracy, including horizontal positioning accuracy and vertical positioning accuracy, refers to the statistical value of the difference between the position provided by BDS and the real position under the specified user conditions. The timing accuracy refers to the statistical value of the offset between the time provided by BDS and the UTC under the specified user conditions.

The service availability refers to the ratio of actual service time to the expected service time, and includes PDOP availability and positioning service availability. The service time refers to the time when the service accuracy meets the specified performance criteria in a given area. The PDOP availability refers to the percentage of the time when the PDOP value in the specified service area is within the PDOP limit under the specified conditions. The position service availability refers to the percentage of the time when the horizontal and vertical positioning accuracy values in the service area meet the specified requirements within the specified time under the specified conditions

# Chapter 2
# Overview of Navigation Satellite Systems

The navigation satellite system is a space-earth integrated system where the satellite transmits the radio navigation signals and provides the PNT services to various users. In 1957, the Soviet Union successfully launched the first artificial earth satellite. In 1958, the United States began to develop the "Transit" navigation satellite system. Undergoing nearly 60 years of development, navigation satellite systems, including Global Navigation Satellite System (GNSS) and Regional Navigation Satellite System (RNSS), have become the important space infrastructure that people rely on.

At present, the navigation satellite systems in the world that have been constructed or are under construction include the US GPS system, the Russian GLONASS system, the Chinese BDS system, the European Union's Galileo system, the Japanese QZSS system, the Indian NAVIC system, etc. These systems can provide global or regional satellite navigation signals.

Most of the global navigation satellite systems have chosen the medium-orbit "Walker" constellation to ensure the global coverage of navigation signals and the average plane distribution according to the principles of satellite PNT and radio signal transmission attenuation, and to reduce the costs of space technology development and construction. Among them, the GPS adopts six orbital planes, while the GLONASS or Galileo uses three orbital planes. The BDS adopts, in addition to three orbital planes, a hybrid constellation system including high-orbit satellites to integrate navigation and augmentation services. RNSS generally utilize high-orbit GEO or IGSO satellite constellation to ensure signal coverage in the service areas.

Different countries and organizations develop different satellite navigation infrastructures with different strategies according to their needs, objectives, and constraints. In general, the current navigation satellite systems mainly provide the following three services: active PNT service, passive PNT service, and satellite-based augmentation service. The QZSS system is expected to provide passive PNT and satellite-based augmentation services



J. Xie et al., *Satellite Navigation Systems and Technologies*, Space Science and Technologies, https://doi.org/10.1007/978-981-15-4863-5_2

after the system upgrading. The BDS has applied unique innovative technologies to the third step (BDS-3) according to the principle of three-step construction, in which active PNT, passive PNT, augmentation services, and global short message services are integrated into one system.
This chapter will elaborate on the basic conditions, history, and development trend of GNSS and RNSS systems, including four GNSS systems, QZSS, and NAVIC. The main navigation satellite systems are shown in Fig. 2.1.

## 2.1 USA GPS

Before the development of GPS, the US Navy completed the building of "Transit" navigation satellite system in 1964. However, the number of "Transit" satellites was small, and the user observation interval was about one hour, so continuous positioning service couldn't be provided. According to the Doppler frequency shift measurement used in the Transit system, the system had a single positioning accuracy of about 100 m. As a result, the "Transit" system couldn't provide continuous, high-precision, and high-dynamic positioning services. Its users were mainly ships and submarines. In order to provide continuous, high-dynamic and high-precision positioning and navigation services to the users, in 1973, the US Department of Defense (DoD) approved the parallel development of the satellite positioning system based on the ranging principle, namely GPS system.

In 1993, the initialized operational capacity of GPS was achieved, and the Standard Positioning Service (SPS) based on full constellation was provided. And in 1994, the constellation composed of 24 GPS satellites was deployed with a global coverage

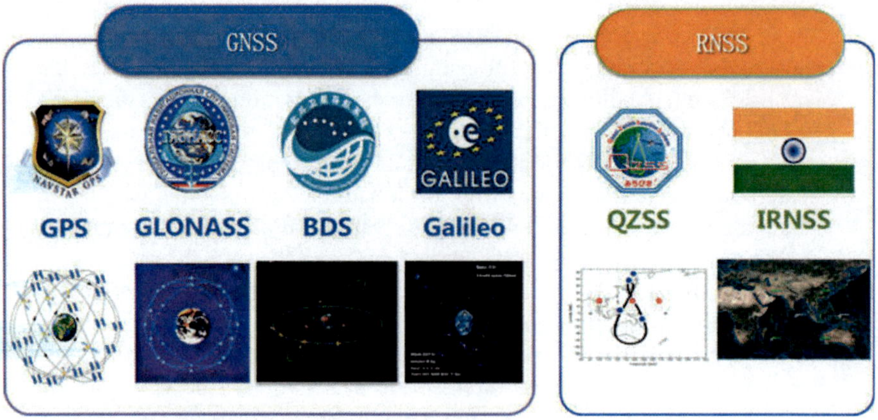


**Fig. 2.1** Navigation satellite system

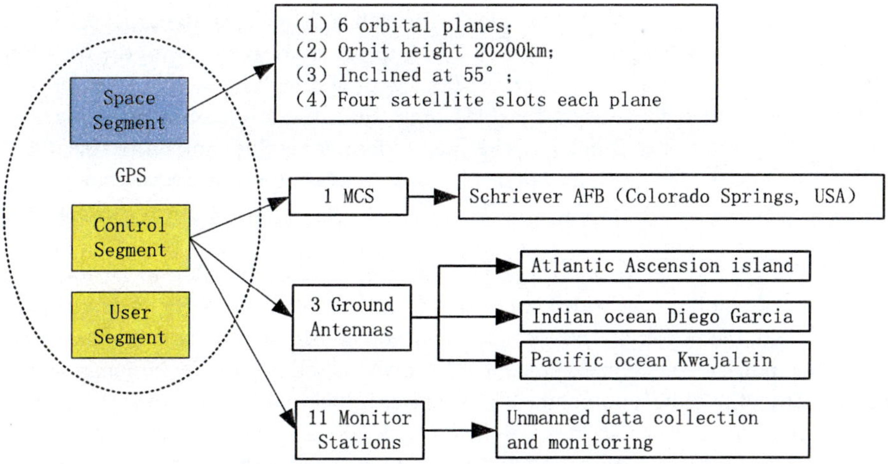


**Fig. 2.2** GPS system architecture

rate of 98%. The GPS consists of space segment, ground segment, and user segment, as shown in Fig. 2.2.

The standard constellation of the GPS space segment consists of 24 MEO satellites, which are unevenly distributed on each of the six orbital planes, with an angle of 55° and an orbital altitude of 20,200 km. Each satellite broadcasts navigation signals to the Earth center through a saddle-shape antenna array, and most of the signal energy is within ±22.5° of the signal beam.

The GPS ground segment consists of a master control station, ground antennas, and monitoring stations. Located at the Consolidated Space Operation Center (CSOC) in Colorado Springs, Colorado, USA, the master station provides services such as data collection, computing, transmission, and diagnostics based on large electronic computers. Ground antennas are located in three US military bases, namely Ascension Island in the Atlantic Ocean, Diego Garcia in the Indian Ocean, and Kwajalein in the Pacific Ocean. The monitoring stations are equipped with GPS receivers, environmental data meters, atomic frequency standards, and processors.

Several generations of GPS satellites have been developed to balance the relationship between inheritance and development. After the start-up of GPS in 1973, the first Block I satellite was launched in 1978, as the experimental satellite at the initial stage of GPS. From 1978 to 1985, 11 satellites of this type were developed. Each of the Block I satellites weighs 760 kg, with the power of 400 W and the designed onboard life of 4–5 years—but some of them have an actual life of 10 years.

The Block II satellites are the first fully-functional satellites. Their improved version is Block IIA, where the letter "A" is the first letter of "advanced". So far, 9 Block II satellites and 19 Block IIA satellites have been developed and launched between 1990 and 1997. The Block II series weighs 1160 kg, with the power of 710 W, and the designed life of 7.5 years. The Block IIA series weighs 1816 kg, with the power of 1100 W, and the designed life of 7.5 years. Each satellite is equipped with 2 cesium clocks and 2 rubidium clocks. In each launch, the Delta-2 rocket carries only one Block IIA satellite. The Block II/IIA series are designed with the function of Selective Availability (eliminated in 2000) to deliberately reduce the positioning accuracy of civilian GPS users by disturbing the satellite clock information and navigation message. The Block IIA satellites with inter-satellite link utilize the ultra-high frequency (UHF) (250–290 MHz) band to increase the spaceborne navigation message data that can help achieve a 180-day technical demonstration of independent operation.

The Block IIR series serves as the "complement" to Block II/IIA. In 1997–2005, 13 Block IIR satellites were developed and launched. The Block IIR series weighs 2032 kg, with the power of 1600 W and the designed life of 7.5 years. It is provided with UHF inter-satellites link, nuclear explosion detector, electronic reconnaissance, and other functions, and equipped with 2 rubidium clocks and 1 cesium clock. In each launch, the Delta-2 rocket carries only one Block IIR satellite. This series is capable of autonomous orbit determination and autonomous navigation message generation. It can operate independently in orbit for more than half a year to ensure that the accuracy of message broadcasting is acceptable.

The Block IIM series is the "modernized" and improved version of the Block IIR series. In 2005–2010, 8 Block IIM satellites were developed and launched. The Block IIM series weights 2060 kg, with the power of 1960 W, and the designed life of 10 years. It is provided with UHF inter-satellite link, nuclear explosion detector, and other functions, and equipped with 2 rubidium clocks and 1 cesium clock. In each launch, the Delta-2 rocket carries only one Block IIM satellite. This series is capable of providing 7 dB power enhancement service to designated areas, and of delivering higher anti-interference and navigation performance. Meanwhile, in order to improve the anti-interference and anti-spoofing abilities of GPS military signals and the high-precision application performance of civilian signals, the L1M and L2M military signals and L2C civilian signal have been introduced to Block IIM to initiate the first phase of GPS modernization plan.

The Block IIF series is the "upgraded" version of Block IIM series. The letter "F" means "Follow on". In 2010–2016, 12 Block IIF satellites were developed and launched. The Block IIF series weights 1630 kg, with the power of 2440 W, and the designed life of 15 years. It is provided with UHF inter-satellites link, nuclear explosion detector, 7 dB power enhancement, and other functions, and equipped with 2 rubidium clocks and 1 cesium clock. In each launch, the Evolved Expendable Launch Vehicle (EELV) carries only one Block IIM satellite. The L5C civilian signal in the L5 band and the military M code signal with power enhancement have been introduced to Block IIF to, initiate the second phase of the GPS modernization plan.

By November 2017, the Block I, Block II, IIA, IIR, IIM, and IIF series of GPS satellites have been developed, and more than 70 satellites launched. Currently, the GPS constellation has 40 satellites in orbit, of which 31 GPS-IIR satellites, 7 GPS-IIM satellites, and 12 GPS-IIF satellites are operational (see Table 2.1).

Currently, the United States is actively promoting the development of the next generation of GPS, namely Block III project. As the third phase of the GPS modernization plan, the Block III series is about to be launched in 2003. This series is

**Table 2.1** GPS satellites in orbit (by the end of november 2017)

| Plane | Slot | Block | Launch time (UTC) | Status |
|---|---|---|---|---|
| A | A1 | GPS-IIF | 4th October 2012 | Operational |
| | A2 | GPS-IIM | 25th September 2006 | Operational |
| | A3 | GPS-IIF | 21st February 2014 | Operational |
| | A4 | GPS-IIM | 15th March 2008 | Operational |
| B | B1 | GPS-IIR | 29th January 2003 | Operational |
| | B2 | GPS-IIF | 28th May 2010 | Operational, first 2F Satellite |
| | B3 | GPS-IIR | 16th July 2000 | Operational |
| | B4 | GPS-IIM | 17th November 2006 | Operational |
| | B5 | GPS-IIF | 25th March 2015 | Operational |
| | B6 | GPS-IIR | 11th May 2000 | In maintenance |
| C | C1 | GPS-IIM | 20th December 2007 | Operational |
| | C2 | GPS-IIF | 15th May 2013 | Operational |
| | C3 | GPS-IIF | 15th July 2015 | Operational |
| | C4 | GPS-IIM | 26th September 2005 | Operational, first 2RM Satellite |
| | C5 | GPS-IIR | 20th March 2004 | Operational |
| D | D1 | GPS-IIR | 6th November 2004 | Operational |
| | D2 | GPS-IIF | 16th July 2011 | Operational |
| | D3 | GPS-IIR | 31st March 2003 | Operational |
| | D4 | GPS-IIF | 17th May 2014 | Operational |
| | D5 | GPS-IIR | 7th October 1999 | Operational |
| E | E1 | GPS-IIF | 29th October 2014 | Operational |
| | E2 | GPS-IIF | 31st October 2015 | Operational |
| | E3 | GPS-IIM | 17th August 2009 | Operational |
| | E4 | GPS-IIR | 30th January 2001 | Operational |
| | E6 | GPS-IIR | 21st December 2003 | Operational |
| F | F1 | GPS-IIR | 10th November 2000 | Operational |
| | F2 | GPS-IIM | 17th October 2007 | Operational |
| | F3 | GPS-IIF | 2nd August 2014 | Operational |
| | F4 | GPS-IIR | 23rd June 2004 | Operational |
| | F5 | GPS-IIF | 5th February 2016 | Operational, last 2F Satellite |
| | F6 | GPS-IIR | 23rd July 1997 | Operational |

designed with further improved anti-interference performance, SAR (Search and Rescue) service performance, and better interoperability. Compared with the GPS II series, the GPS-III is expected to significantly improve the navigation performance in military and civilian fields. Meanwhile, it can quickly turn off the navigation signal in specific service areas as needed.

**Table 2.2** GPS-III Satellite launch plan

| SVN | Satellite | Launch plan | Status |
|---|---|---|---|
| 74 | Block IIIA-01 | December 2018 | Not launched |
| 75 | Block IIIA-02 | April 2019 | Not launched |
| 76 | Block IIIA-03 | October 2019 | Not launched |
| 77 | Block IIIA-04 | 2020 | Not launched |
| 78 | Block IIIA-05 | | Not launched |
| 79 | Block IIIA-06 | 2021 | Not launched |
| 80 | Block IIIA-07 | | Not launched |
| 81 | Block IIIA-08 | 2022 | Not launched |
| 82 | Block IIIA-09 | | Not launched |
| 83 | Block IIIA-10 | 2023 | Not launched |

To quickly respond to the commands and shorten the message updating period, a SAR payload and a 100 Mb/s high-speed inter-satellite link with Ka or V band have been added to the GPS-IIIA, whose designed life is 15 years. The signal transmission mode has been changed to point-to-point transmission. In terms of navigation signals, the coverage of military M code signal will be extended to the ground, and the L1C civil signal compatible with the Galileo system and interoperable in the L1 band will be added.

It is estimated that by 2030, 32 GPS-III satellites will be developed to increase the regional received power by 20 dB and to provide additional SAR services and space environment detection. Compared with GPS II, the performance of GPS-III is expected to be greatly improved, with the positioning accuracy to be improved by 4 times, the timing accuracy to be doubled, and the availability to be increased by an order of magnitude. The GPS-III provides users with safety assurance for the first time, promises to meet the integrity requirements of non-precision approach, and strives to meet the integrity requirements of Class-I precision approach. At the same time, it will greatly improve the anti-jamming capability of the system, ensuring the positioning accuracy in a power enhancement environment with strong interference.

The first Block IIIA satellite was originally planned to be launched in 2014. After being postponed for several times, it was finally launched in 2018. At that time, a spot beam (an area with a diameter of about 600 miles) was used for the in-orbit demonstration and verification of broadcasting navigation. The US GPS-III satellites launch plan is shown in Table 2.2.

## 2.2 Russian GLONASS

The GLONASS is a global navigation system developed by the Soviet Union on the basis of the low-orbit satellite navigation system "Tsikada". It has been deployed in parallel with the US GPS. The first GLONASS satellite was launched in 1982, after

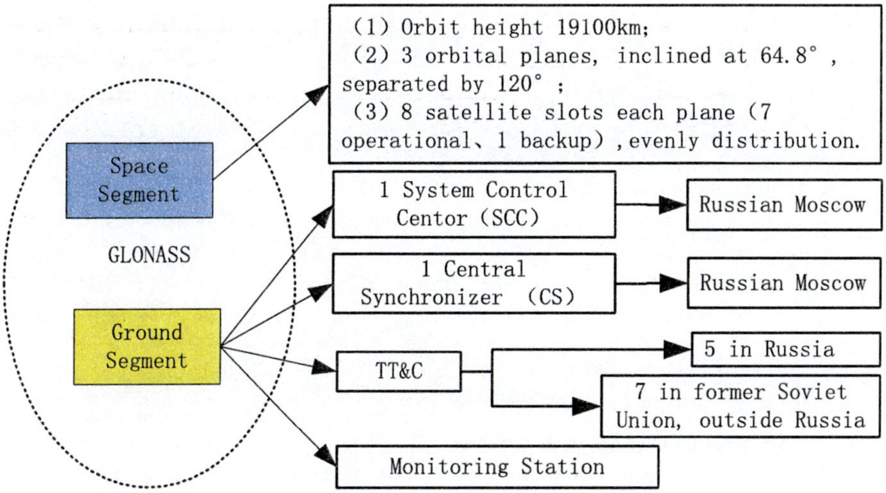


**Fig. 2.3** GLONASS system architecture

10 years of research and development. The GLONASS system is different from GPS in constellation, signal, and satellite design.

The GLONASS consists of space segment, ground segment, and user segment. The system architecture is shown in Fig. 2.3.

The GLONASS constellation consists of 24 MEO satellites distributed on three orbital planes, which have an inclined angle of 64.8° and an altitude of 19,100 km. On each orbital plane, eight satellites are evenly distributed. Each satellite broadcasts the navigation signals to the Earth center through a saddle-shape antenna array. Due to the higher geographic latitude of the northern hemisphere, the GLONASS orbital inclination is designed to be 64.8° in order to optimize the local coverage performance of satellite navigation signals.

The GLONASS ground segment is composed of a system control center, a central synchronizer, TT&C (Telemetry, Tracking, and Command) centers, and monitoring stations. The main functions of the ground segment are satellite orbit measurement, time measurement, navigation message generation, and telemetry and remote control, as shown in Fig. 2.4.

The GLONASS time ($t_{\text{GLONASS}}$) is provided by the hydrogen atomic clock of central synchronizer, whose daily stability is better than 1–5 $\times$ $10^{-14}$. The deviation between the GLONASS time and the Universal Time Coordinated (UTC (SU) is controlled within 1 ms. Periodic integer corrections are performed once a year or once a year and a half. The time of each GLONASS satellite is periodically calibrated to ensure that the time difference between satellite and GLONASS system is within 10 ns. The space stability of the spaceborne atomic clock is better than 5 $\times$ $10^{-13}$ (on GLONASS satellite) and 1 $\times$ $10^{-13}$ (on GLONASS-M satellite) to ensure that the measurement time difference between satellites is less than 20 ns (between GLONASS satellites) and 8 ns (between GLONASS-M satellites).

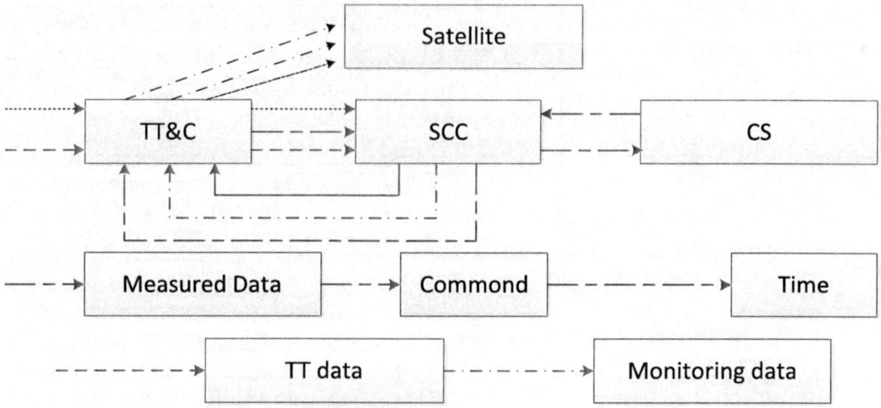


**Fig. 2.4** Diagram of GLONASS ground segment

The GLONASS satellite began to be developed in the Soviet Union period. Now it has evolved into several series, including prototype verification series, GLONASS series, GLONASS-M series, GLONASS-K1 series, GLONASS-K2 series, and GLONASS-KM series. Different from the GPS that broadcasts the code division multiple access (CDMA) signals, the GLONASS broadcasts the frequency division multiple access (FDMA) signals to the Earth center so that the error rate caused by multi-path frequency selective channel is effectively reduced. Moreover, the distribution of different satellites in different signal bands is helpful for the anti-jamming performance of GLONASS. With the development of receiver technology, the advantages of the FDMA technique at the receiver end have gradually disappeared. Instead, the problems of low spectrum utilization and large resource consumption found in multimode receivers have become increasingly prominent. Therefore, the GLONASS modernization project has started with the CDMA signal transmission by GLONASS-K1 satellite.

In 1982–1985, the prototype verification satellites were launched, and the demonstration of a test system consisting of four satellites was completed.

As the first generation of GLONASS family, the GLONASS satellites were launched from 1985 to 2005. The first group weights 1250 kg, with the power of 1000 W, and the designed life of 1 year. The follow-up group weights 1415 kg, with the power of 1000 W, and the designed life of 3 years. So far, a total of 78 satellites have been developed. They were launched by the Proton rocket, which carries three satellites in each launch. The GLONASS series takes a cesium clock as spaceborne frequency reference. Compared with the prototype verification series, the time-space reference performance of this series (like frequency stability) is significantly improved. This series is equipped with a propulsion device that can adjust the position in orbit.

Due to the short designed life of GLONASS satellites and the Russian economy depression in the late 1990s, the satellite replenishment was delayed, making it difficult to provide continuous stable services after 1995. At the end of 2001, only six satellites were in orbit, including four satellites that were able to provide navigation signals normally.

In order to change this situation, the Russian government approved the "GLONASS 2002.2011 Development Plan" in August 2001, and started the GLONASS modernization plan. The GLONASS-M, in which the letter "M" means "Modernized", is the product of GLONASS modernization, also known as the second generation of GLONASS satellite.

As the first step of the modernization plan, the GLONASS-M satellite signal was improved. In the G2 band (1246 + MHz), the second civilian signal L2OF was added to provide duel frequency service so that the interference with radio astronomy service was mitigated. Since 2014, the GLONASS-M satellites have been broadcasting the CDMA civil signal L3OC in the G3 band (1202.025 MHz). At the same time, a laser reflector near the transmitting antenna and an inter-satellite link was added to enable the satellites to achieve more accurate satellite orbit determination through laser measurement, and to demonstrate their ability to operate autonomously.

With the launch of GLONASS-M satellites, the GLONASS recovered fully at the end of 2011. The GLONASS-M is the current main GLONASS satellite type in orbit, with a weight of 1480 kg, the power of 1270 W, and the designed life of 7 years. So far, a total of 45 satellites have been developed and launched by the Proton rocket carrying three satellites per launch.

The GLONASS-K is called the third generation of GLONASS satellites. The launch of two GLONASS-K1 satellites in February 2011 and December 2014, respectively, officially started the second phase of the GLONASS modernization plan. The GLONASS-K is designed with a weight of 750 kg, the power of 1400 W, and the service life of 10–12 years, and is equipped with an inter-satellite link. It is planned to be launched by the Proton rocket carrying 6 satellites per launch or the Soyuz-2 rocket carrying one satellite per launch. Cesium clock is used as its frequency reference. Compared with the second generation, the weight of this satellite has been reduced by half, while the SAR payload added. The GLONASS-K satellites fall into two types: GLONASS-K1 and GLONASS-K2. The GLONASS-K1 broadcasts CDMA civil navigation signal L3OC in the G3 band (1202.025 MHz). The GLONASS-K2 broadcasts CDMA civil navigation signals L1OC and L2OC and military signals L1SC and L2SC in the G1 (1600.995 MHz) and G2 bands (1248.06 MHz), in addition to the signals broadcasted by GLONASS-K1, thereby enabling the GLONASS system to provide CDMA services in civil tri-band and military dual-band.

By August 2019, Russia's GLONASS constellation has embraced 27 satellites in orbit, including 23 GLONASS-M satellites and 2 GLONASS-K1 satellites, 24 of which are operational (see the Table 2.3).

The GLONASS modernization plan is still in progress towards GLONASS-KM, the next generation of GLONASS satellites.

The GLONASS-KM is expected to have new features and improved performance while inheriting the GLONASS-K2's signals in the G1 (1575.42 MHz), G2 (1207.14 MHz), and G3 (1176.45 MHz) bands. In addition, three CDMA navigation signals L1OCM, L2OCM, and L3OCM will be transmitted to improve the interoperability with other GNSS systems.

**Table 2.3** GLONASS satellites in orbit (by the end of august 2019)

| Plane | Slot | Block | Launch time (UTC) | Status |
|---|---|---|---|---|
| I | 1 | GLONASS-M | 14th December 2009 | Operational |
| | 2 | GLONASS-M | 26th April 2013 | Operational |
| | 3 | GLONASS-M | 4th November 2011 | Operational |
| | 4 | GLONASS-M | 2nd October 2011 | Operational |
| | 5 | GLONASS-M | 17th June 2018 | Operational |
| | 6 | GLONASS-M | 14th December 2009 | Operational |
| | 7 | GLONASS-M | 4th November 2011 | Operational |
| | 8 | GLONASS-M | 4th November 2011 | Operational |
| II | 9 | GLONASS-K1 | 30th November 2014 | Operational |
| | 10 | GLONASS-M | 25th December 2006 | Operational |
| | 11 | GLONASS-M | 29th May 2016 | Operational |
| | 12 | GLONASS-M | 27th May 2019 | Operational |
| | 13 | GLONASS-M | 25th December 2007 | Operational |
| | 14 | GLONASS-M | 22nd September 2017 | Operational |
| | 15 | GLONASS-M | 3th November 2018 | Operational |
| | 16 | GLONASS-M | 2nd September 2010 | Operational |
| III | 17 | GLONASS-M | 7th February 2016 | Operational |
| | 18 | GLONASS-M | 23rd March 2014 | Operational |
| | 19 | GLONASS-M | 26th October 2007 | Operational |
| | 20 | GLONASS-K1 | 26th February 2011 | First GLONASS K1, flight testing |
| | | GLONASS-M | 26th October 2007 | Operational |
| | 21 | GLONASS-M | 14th June 2014 | Operational |
| | 22 | GLONASS-M | 1st March 2010 | Operational |
| | 23 | GLONASS-M | 1st March 2010 | Operational |
| | 24 | GLONASS-M | 1st March 2010 | Operational |

## 2.3 EU Galileo

Developed by the European Union (EU) and the European Space Agency (ESA), the Galileo navigation satellite system was officially launched in March 2002. In December 2016, its global early operability was announced.

The composition of the Galileo system is shown in Fig. 2.5.

The Galileo constellation consists of 30 MEO satellites, including 24 operational satellites, and 6 backup satellites. They are evenly distributed on three orbital planes at an inclined angle of 56° and an altitude of 23,322 km. Each satellite broadcasts navigation signals to the Earth center through a saddle-shape antenna array.

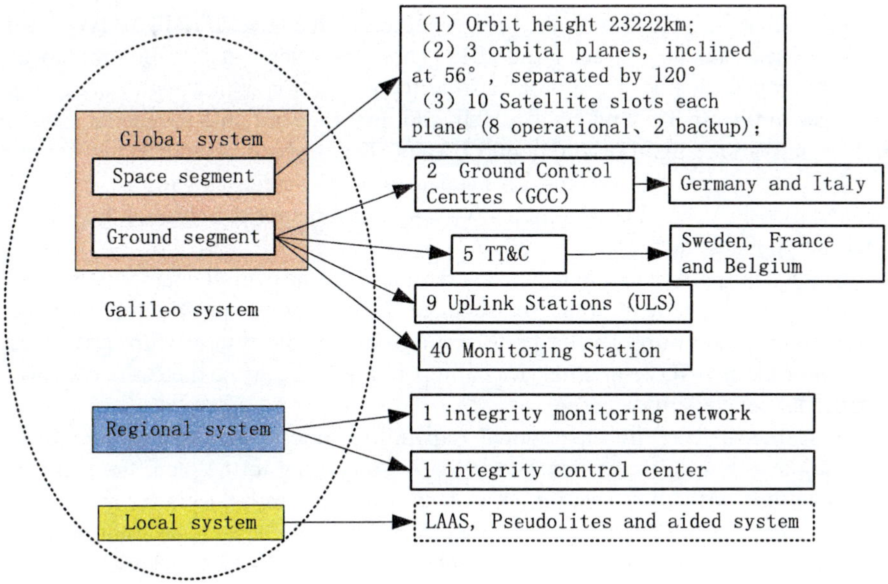


**Fig. 2.5** Galileo system architecture

Due to a higher altitude than GPS and GLONASS, the Galileo has a better signal coverage so that at any point on the Earth, at least 6 satellites can be seen at any time with an elevation angle greater than 10°.

The ground segment of the Galileo system consists of two ground control centers, five TT&C stations, 10 mission data uplink stations (ULS), and monitoring stations distributed around the world. TT&C stations are mainly used to transmit remote control commands and serve as a backup for mission data uplink service by transmitting the navigation message data of Open Service (OS), Life-of-Life (SoL) and Public Regulated Service (PRS) to the satellites. ULS stations are used to transmit six types of data to the satellites, including the data of OS, SoL, External Region Integrity Systems (ERIS), public authorization services, Commercial Services (CCS), and Search and Rescue services (SAR).

The Galileo test satellites include the GIOVE-A launched in 2005 and the GIOVE-B launched in 2008. The GIOVE-A is designed with a weight of 600 kg, the power of 700 W, and the life of 2 years, and is equipped with two cesium clocks. The GIOVE-B has a total weight of 523 kg and supplies 700 W of power. It is launched by the Soyuz-FG rocket carrying one satellite per launch. The GIOGVE-B, whose designed life is 2 years, is equipped with 2 cesium clocks and a passive hydrogen clock. The daily stability of its atomic clocks has been improved from 10 to 1 ns. The GIOVE-A and GIOVE-B were developed to demonstrate the broadcasting performance of the Galileo satellite family and its ability to monitor the MEO orbit. The navigation signals they broadcast contain no navigation messages.

The Galileo-IOV (In Orbit Validation) satellites include the PFM, FM2, FM3, and FM4 satellites launched in 2011 and 2012. These satellites carry 2 rubidium clocks and 2 hydrogen clocks, while broadcasting the navigation signals with navigation messages in the E1, E5, and E6 bands. In addition, they are able to receive ground SAR signals in the band of 406.0–406.1 MHz. The payload of SAR signal is sent to the rescue center through the L6 band to provide search and rescue services. Each Galileo-IOV satellite is equipped with a laser reflector to achieve higher orbit determination accuracy.

The main purpose of Galileo-IOV is to demonstrate the overall operation of space segment, ground segment, and user segment, as well as the positioning, velocity measurement, and timing service performance of the whole system. Currently, three of the four Galileo-IOV satellites are still operational as part of the Galileo's early operational constellation.

The Galileo-FOC (Full Operational Capability) series is the formal satellite of the Galileo system initially launched in 2014. According to the plan, the Galileo-FOC satellites will fill the entire constellation on three orbital planes to provide global navigation and positioning services. The Galileo-FOC series weights 680 kg and supplies 1600 W of power. It is launched by the Soyuz-STB rocket carrying 2 satellites per launch, or by the Ariane 5 rocket carrying 4 satellites per launch. The designed life of Galileo-FOC is 12 years.

By November 2017, the European Union's Galileo constellation has embraced 17 satellites in orbit, 15 of which are operational, including 3 Galileo-IOV satellites and 12 Galileo-FOC satellites (see Table 2.4).

Owing to the use of the hydrogen atomic clock as the time-frequency reference of space segment and the modernized design of navigation signals, the signal transmission quality of Galileo is better than GPS and GLONASS.

## 2.4 Japanese QZSS

For the users in the countries and regions with urban canyons and mountainous terrain like Japan, it is very important to observe satellites at high elevation angles. Through the cooperation between the Japanese government and private companies, the Quasi-Zenith Satellite System (QZSS) has been developed to solve the problem of satellite performance declination due to the obstacles in radio signal transmission and the lack of an effective number of GPS satellites.

The QZSS system was originally defined as a multi-tasking satellite system that provides communication, broadcasting, and locating services to mobile phone users at high viewing angles in a designated area through the enhanced GPS. On the other hand, when the GPS signal is interrupted, Japan can still rely on the independent navigation and positioning capabilities of QZSS.

The block diagram of the Japanese QZSS system is shown in Fig. 2.6.

In 2010, the first QZSS satellite "QZS-1" was developed and launched. After one year of in-orbit testing, the QZS-1 successfully completed the task of technical

**Table 2.4** EU Galileo satellites in orbit (by the end of november 2018)

| Plane | Slot | Block | Launch time (UTC) | Status |
|---|---|---|---|---|
| A | A1 | Galileo-FOC | 12th December 2017 | Operational |
| | A2 | Galileo-FOC | 24th May 2016 | Operational |
| | A3 | Galileo-FOC | 12th December 2017 | Operational |
| | A4 | Galileo-FOC | 12th December 2017 | Operational |
| | A5 | Galileo-FOC | 11th September 2015 | Operational |
| | A6 | Galileo-FOC | 24th May 2016 | Operational |
| | A7 | Galileo-FOC | 12th December 2017 | Operational |
| | A8 | Galileo-FOC | 11th September 2015 | Operational |
| B | B1 | Galileo-FOC | 25th July 2018 | Commissioning |
| | B2 | Galileo-FOC | 25th July 2018 | Commissioning |
| | B3 | Galileo-FOC | 27th March 2015 | Operational |
| | B4 | Galileo-FOC | 25th July 2018 | Commissioning |
| | B5 | Galileo-IOV | 21st October 2011 | Operational |
| | B6 | Galileo-IOV | 21st October 2011 | Operational |
| | B7 | Galileo-FOC | 25th July 2018 | Commissioning |
| | B8 | Galileo-FOC | 27th March 2015 | Operational |
| C | C1 | Galileo-FOC | 17th November 2016 | Operational |
| | C2 | Galileo-FOC | 17th December 2015 | Operational |
| | C3 | Galileo-FOC | 17th November 2016 | Operational |
| | C4 | Galileo-IOV | 12th October 2012 | Operational |
| | C6 | Galileo-FOC | 17th November 2016 | Operational |
| | C7 | Galileo-FOC | 17th December 2015 | Operational |
| | C8 | Galileo-FOC | 17th November 2016 | Operational |

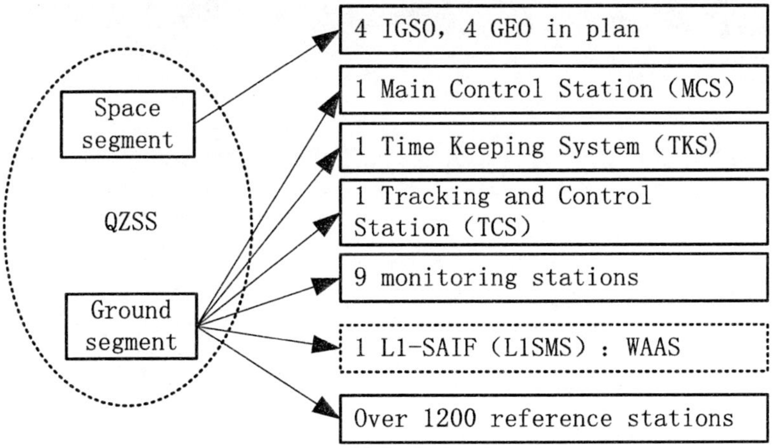


**Fig. 2.6** QZSS architecture

confirmation and application demonstration. In July 2017, Japan launched the fourth QZSS satellite so that the QZSS was completely established. Currently, Japan is planning to increase the number of QZSS satellites to seven in order to upgrade the QZSS into an independent navigation and augmentation system similar to BDS-2.

The QZSS consists of a number of highly inclined geosynchronous orbit satellites with the same ground trace. By selecting an appropriate eccentricity and inclination, the minimum elevation angle can exceed 70° in the service area in 24 h. This means that at any time, the user can receive at least one satellite signal from QZSS at a high elevation angle in the zenith direction. This is why the system is called "Quasi-Zenith".

The QZSS is capable of transmitting the signals compatible with the current and future GPS L1, L2, and L5 civil signals (in terms of center frequency, bandwidth, PRN code, and information structure). These signals are transmitted from the QZSS satellites to the entire visible area on the Earth's surface. The service area covers not only Japan, but also East Asia and Oceania. The basic navigation area of QZSS is 125°–146° East longitude and 25°–45° North latitude, covering Japan and neighboring regions (see Fig. 2.7).

The satellite-based augmentation service complements and enhances the GPS through the facilities of QZSS satellite and ground segments. The service area, which is closely related to the layout of ground segment is 129°–146° East longitude and 26°–45° North latitude, covering Japan and neighboring regions (see Fig. 2.8).

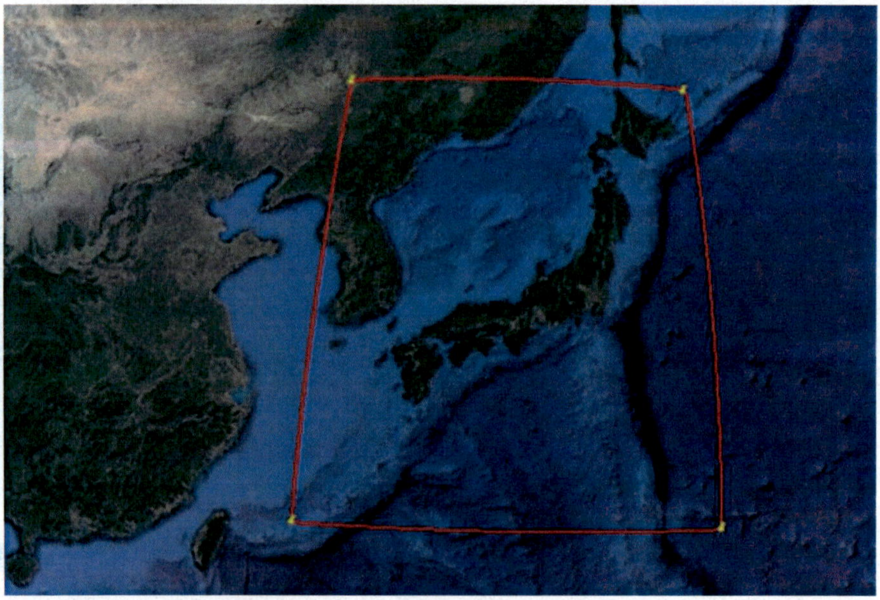

**Fig. 2.7** Basic navigation service area of QZSS

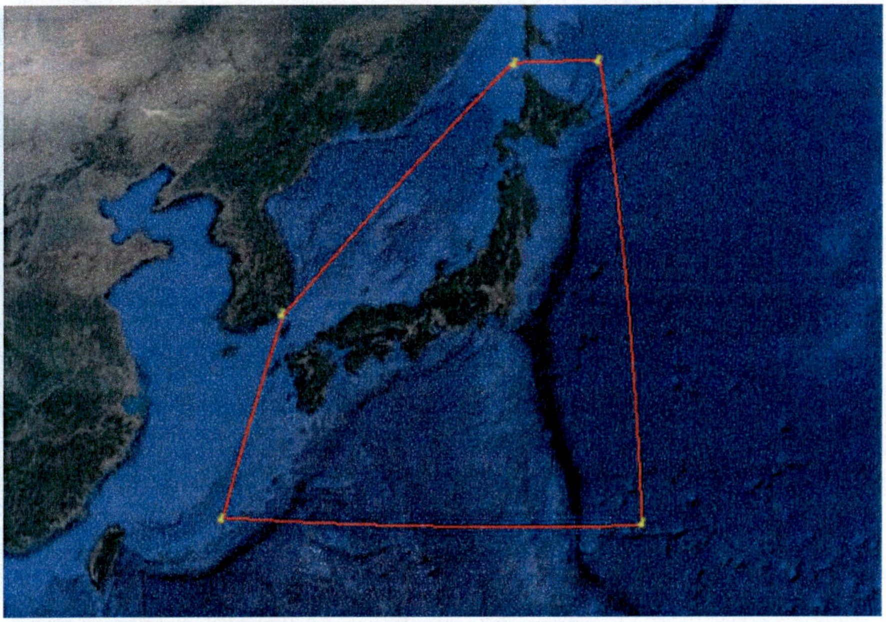

**Fig. 2.8** SBAS area of QZSS

## 2.5 Indian NAVIC

NAVIC (Navigation with Indian Constellation), the former IRNSS (Indian Regional Navigation Satellite System), is India's independent regional navigation satellite system. The first IRNSS-1A satellite was launched in July 2013. By April 2018, a total of seven NAVIC satellites have been operational in orbit. The block diagram of NAVIC system is shown in Fig. 2.9.

The NAVIC constellation consists of seven satellites, including three GEO satellites and four IGSO satellites. The International Telecommunication Union has accepted the Indian government's application for increasing satellite orbits in the constellation. The three GEO satellites are fixed at east longitude 34°, 83°, and 131.5°, respectively. The four IGSO satellites are distributed on two orbital planes. Their ground traces form two "8" patterns where the intersections are at east longitude 55° and 111.5° respectively. The IGSO satellites I1 and I3 are on one orbital plane, while I2 and I4 are on the other orbital plane. The phase difference between the two planes is 56°. The service area is 40°–140° East longitude and 40° North and South latitude. The satellites broadcast single-band and dual-band navigation signals, with the positioning accuracy of standard service better than 20 m.

The NAVIC ground segment includes satellite control center, monitoring station, range and integrity monitoring stations, timing center, CDMA ranging station, laser ranging station, navigation center, and data communication network. Among them, the main function of NAVIC monitoring station is to receive data from GEO and

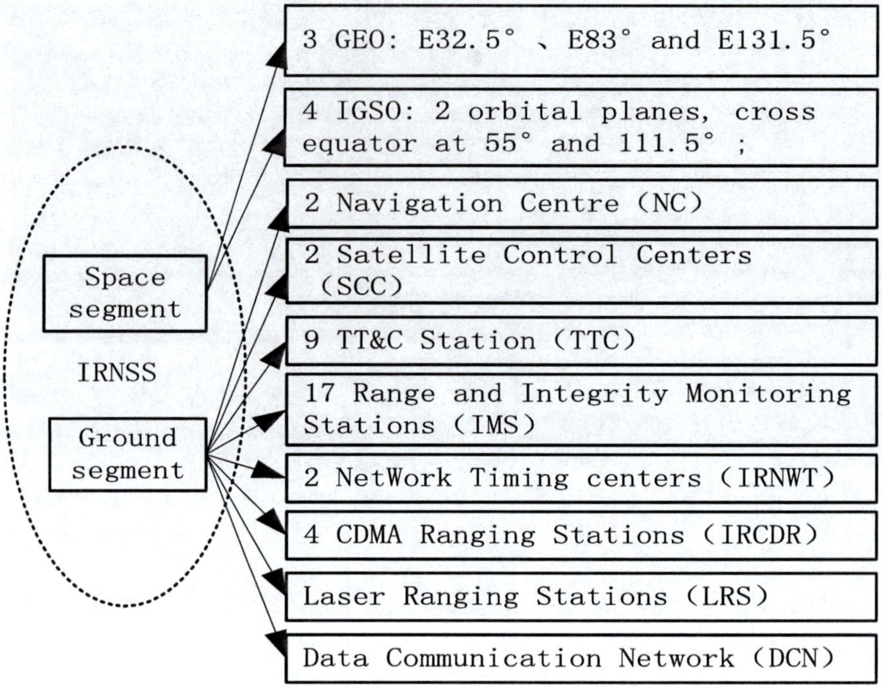


**Fig. 2.9** IRNSS architecture

IGSO satellites, correct the received ranging data, and transmit the original data and correction data to the control center.

The main function of the navigation center is to calculate the satellite ephemeris, satellite clock correction parameters, ionospheric delay correction data and the corresponding integrity information, transmit the calculated results to the uplink station, and then broadcast them to the user through the GEO satellite. The satellite control center is mainly responsible for managing, controlling, and maintaining the normal operation of the satellites in orbit.

The CDMA ranging station and laser ranging station is responsible for collecting the ranging information from NAVIC satellites and then transmitting the correction data to the navigation center.

The NAVIC satellites broadcast civilian SPS signals and authorized signals in the L5 (1176.45 MHz) and S (2492.028 MHz) bands, providing dual-band services for civilian and military uses.

The NAVIC satellites in orbit by November 2018, are shown in Table 2.5.

**Table 2.5** NAVIC satellites in orbit (by the end of november 2018)

| Satellite | Orbit | Launch time (UTC) | Status |
|---|---|---|---|
| 1A | IGSO | 1st July 2013 | Failed in orbit |
| 1B | IGSO | 4th April 2014 | Operational |
| 1C | GEO | 15th October 2014 | Operational |
| 1D | IGSO | 28th March 2015 | Operational |
| 1E | IGSO | 20th January | Operational |
| 1F | GEO | 10th March 2016 | Operational |
| 1G | GEO | 28th April 2016 | Operational |
| 1I | IGSO | 12th April 2018 | Operational |

## 2.6 BeiDou Navigation Satellite System (BDS)

BDS is an independent and self-constructed navigation satellite system that focuses on the needs for national security and economic and social development. It is an important national space infrastructure providing all-weather all-day high-precision positioning and navigation services to the users all over the world.

In the late twentieth century, China began to explore a development path of navigation satellite system suitable for national conditions, and gradually developed a "three-step" development strategy.

**The first step** is to build the BDS-1 system (also known as Beidou navigation satellite test system). In 1994, the BDS-1 system project was started. In 2000, two GEO satellites were successfully launched. Thus, the system was completely built and put into operation with active RDSS to provide positioning, timing, and short message communication services for users in China. In 2003, the third GEO satellite was launched to enhance the system performance and expand the application.

**The second step** is to build the BDS-2 system. The construction of the BDS-2 system was started in 2004. By the end of 2012, 14 satellites (including 5 GEO satellites, 5 IGSO satellites, and 4 MEO satellites) had been launched. The BDS-2 system provided passive PNT services such as GPS, in addition to wide-area differential service and short message communication service. Furthermore, its service area was expanded to the Asia Pacific region.

**The third step** is to build the BDS-3 system. The construction of the BDS-3 system was started in 2009. The BDS-3 system provides services such as passive PNT, wide-area differential, and short message communication. In addition, by 2018, it has provided basic services to the countries along the "Belt and Road" and in neighboring regions. By 2020, 35 satellites are expected to be launched to provide services to the global users.

The BDS consists of three segments: space segment, ground segment, and user segment. The space segment includes a number of GEO satellites, IGSO satellites, and MEO satellites. The ground segment includes several ground stations such as master control station, time synchronization/uplink station, and monitoring station. The user segment is composed of various products, such as chips, modules and antennas, terminals, and application systems and services compatible with other navigation satellite systems.

### 2.6.1 BDS-1 System

BDS-1 is a two-way active ranging navigation system. It relies on two GEO satellites and ground control stations to deliver the radio determination satellite services (RDSS) to the service area, including positioning, timing, and location reporting/short message communication services. In the RDSS services, the positioning and timing services are provided by calculating the position and relative radio navigation parameters through the ground control system other than the user, and then forwarding the position information to the user through satellite. Since the user is required to apply and transmit the navigation signal provided by RDSS to the satellite, this system is also called an active navigation system.

In January 1994, China approved the "Double Star Navigation and Positioning System" project and renamed it "Beidou No. 1". From then on, the development of China's navigation and positioning satellite system began. Several years later, the first Beidou-1 satellite was successfully launched on October 31st, 2000. Shortly thereafter, the second Beidou-1 satellite was launched on December 21st, 2000. The third Beidou-1 satellite was launched on May 25th, 2003, as a backup satellite. After the launch of Beidou-1 into the orbit, a large number of tests and demonstrations for system time, system coordinates, and information flow have been carried out in the satellite system and ground control system. The official announcement of BDS-1 operation on December 15th, 2003, means that the RDSS has been implemented for the first time in the world, and that China, has become the third country in the world to provide PNT services through the satellite system. It is also the first milestone on the road of navigation satellite system construction for China.

The working principle and process of BDS-1 system are as follows:

(1) The ground control segment conducts the satellite orbit determination and ionospheric correction, and then transmits the results to the satellite at a specific signal frequency as required.
(2) The satellite broadcasts the navigation signals to all the users in the corresponding areas within the antenna beam coverage.
(3) When the user needs the positioning service, the user terminal sends a signal to the ground control segment via satellite.
(4) After receiving the user's request signal, the ground center demodulates the information sent by the user and measures the distances from the user to the two satellites.
(5) The ground center sends the positioning information to the user via satellite.

The working principle and process of the BDS-1 system are shown in Fig. 2.10.

The BDS-1 has implemented and demonstrated lots of PNT technologies, including RDSS active positioning, one-way/two-way timing, navigation and

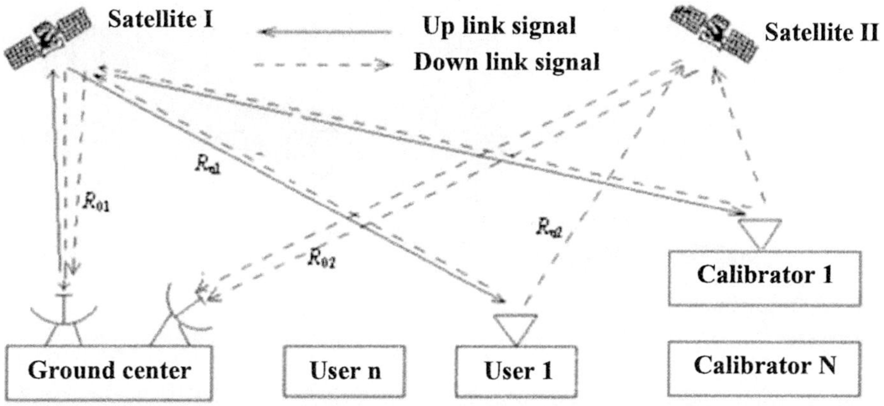


**Fig. 2.10** Principle diagram of BDS-1

communication integrated signal, user random access based on burst signals, and ID identification based on time division/code division.

The BDS-1 is capable of providing the 20 cm positioning accuracy, 10 ns two-way timing accuracy, two-way reporting within the 1 s delay, and 120 Chinese characters (240 bytes). Its services cover China and neighboring areas in the 70°–145° East longitude and 5°–55° North latitude. Practically, the users use multiple receivers for parallel processing in order to communicate beyond 120 Chinese characters. Moreover, because the BDS-1 actually has three GEO satellites in orbit, the users have also innovatively created the three-satellite passive positioning mode.

In the RDSS application, the user equipment is required to not only receive the request signal from the ground control segment via the GEO satellite, but also transmit the acknowledgment signal to the ground via satellite. The service capacity of BDS-1 depends on the tolerated channel blocking rate, ground segment calculation capacity, and channel capacity. As a result, the number of simultaneous users is limited, and the positioning of the fast-moving objects is unsatisfied. In addition, since the user equipment needs to transmit signals, it is difficult for the equipment to achieve weight reduction and miniaturization.

The BDS-1 satellite is equipped with an inbound signal transponder (to the station), outbound signal transponder (from the station), inter-station time synchronization and data transmission transponder (C/C transponder), high-power amplifiers, and high-gain reflector antennas. The satellite transparently forwards the time information generated by the ground segment, while implementing and demonstrating various PNT techniques including the transparent forwarding of positioning measurement, the technique of multi-beam large-aperture deployable antenna, the micro-discharge control of high-power components, and the electromagnetic compatibility with complex multi-frequency payload.

The BDS-1 satellite has a DFH-3 platform with three-axis stability, and a 2000 mm × 1720 mm × 2200 mm box-shaped structure composed of a service bay, a propulsion bay, and a payload bay. It weighs 2300 kg, with the designed life of 8 years.

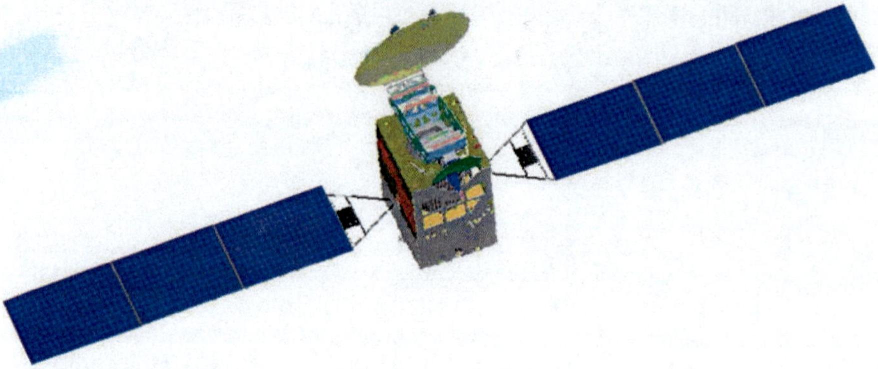

**Fig. 2.11** BDS-1 satellite

**Table 2.6** Launch time and in-orbit status of BDS-1 satellites

| No. | Satellite | Launch time (Beijing time) | Payload and function | Status |
|---|---|---|---|---|
| 1 | BDS-1-01 | 31st October 2000 | RDSS, C/C transponder | GEO: 140 °E, Retired |
| 2 | BDS-1-02 | 21st December 2000 | RDSS, C/C transponder | GEO: 80 °E, Retired |
| 3 | BDS-1-03 | 25th May 2003 | RDSS, C/C transponder | GEO: 110.5 °E, Retired |
| 4 | BDS-1-04 | 3rd February 2007 | RDSS, C/C transponder | GEO: 140 °E. \ |

The satellite system is composed of multiple subsystems for payloads (transponder, antenna), power supply, TT&C, attitude and orbit control, propulsion, thermal control, structure, etc. (see Fig. 2.11).

The conditions of the BDS-1 satellites are shown in Table 2.6.

## 2.6.2 *BDS-2 System*

The BDS-2 is the second step of BDS's "three-step development" strategy, where some of the BDS-1 technologies can still be found. However, it carries high-precision atomic clocks. The BDS-2 satellite generates the navigation signal with a message while keeping itself synchronized with the BDS time by two-way ranging between satellite and ground segment. Besides the RDSS service inherited from BDS-1, the BDS-2 also provides the RNSS (Radio Navigation Satellite Service) passive positioning, GPS, and augmentation services.

The working principle and process of BDS-2 RNSS are as follows:

(1) The ground control segment completes the satellite orbit determination, ephemeris measurement, time comparison, ionospheric correction, etc.

(2) The ground control segment uplinks the signals, including navigation messages, to satellites.
(3) The satellites receive the uplinked signals, complete the satellite-ground distance measurement by the spaceborne receivers, generate the downlink navigation signals, and broadcast them to the user.
(4) The user receives the navigation signals transmitted by more than 4 satellites at the same time, measures the distances between users and satellites, calculates the spatial coordinates of those satellites, and obtains the positioning and timing information.

In August 2004, the construction of the BDS-2 project was approved by the Chinese government. On April 14th, 2007, the first BDS-2 satellite was successfully launched to demonstrate the system's technical principle, key technologies, and development plan feasibility. In the space segment, ground segment, and user terminal of this satellite, the RDSS and RNSS were integrated, and the difficulties in compatibility coordination, transceiver isolation, and terminal miniaturization were conquered. The space environment of MEO orbit was also studied to occupy the applied frequency and orbit resources in time.

In 2010–2012, 14 BDS-2 satellites were successfully launched to form a hybrid constellation of "5GEO + 5IGSO + 4MEO". At the same time, the construction of the ground operation and control system was completed. On December 27th, 2012, China officially announced the start of BDS-2 services in the Asia Pacific region. This indicated that the project perfectly balanced the relationships among construction, demonstration, and application. In March and June 2016, two BDS-2 backup satellites were successfully launched (including 1 IGSO satellite and 1 GEO satellite). The in-orbit backup satellites could effectively ensure the continuity, stability, and reliability of system services.

The BDS-2 adopts a hybrid constellation composed of GEO, IGSO, and MEO satellites (Fig. 2.12). The combination of GEO and IGSO satellites plays a dominant role in the constellation and provides an improved coverage in the service region. The MEO satellites play a secondary role and are mainly used for the technical demonstration of BDS-3 satellites. This constellation design has outstanding performance in low-latitude areas. In the countries along the "Belt and Road", the number of visible satellites in this constellation is kept at 7–9.

The BDS-2 ground segment consists of 1 master control station, 27 monitoring stations (7 primary stations and 20 secondary stations), and two-time synchronization/uplink stations. The master control station is the operation and control center of the ground segment and space segment. Its main tasks are: to collect original data from navigation signal monitoring and time synchronization observation and comparison; to complete the system time synchronization; and to carry out the satellite clock forecasting, precise satellite orbit determination and ephemeris forecasting, ionospheric correction, wide-area differential correction, system integrity monitoring, and other information processing processes. It is also responsible for task planning and scheduling, and for system operation management and control. At the same time, the master station also needs to observe and compare the time of all the

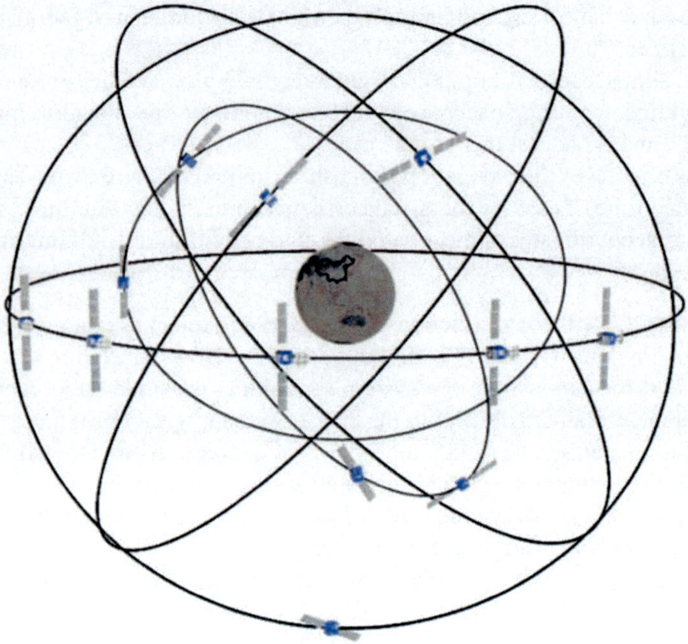

**Fig. 2.12** BDS-2 constellation

satellites, observe and compare the inter-station time or the time of time synchronization/uplink stations, and uplink the navigation message parameters and broadcast information into satellites through the time synchronization/uplink stations. The monitoring stations continuously monitor the navigation signals broadcasted by each satellite (to be specific, monitor the receivers), and provide real-time observation data for precise orbit determination, ionospheric correction, wide-area differential correction, and integrity determination. The primary monitoring stations are mainly used for satellite orbit determination and ionospheric correction. The secondary monitoring stations are mainly used for wide-area differential correction and integrity monitoring. Three of the primary monitoring stations are built together with the master control station and the two-time synchronization/uploading stations. The other monitoring stations are distributed evenly in China's land area and built independently. The time synchronization/uplink stations cooperate with the master station to observe and compare the satellite-ground time and uplink the navigation message parameters into satellites

The BDS-2 GEO satellite has an improved DFH-3 platform, which is 400 mm higher than the classic DFH-3 platform. The satellite has a 2400 mm $\times$ 1720 mm $\times$ 2200 mm box-shaped structure composed of a service bay, a propulsion bay, and a payload bay. It weighs 3000 kg, with the designed life of 8 years. The satellite system is composed of multiple subsystems for navigation, antenna, power supply, TT&C, attitude and orbit control, propulsion, thermal control, and structure. The

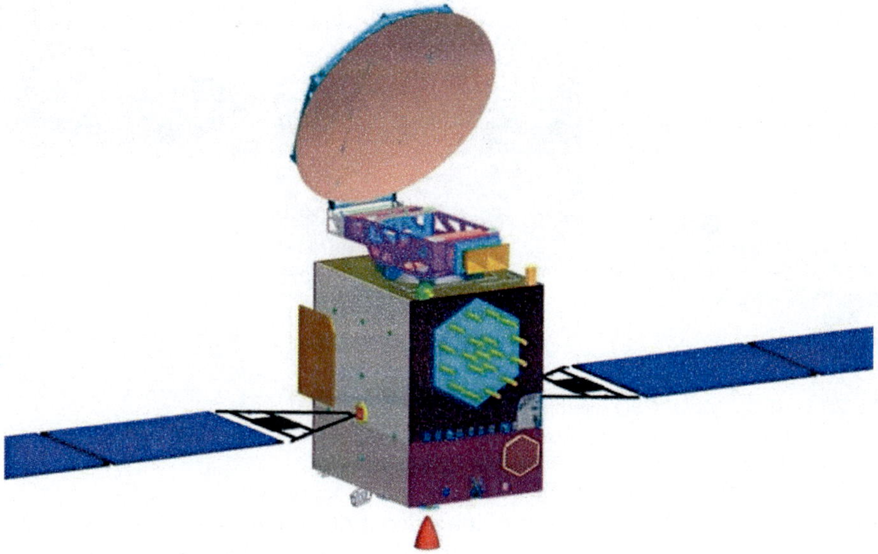

**Fig. 2.13** BDS-2 GEO satellite

spaceborne payloads include RDSS payload, inter-station time synchronization and data transmission transponder, uplink and precision ranging receiver, RNSS payload, laser reflectors, etc. The appearance of the BDS-2 GEO satellite is shown in Fig. 2.13.

The BDS-2 MEO/IGSO satellite has a DFH-3 platform with three-axis stability. The satellite has a 2000 mm × 1720 mm × 2200 mm box-shaped structure composed of a service bay, a propulsion bay, and a payload bay. It weighs 2160/2300 kg, with the designed life of 8 years. The MEO satellite is launched by the CZ-3B rocket carrying two satellites per launch, and the IGSO satellite launched by the C-3A rocket carrying one satellite per launch. This satellite system is composed of multiple subsystems for navigation, antenna, power supply, TT&C, attitude and orbit control, propulsion, thermal control, and structure. The spaceborne payloads include RDSS payload, inter-station time synchronization, and data transmission transponder, uplink & precision ranging receiver, RNSS payload, laser reflectors, etc. The appearance of the BDS-2 MEO/IGSO satellite is shown in Fig. 2.14.

The payloads on the MEO/IGSO satellite include high-stability atomic clock, uplink and precision ranging receiver, navigation signal broadcasting payload, and high-power saddle-shape antenna array, etc.

The 10 MHz signal with high accuracy and stability provided by the spaceborne rubidium clock serves as the time-frequency reference of the entire satellite system. The thermal control subsystem provides lifelong temperature control for the satellite, especially for spaceborne clocks and high-power amplifiers, whose temperature can be kept within ±0.3 °C/orbital period.

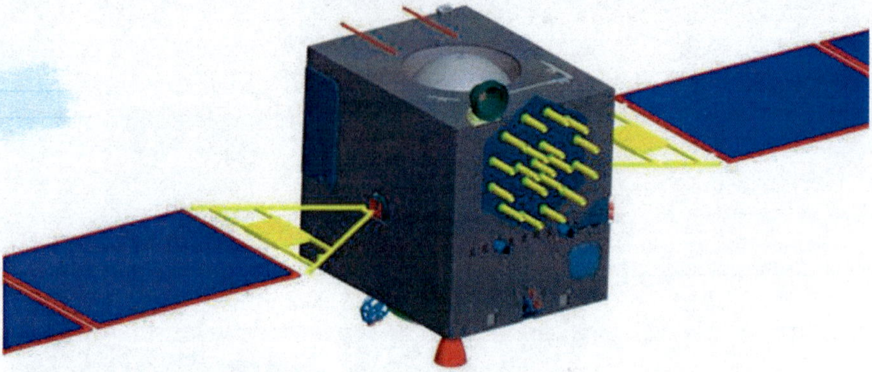

**Fig. 2.14** BDS-2 MEO/IGSO satellite

The spaceborne receiver adopts the high-gain coding and interference suppression technique based on continuous bidirectional ranging to achieve the measurement accuracy of 1 ns under a high signal dynamic level and a serious interference condition.

The system-level electromagnetic compatibility (EMC) technique, the high-power micro-discharge mitigation technique, and the heat dissipation measures are introduced to MEO/IGSO to achieve the 600 W navigation broadcasting capacity and the L/S/C frequency transmission compatibility. To solve the problem of signal delay variation in the space environment, the navigation signal broadcasting payload combines pre-compensation techniques (including baseband processing) with RF channel and antenna to achieve the transmission delay variation better than 0.5 ns/day.

In order to obtain a stable phase center of navigation antenna at different elevation angles and different temperatures, each satellite is designed with a special antenna network to ensure that the deviation of pseudorange measurement at each elevation angle is better than 3 cm at the temperature of ±150 °C. The navigation antenna is a multi-band compatible antenna design that achieves low PIM through passive intermodulation (PIM) mitigation. Furthermore, with the intelligent control technique in the beam direction of the navigation signal transmitting antenna that is realized through the joint design of software and hardware, the regional power of navigation signal has been increased.

### *2.6.3 BDS-3 System*

The BDS-3 is the third step of BDS's "three-step development" strategy. It provides RNSS services and short message communication worldwide, while delivering RDSS, satellite-based enhancement, power enhancement, and other special services in China and neighboring regions. During the BDS-3 design, great modification and improvement are made, as more requirements are considered such as space service

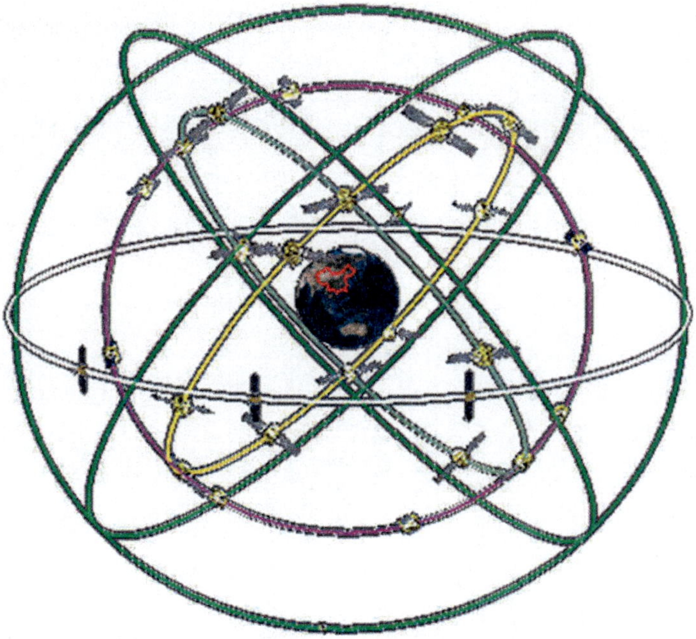

**Fig. 2.15** BDS-3 constellation

volume (SSV) [27, 28], multi-system compatibility, and interoperability, etc. Furthermore, the BDS-3 system also carries on some of the BDS-2 services to enable smooth transition and upgrading on the user side.

The BDS-3 adopts a hybrid constellation composed of "3GEO + 3IGSO + 24MEO" satellites (see Fig. 2.15). With the modernized signal design, inter-satellite links, and improved spaceborne atomic clocks, the BDS-3 can provide high-performance stable interoperable services worldwide.

The BDS-3 project was started in 2009. In 2015, five BDS-3 test satellites were launched to demonstrate new signals, new key technologies, and new products. On November 5, 2017, the first two MEO service satellites developed by the China Academy of Space Technology (CAST) were launched together at the Xichang Satellite Launch Center. By 2018, the BDS-3 system has provided basic services to the countries along the "Belt and Road" and in neighboring regions. It is expected to provide services to the global users by 2020.

The launch time and in-orbit condition of Beidou satellites are shown in Table 2.7.

The modernized design introduced to the BDS-3 navigation signal, including signal spectrum separation, pilot/data orthogonality, and power rate adjustability, as well as new modulation method like QMBOC, has improved the compatibility, interoperability, efficiency, and performance of this signal.

In order to solve the problem of observation period insufficiency, an inter-satellite link has been established in BDS-3 to improve the result of orbit determination and

**Table 2.7** Launch time and in-orbit condition of BDS-2 and BDS-3 satellites (by the end of december 2019)

| No. | Satellite | Launch time (Beijing time) | Payload and function | Status |
|---|---|---|---|---|
| 1 | BDS-2.M-S | 14th April 2007 | Uplink data uploading and ranging, RNSS | MEO, retired |
| 2 | BDS-2.G2 | 15th April 2009 | RDSS, C/C transponder, Uplink data uploading and ranging, RNSS | GEO, / |
| 3 | BDS-2.G1 | 17th January 2010 | RDSS, C/C transponder, Uplink data uploading and ranging, RNSS | GEO: 140 °E, Operational |
| 4 | BDS-2.G3 | 2nd June 2010 | RDSS, C/C transponder, Uplink data uploading and ranging, RNSS | GEO: 110.5 °E, Operational |
| 5 | BDS-2.I1 | 1st August 2010 | RDSS, C/C transponder, Uplink data uploading and ranging, RNSS | IGSO, Operational |
| 6 | BDS-2.G4 | 1st November 2010 | RDSS, C/C transponder, Uplink data uploading and ranging, RNSS | GEO: 160 °E, Operational |
| 7 | BDS-2.I2 | 18th December 2010 | RDSS, C/C transponder, Uplink data uploading and ranging, RNSS | IGSO, Operational |
| 8 | BDS-2.I3 | 10th April 2011 | RDSS, C/C transponder, Uplink data uploading and ranging, RNSS | IGSO, Operational |
| 9 | BDS-2.I4 | 27th July 2011 | RDSS, C/C transponder, Uplink data uploading and ranging, RNSS | IGSO, Operational |
| 10 | BDS-2.I5 | 2nd December 2011 | RDSS, C/C transponder, Uplink data uploading and ranging, RNSS | IGSO, Operational |
| 11 | BDS-2.G5 | 25th February 2012 | RDSS, C/C transponder, Uplink data uploading and ranging, RNSS | GEO: 58.75 °E Operational |
| 12 | BDS-2.M3 | 30th April 2012 | Uplink data uploading and ranging, RNSS | MEO, Operational |
| 13 | BDS-2.M4 | 30th April 2012 | Uplink data uploading and ranging, RNSS | MEO, Operational |
| 14 | BDS-2.M5 | 19th September 2012 | Uplink data uploading and ranging, RNSS | MEO, / |
| 15 | BDS-2.M6 | 19th September 2012 | Uplink data uploading and ranging, RNSS | MEO, Operational |
| 16 | BDS-2.G6 | 25th October 2012 | RDSS, C/C transponder, Uplink data uploading and ranging, RNSS | GEO: 84 °E Operational |
| 17 | BDS-3-I1-S | 30th March 2015 | Uplink data uploading and ranging, RNSS (B1), C/C transponder | IGSO, Operational |

(continued)

**Table 2.7** (continued)

| No. | Satellite | Launch time (Beijing time) | Payload and function | Status |
|---|---|---|---|---|
| 18 | BDS-3-M1-S | 25th July 2015 | Uplink data uploading and ranging, RNSS, Inter-sat link, RNSS(S) | MEO, Operational |
| 19 | BDS-3-M2.S | 25th July 2015 | Uplink data uploading and ranging, RNSS, Inter-sat link | MEO, Operational |
| 20 | BDS-3-I2.S | 29th September 2015 | Uplink data uploading and ranging, RNSS, Inter-sat link, RNSS(S), Hydrogen atomic clock | IGSO, Operational |
| 21 | BDS-3-M3-S | 1st February 2016 | Uplink data uploading and ranging, RNSS, Inter-sat link, Hydrogen atomic clock | MEO, Operational |
| 22 | BDS-2.I6 | 30th March 2016 | RDSS, C/C transponder, Uplink data uploading and ranging, RNSS | IGSO, Operational |
| 23 | BDS-2.G7 | 12th June 2016 | RDSS, C/C transponder, Uplink data uploading and ranging, RNSS | GEO: 144.5 °E Operational |
| 24 | BDS-3-M1 | 5th November 2017 | Uplink data uploading and ranging, RNSS, Inter-sat link | MEO, Operational |
| 25 | BDS-3-M2 | 5th November 2017 | Uplink data uploading and ranging, RNSS, Inter-sat link | MEO, Operational |
| 26 | BDS-3-M3 | 12th February 2018 | Uplink data uploading and ranging, RNSS, Inter-sat link | MEO, Operational |
| 27 | BDS-3-M4 | 12th February 2018 | Uplink data uploading and ranging, RNSS, Inter-sat link | MEO, Operational |
| 28 | BDS-3-M5 | 29th July 2018 | Uplink data uploading and ranging, RNSS, Inter-sat link | MEO, Operational |
| 29 | BDS-3-M6 | 29th July 2018 | Uplink data uploading and ranging, RNSS, Inter-sat link | MEO, Operational |
| 30 | BDS-3-M7 | 12th January 2018 | Uplink data uploading and ranging, RNSS, Inter-sat link, Hydrogen atomic clock | MEO, Operational |
| 31 | BDS-3-M8 | 12th January 2018 | Uplink data uploading and ranging, RNSS, Inter-sat link, Hydrogen atomic clock | MEO, Operational |
| 32 | BDS-2.I7 | 10th July 2018 | Uplink data uploading and ranging, RNSS | IGSO, Operational |
| 33 | BDS-3-M9 | 30th March 2018 | Uplink data uploading and ranging, RNSS, Inter-sat link, Hydrogen atomic clock | MEO, Operational |
| 34 | BDS-3-M10 | 30th March 2018 | Uplink data uploading and ranging, RNSS, Inter-sat link, Hydrogen atomic clock | MEO, Operational |

(continued)

**Table 2.7** (continued)

| No. | Satellite | Launch time (Beijing time) | Payload and function | Status |
|---|---|---|---|---|
| 35 | BDS-3-M11 | 25th August 2018 | Uplink data uploading and ranging, RNSS, Inter-sat link, Hydrogen atomic clock, Short Message Communication (SMC) | MEO, Operational |
| 36 | BDS-3-M12 | 25th August 2018 | Uplink data uploading and ranging, RNSS, Inter-sat link, Hydrogen atomic clock, SMC | MEO, Operational |
| 37 | BDS-3-M13 | 19th September 2018 | Uplink data uploading and ranging, RNSS, Inter-sat link, MEO SAR, SMC | MEO, Operational |
| 38 | BDS-3-M14 | 19th September 2018 | Uplink data uploading and ranging, RNSS, Inter-sat link, MEO SAR, SMC | MEO, Operational |
| 39 | BDS-3-M15 | 15th October 2018 | Uplink data uploading and ranging, RNSS, Inter-sat link, Hydrogen atomic clock SMC | MEO, Operational |
| 40 | BDS-3-M16 | 15th October 2018 | Uplink data uploading and ranging, RNSS, Inter-sat link, Hydrogen atomic clock SMC | MEO, Operational |
| 41 | BDS-3-G1 | 1st November 2018 | RDSS, C/C transponder, Uplink data uploading and ranging, RNSS, Inter-sat link, Hydrogen maser clock | GEO: 140 °E, Operational |
| 42 | BDS-3-M17 | 19th November 2018 | Uplink data uploading and ranging, RNSS, Inter-sat link, SMC | MEO, Operational |
| 43 | BDS-3-M18 | 19th November 2018 | Uplink data uploading and ranging, RNSS, Inter-sat link, SMC | MEO, Operational |
| 44 | BDS-3-I1 | 20th April 2019 | Uplink data uploading and ranging, RNSS, Inter-sat link, Hydrogen atomic clock | IGSO, Operational |
| 45 | BDS-2.G8 | 17th May 2019 | RDSS, C/C transponder, Uplink data uploading and ranging, RNSS | GEO: 80 °E, Operational |
| 46 | BDS-3-I2 | 25th June 2019 | Uplink data uploading and ranging, RNSS, Inter-sat link, Hydrogen atomic clock | IGSO, Operational |
| 47 | BDS-3-M23 | 23th September 2019 | Uplink data uploading and ranging, RNSS, Inter-sat link, MEO SAR, SMC | MEO, Operational |
| 48 | BDS-3-M24 | 23th September 2019 | Uplink data uploading and ranging, RNSS, Inter-sat link, MEO SAR, SMC | MEO, Operational |
| 49 | BDS-3-I3 | 5th November 2019 | Uplink data uploading and ranging, RNSS, Inter-sat link, Hydrogen atomic clock | IGSO, Operational |
| 50 | BDS-3-M21 | 23th November 2019 | Uplink data uploading and ranging, RNSS, Inter-sat link, MEO SAR, SMC | MEO, Operational |
| 51 | BDS-3-M22 | 23th November 2019 | Uplink data uploading and ranging, RNSS, Inter-sat link, MEO SAR, SMC | MEO, Operational |
| 52 | BDS-3-M19 | 16th December 2019 | Uplink data uploading and ranging, RNSS, Inter-sat link, SMC | MEO, Operational |
| 53 | BDS-3-M20 | 16th December 2019 | Uplink data uploading and ranging, RNSS, Inter-sat link, SMC | MEO, Operational |

time synchronization by measuring the distance between satellites and calculating and correcting the error. Through the inter-satellite link, the BDS-3 has accomplished global TT&C, precise constellation orbit determination, and 60-day autonomous navigation with an accuracy of 15 m, by relying on only domestic stations.

### *2.6.4 BDS-3 Satellite*

The BDS-3 GEO satellite has an improved DFH-3B platform. The satellite has a 2360 mm × 2100 mm × 3600 mm box-shaped structure composed of a service bay, a propulsion bay, and a payload bay whose north and south plates are extended to 2960 mm in the $\pm x$ direction. It weighs 5400 kg, with the designed life of 12 years. The BDS-3 GEO satellite is launched by the CZ-3B rocket and enters into the orbit by orbital transfer. The satellite platform contains 9 subsystems for integrated electronics, orbit control, propulsion, TT&C, power supply, integrated layout, autonomous operation, thermal control, and structure, respectively. The spaceborne payload contains 3 subsystems for navigation, antenna, and transponder respectively.

The BDS-3 IGSO satellite has a DFH-3B platform. The satellite has a 2360 mm × 2100 mm × 3600 mm box-shaped structure composed of a service bay, a propulsion bay and a payload bay. It weighs 5400 kg, with the designed life of 12 years. The BDS-3 IGSO satellite is launched by the CZ-3B rocket and enters into the orbit by orbital transfer. The satellite platform contains 9 subsystems for integrated electronics, orbit control, propulsion, TT&C, power supply, integrated layout, autonomous operation, thermal control, and structure, respectively. The spaceborne payload contains 2 subsystems for navigation and antenna, respectively.

The BDS-3 MEO satellite has a new platform designed for navigation satellites, which has a truss type structure and consists of a single propulsion system, an improved power supply, and an integrated electronic system. The satellite has a 1804 mm × 1224 mm × 2300 mm box-shaped structure composed of, a service bay, a propulsion bay, and a payload bay. It weighs 1060 kg, with the designed life of 10 years. The BDS-3 MEO satellite is launched by the CZ-3B rocket and enters into the orbit via the YZ-1 upper stage. The satellite platform contains 9 subsystems for integrated electronics, orbit control, propulsion, TT&C, power supply, integrated layout, autonomous operation, thermal control, and structure, respectively. The spaceborne payload contains 2 subsystems for navigation and antenna, respectively.

The definitions and functions of the 12 subsystems mentioned above will be described briefly as follows:

(1) Integrated electronic subsystem

Through the 1553B bus and an integrated computer, an information network is established to help the integrated electronic subsystem execute the attitude control and satellite management tasks, including energy management, autonomous navigation calculation, inter-satellite routing, antenna pointing calculation, and autonomous health management. This subsystem consists of a central management unit, a data processing and routing unit, and an integrated service unit.

(2) Orbit control subsystem

The orbit control subsystem works with the integrated electronic subsystem to control the satellite attitude and orbit after the satellite-rocket separation. The components in this subsystem include: reaction wheel, magnetic torque generator, solar array drive assembly (SADA), floated gyroscope, fiber optic gyroscope, earth sensor, sun sensor, star sensor, etc.

(3) Propulsion subsystem

The propulsion subsystem provides the integrated electronic subsystem with the thrust required for attitude and orbit control after the satellite-rocket separation. The propulsion subsystems of GEO and IGSO satellites are based on a bipropellant engine assembly that is mainly composed of a 490 N engine, a 10 N thruster, and a propellant tank. In contrast, the propulsion subsystem of the MEO satellite is based on a monopropellant engine assembly that is mainly composed of an engine, a 5 N thruster, and a propellant tank.

(4) TT&C subsystem

The TT&C subsystem provides functions such as telemetry, remote control, ranging, and velocity measurement. It is designed with S-band multiple signals in order to satisfy the requirements of ranging, data transmission, and anti-jamming.

(5) Power supply subsystem

The Power supply subsystem is a fully regulated single-bus system consisting of GaAs solar array, lithium battery pack, power controller unit (PCU), battery balance manager, and other components. The power supplies of GEO and IGSO satellites can provide 100 V power, while that of the MEO satellite can provide 42 V.

(6) Integrated layout subsystem

The integrated layout subsystem is the integration of the initiation device management module, power distribution module, cable network, and bus conversion adapter. It provides the functions of initiation device management, power supply and distribution, bus conversion, and low-frequency signal transmission.

(7) Autonomous operation subsystem

This subsystem enables the autonomous operation of navigation constellation by using the phased array antenna and autonomous navigation software for inter-satellite routing and antenna pointing calculation. The inter-satellite links between the satellites in the constellation increase the satellite observations, improve the accuracy of orbit determination, raise the updating frequency of satellite ephemeris and clock information, and realize the measurement and control of overseas satellites. Therefore, the management capacity and navigation performance of the whole system has been improved.

(8) Thermal control subsystem

The task of the thermal control subsystem is to ensure that the temperature of all the spaceborne devices, equipments, and components is kept within the required range in the launch, orbit transfer, and in-orbit phases. This subsystem is mainly composed of thermal control coating, multi-layer insulation component, heating pipe, temperature sensor, heater, heat-conducting filler, heat expansion plate, etc.

(9) Structural subsystem

The structural subsystem is mainly to receive and pass on various external forces generated during launching, flight, and ground operations throughout the satellite life. It provides support to the spaceborne equipments in order to meet the requirements of position and accuracy during the instrument installation.

(10) Navigation subsystem

The navigation subsystem is mainly to complete the uplink signal reception, navigation message/signal generation, and navigation signal broadcasting. It is equipped with atomic clocks that have been calibrated and synchronized with ground frequency and time reference through bidirectional comparison. The atomic clocks are capable of integrity monitoring and smooth switching, which makes the synchronization of multiple atomic clocks possible. When the operational clock fails, the satellite can diagnose and smoothly switch to the backup clock to ensure the continuity of time-frequency signals. Thus, the reliability and integrity of navigation service can be greatly improved.

(11) Antenna subsystem

The antenna subsystem receives the L-band uplink signal from the user and the B1, B2, B3 signals from the navigation subsystem, and broadcasts the L-band navigation signal. The antenna subsystem of the GEO satellite receives the L-band uplink signal from the user, and broadcasts the S-band downlink signal to the user to provide the RDSS service. The GEO satellite antenna subsystem receives the C-band uplink signal and broadcasts the C-band downlink signal to complete the inter-station time synchronization and communication.

(12) Transponder subsystem

The transponder subsystem of the GEO satellite includes a fixed beam repeater, a spot beam repeater, and an inter-station time synchronization and communication repeater (C/C) that serve the fixed beam signals, spot beam signals, and C/C signals, respectively. Among them, the fixed beam repeater forwards the inbound and outbound RDSS signals to achieve the user positioning and message communication so as to mainly meet the demand for RDSS in China and neighboring regions. The spot beam repeater forwards the inbound and outbound signals for short message communication and provides RDSS services to the users that are beyond the coverage of a fixed spot beam. The inter-station time synchronization and data transmission repeater (C/C) forwards the station-specific time signals and data to achieve inter-station time synchronization and data transmission among Chinese ground stations.

# Chapter 3
# Satellite Navigation Uplink and Reception Technology


Uplink is critical for a satellite navigation system as an important means for the ground segment to upload the information. The uplink channels include a remote control channel for satellite TT&C and satellite operation management, as well as an information upload channel, for managing and controlling the spaceborne payload. The technology of TT&C channel in the navigation satellite system is similar to that in a conventional satellite system, and therefore, will not be elaborated on in this book.

The satellite navigation uplink system mainly provides the uplink service related to navigation information. The uplink is responsible for updating and maintaining navigation information such as navigation satellite ephemeris, satellite clock offset, and satellite almanac, thereby providing an important guarantee for the navigation system to stably broadcast the signals. At the same time, the BDS ground station makes two-way comparison between the downlink and uplink pseudorange measurement data to realize the satellite time synchronization.

This chapter introduces the function and design of navigation satellite uplink, and discusses the clock time synchronization technology, the technology of uplink signal reception and precise ranging, the anti-interference technology of uplink signal, and the type of uplink information.


## 3.1 Function and Design of Satellite Navigation Uplink

### *3.1.1 Function of Uplink*

The main functions of satellite navigation uplink are to upload and measure the satellite-ground time offset data. For a navigation satellite system, uplink is a bridge between the ground control segment and space segment. It is responsible for updating




J. Xie et al., *Satellite Navigation Systems and Technologies*, Space Science and Technologies, https://doi.org/10.1007/978-981-15-4863-5_3

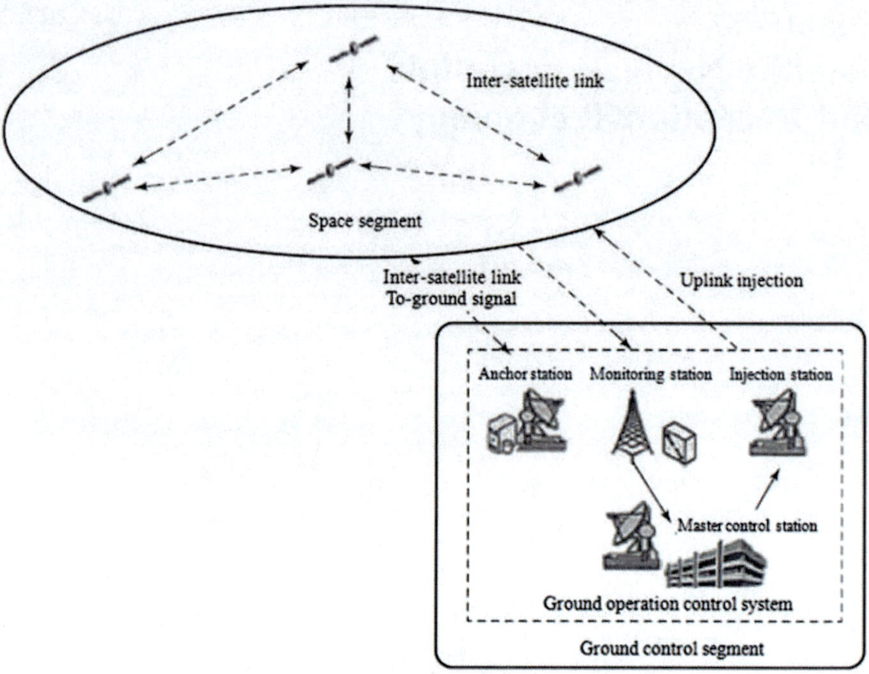


**Fig. 3.1** Uplink of Beidou navigation satellite system

the navigation messages such as satellite ephemerides, satellite clock offsets, and satellite almanacs. The uplink diagram of the Beidou navigation satellite system is shown in Fig. 3.1.

The data uplinking is completed by the ground control segment when the satellite is directly visible. The ground control system consists of a master control station, monitoring stations, and upload stations. The monitoring station receives the navigation signal and generates the downlink measurement data. The master control station processes in a centralized way the downlink measurement data transmitted by the satellite, and generates the uplink data such as ephemeris and satellite clock offsets, which will be uploaded to the satellite by the upload station, as shown in Fig. 3.2.

The main functions of BDS uplink include:

**Satellite-ground time synchronization**: The satellite payload receives the uplink signal from the ground operation control system and completes the high-precision measurement to obtain the uplink measurement data. At the same time, the ground operation control system receives the downlink B3-frequency navigation signal to obtain the downlink measurement data. Then the uplink and downlink measurement data are compared to realize satellite-ground time synchronization.

**Satellite data upload**: The navigation data processed and generated by the master station, such as broadcast ephemeris, satellite almanac, and clock difference, are

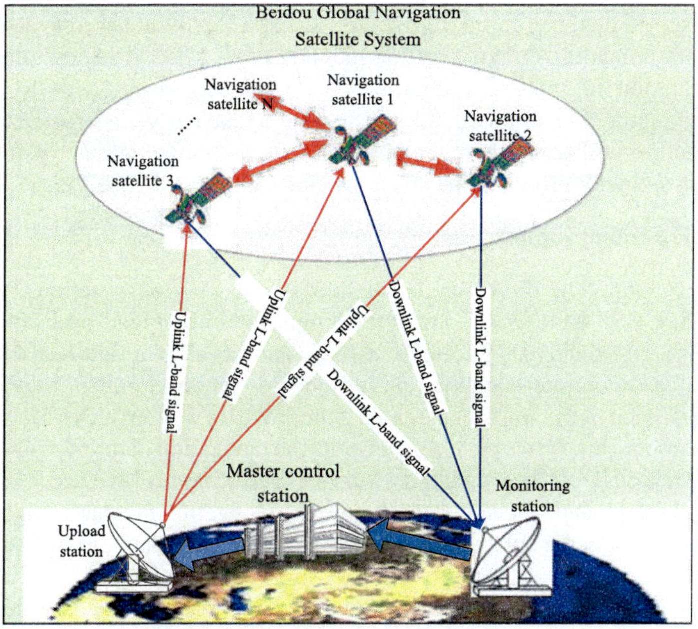


**Fig. 3.2** Uplink of the ground control segment

injected into the satellite via the uplink. Meanwhile, the uplink is used for uploading the telecommands.

### *3.1.2 General Design of Uplink*

1. Design of GPS uplink

The GPS is transitioning from the GPS II to the GPS III, while its uplink mode is being transformed from direct single-satellite upload by global stations to inter-satellite upload by local stations. At the beginning of the GPS–II era, the uplink was only used for uploading basic navigation messages. Everyday, the four upload stations established around the world uploaded the ephemeris and time offset information regularly predicted and updated by the master control station to the visible single satellite through an *S*-band antenna.

With the continuous expansion of GPS system applications, civil aviation users are increasingly demanding single-frequency ionosphere correction information and integrity information, which, however, fails to meet the requirements of the aviation class-I precision approach. Therefore, the United States established a Wide Area Augmentation System (WAAS) to provide the GPS augmentation service by using a GEO satellite to broadcast the integrity and ionospheric correction parameters to regional users. Thus, the uplink of the GPS II has actually undergone a functional change from uploading only navigation messages to a MEO satellite, to uploading the augmentation information, in addition to navigation messages, to a GEO satellite. Different from the *S*-band uplink frequency used by the early GPS system, the GPS III in the future will adopt the *C*-band to improve its remote telemetry and upload capabilities and will introduce the "Contact One Contact All" method.

2. Design of Galileo uplink

The Galileo adopts the direct single-satellite upload by global stations, while the GLONASS, QZSS and NAVIC adopt the single-satellite upload by local stations. For the GNSS gradually upgrading, planning, and developing inter-satellite links, the upload based on inter-satellite link has become its development trend. Galileo has been designed in a main-backup mode to improve the system redundancy [36]. In the general mode, the system mainly transmits the navigation data for public service, life safety, regional expansion integrity service, public control service, commercial service, and search and rescue service, where the navigation message length is 12 h. In the backup mode, the system uploads navigation service data through the remote telemetry link.

Different from GPS III, the Galileo applies the one-to-one upload technique only to visible satellites. Like GPS III, it also relies mainly on *C*-band, a legal band used by the ITU's air services with simple service types and less malicious interference. The uplink signal in this system is DS-CDMA in the O-QPSK mode, where both the components *I* and *Q* undertake the data transmission tasks by using convolutional codes. In the backup mode, TT&C frequency band is used.

3. Design of BDS uplink

The BDS uplink is working in three modes: *L*-band, *S*-band, and Ka-band. The *L*-band is the main signal used for uploading the satellite navigation messages. The main design feature of BDS is to achieve satellite-ground time synchronization through the bidirectional comparison between *L*-band uplink and B3 *L*-band downlink. The path delay between channels can be canceled to improve time synchronization accuracy. Furthermore, a single station is enough for bidirectional time synchronization so that the requirements for the number and distribution of monitoring stations can be relaxed. To reduce potential interference effect of aeronautical radio navigation radars and early warning radars on the ground on *L*-band, the signal uplink to BDS satellites is designed with a specific anti-interference feature to ensure reliable information upload and high two-way comparison accuracy. The signal uplink is composed of a pilot channel and a data channel, thereby achieving both data transmission and signal tracking. The uplink signal is transmitted in the DS-CDMA mode, using precision ranging code as its spreading code. A variety of error detection and correction techniques are also used to improve the reliability of data transmission in complex electromagnetic environments.

The TT&C link in the *S*-band and the inter-satellite link (to-ground) in the Ka-band are the backups of the uplink in the *L*-band. With direct upload and inter-satellite link, the system realizes information distribution in the whole network. Similar to the "Contact One Contact All" design of GPS III, this function is able to improve the efficiency of uplink operation and control.

### 3.1.3 Design of Uplink Payload in BDS Satellite

The BDS uplinking includes: the transmission of *L*-band signals to the satellite through the ground control segment, and the signal reception, measurement, and demodulation by the uplink reception payload in the space segment.

The block diagram of the uplink reception payload is shown in Fig. 3.3. The payload is mainly composed of two parts: *L* antenna and uplink receiver. The uplink receiver includes an *L* preselected coupler, an *L* modulator, a switch, and an *L*-band ranging receiver.

1. *L*-band antenna

Since the designed uplink signal frequency is in the same band with the downlink navigation signal, the antenna reception of the uplink signal from the ground will be affected by the downlink navigation signal. Therefore, it is necessary to provide the uplink antenna with a low-sidelobe design, of which the beam performance directly supports the uplink signal reception (Fig. 3.4).

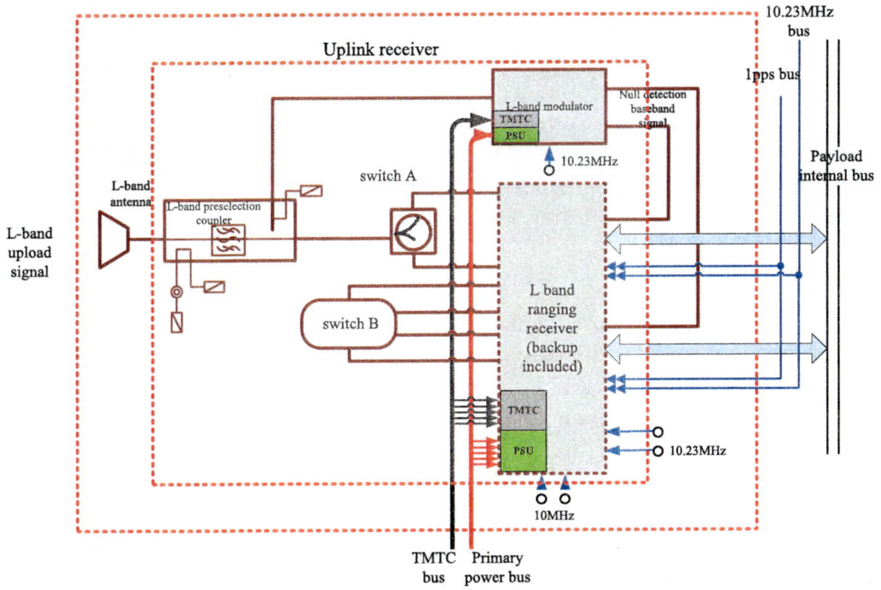


**Fig. 3.3** Block diagram of typical spaceborne uplink reception payload

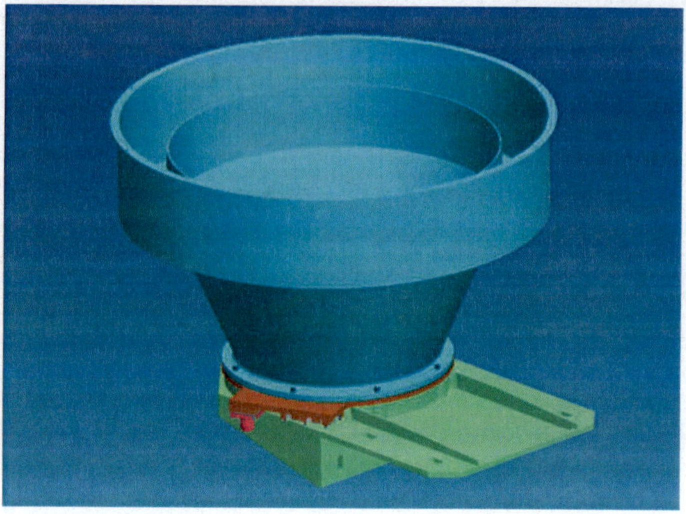

**Fig. 3.4** Typical uplink antenna

The antenna with choke ring is adopted to obtain low side and back lobes. The choke ring is designed with a circular groove to suppress the energy diffracted from the antenna edge and reduce lateral and backward radiation. The application of chock ring can effectively reduce side and back lobes to below −30 dB, much lower than the horn design with an equivalent gain, whose side lob is about −17 to −18 dB. Meanwhile, the use of choke ring can reduce the signal cross-polarization, produce a rotationally symmetric radiation pattern, and improve the stability of the phase center in the band. In the absence of choke ring, the fluctuation of the phase center may be as big as nearly 30°. By using the phase-stabilizing technique, the phase center fluctuation of the *L*-band antenna can be kept below 1°, and the low side lobes can be obtained, thus significantly reducing the interference of the downlink signal with the uplink signal and guaranteeing the uplink performance.

2. Uplink receiver

A digital uplink receiver is used to achieve the fast acquisition, stable tracking, anti-interference, low-loss demodulation, and precision measurement of dual-channel uplink signals. In addition, a null detection channel is designed for receiving the channel null correction information continuously. In order to satisfy the precise measurement requirements in uplink reception, the measures of precise delay correction are taken to achieve a stable delay in the long-term orbit operation of satellites.

In addition to the function and performance of the receiver, its weight, volume, and power consumption must be taken into full consideration to realize the miniaturization and low power consumption. Furthermore, it is necessary to adopt the anti-irradiation and anti-Single Event Upset (SEU) design in order to improve the adaptability to the space environment and ensure the long-term reliable operation of the uplink receiver in the space.

## 3.2 Satellite-Ground Time Synchronization Technology

The satellite-ground time synchronization technology refers to the synchronization of spaceborne atomic clocks with the system time maintained by the ground segment. In the navigation satellite system, high-precision PNT service can be achieved only by accurately knowing the clock difference between each spaceborne atomic clock and ground-based atomic clock group, and realizing the satellite-ground time synchronization. Therefore, satellite-ground time synchronization is the key technology of the navigation satellite system and is the directly decisive factor of navigation accuracy.

Generally, satellite-ground time synchronization is mainly achieved by the following methods: radio-based one-way/two-way time synchronization, laser-based time synchronization, reverse positioning, and pseudorange and radar ranging. The accuracy of laser-based time synchronization, up to 50 ps, is the highest, but is very unreliable due to high susceptibility to weather. Currently, it is mainly used in European ACES and T2L2. With high-precision better than 1 ns, the radio-based two-way time synchronization is unsusceptible to weather and can work continuously for a longtime. Therefore, it is the most popular satellite-ground time synchronization technology in the satellite navigation systems, especially BDS and Galileo systems. The inverse positioning method, exclusively used by GPS, has an outstanding advantage. That is, it can identify the satellite position while realizing time synchronization with the accuracy within 10 ns—generally 6 ns and not exceeding 18 ns. The method of pseudorange and radar ranging, which is adopted by GLONASS, is based on $C$-band responsive radar, with the precision of about 10 ns [39].

### *3.2.1 Radio-Based One-Way Satellite-Ground Time Synchronization Technology*

The working principle of one-way time synchronization technology is that the ground system receives the navigation signal from the satellite, takes the measurements, and transmits the measurement data back to the satellite to realize the synchronization between ground-based and spaceborne atomic clocks. It is simple and easy to operate, and its receiver is inexpensive, but it is significantly susceptible to tropospheric delay, ionospheric delay and satellite orbit. Therefore, the synchronization accuracy of this method is relatively low.

The principle of this technology is shown in Fig. 3.5. After being acquired by the receiver, the navigation signal from the satellite will be processed by ground

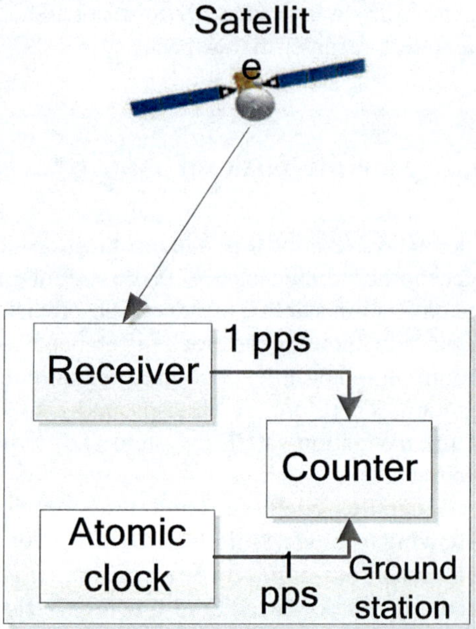


**Fig. 3.5** Block diagram of radio-based one-way satellite-ground time synchronization

station to obtain the pseudo delay TIC(1). Then the difference between spaceborne and ground-based atomic clocks can be expressed as

$$\delta t_i = \text{TIC}(1) - [\text{TX}(2) + \text{SPD} + \text{SCD} + \text{RX}(1)] \quad (3.1)$$

where $\delta t_i$ represents the satellite clock difference ($i$ is the satellite number); TX(2) is the signal transmission delay; SPD is the signal path delay; SCD is the satellite's Sagnac effect correction; and RX(1) is the reception delay of the ground station.

The position of the receiving antenna in the ground station is measured at first, and then the distance from the satellite to the ground station is calculated in accordance with the satellite navigation message to obtain geometric signal path delay (SPD). Based on the satellite parameters and ground station coordinates, the Sagnac effect correction (SCD) of navigation signal can be calculated through the Sagnac model. By referring to the pre-calibrated reception delay of the ground station and the transmission delay of the satellite, the difference between spaceborne and ground-based atomic clocks can be obtained by considering the above known parameters.

In the one-way time synchronization method, the ground station only needs to receive signals rather than transmit signals, so the ground receiver is simple. This method is widely used in civil navigation and positioning. Due to the influence of tropospheric delay, ionospheric delay, satellite clock difference, satellite motion, and other factors, the uncertainty of time synchronization in this method cannot be neglected.

### 3.2.2 *Radio-Based Two-Way Satellite-Ground Time Synchronization Technology*

In the two-way time synchronization method, the two-way (transmission and reception) pairwise mode is adopted. The uplink and downlink paths are correlated with each other to achieve the relevant measurement, and then the measured results are compared. In such a way, the errors of tropospheric and ionospheric delays can be greatly reduced. This is a highly accurate and widely used method. The basic principle of this method is shown in Fig. 3.6.

In the ground station, the 1 pps clock signal is first modulated into an intermediate-frequency carrier signal, which is up-converted to the radio frequency band and then sent to the satellite after being amplified by a power amplifier. After receiving the signal, the satellite first amplifies it with a low-noise amplifier, then down-converts it

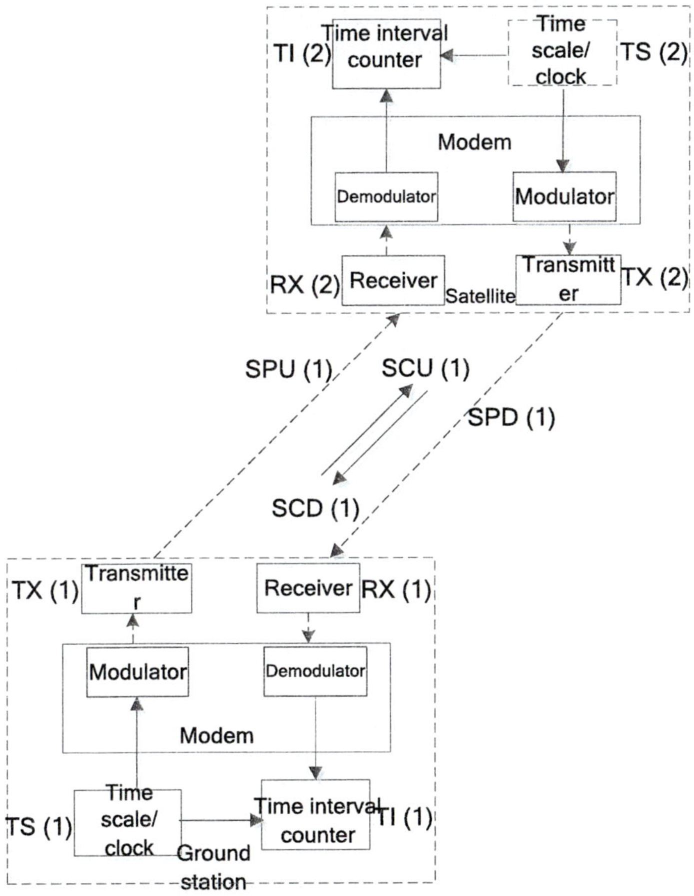


**Fig. 3.6** Block diagram of radio-based two-way satellite-ground time synchronization

to an intermediate frequency, demodulates it, and finally measures the time difference in accordance with the pseudo-code correlation principle. At the same time, the satellite sends a signal to the ground station, which receives the signal and measures the time difference.

In the radio-based two-way satellite-ground time synchronization system, the time transfer relationship between the ground station and satellite can be expressed as follows:

The counter of the ground station is

$$\mathrm{TI}(1) = \mathrm{TS}(1) - \mathrm{TS}(2) + \mathrm{TX}(2) + \mathrm{SPD} + \mathrm{SCD} + \mathrm{RX}(1) \tag{3.2}$$

The counter of the satellite is

$$\mathrm{TI}(2) = \mathrm{TS}(2) - \mathrm{TS}(1) + \mathrm{TX}(1) + \mathrm{SPU} + \mathrm{SCU} + \mathrm{RX}(2) \tag{3.3}$$

where $i = 1$ is for ground station and $i = 2$ for satellite; TS($i$) is the time scale of $i$; TI($i$) is the time interval counter in $i$; TX($i$) is the transmission delay of $i$ (including that of modem); RX($i$) is the reception delay of $i$ (including that of modem); SPU is the propagation delay of the uplink signal transmitted by the ground station; SPD is the propagation delay of satellite downlink signal; SCU is the Sagnac effect correction of the ground station uplink; and SCD is the Sagnac effect correction of the satellite downlink.

By combining Eqs. (3.2) and (3.3), we get

$$\begin{aligned}\delta t_i &= 0.5 \times [\mathrm{TI}(1) - \mathrm{TI}(2)] - 0.5 \times [\mathrm{SCD} - \mathrm{SCU}] + 0.5 \times [\mathrm{SPU} - \mathrm{SPD}] \\ &\quad + 0.5 \times [\mathrm{TX}(1) - \mathrm{RX}(1)] - 0.5 \times [\mathrm{TX}(2) - \mathrm{RX}(2)]\end{aligned} \tag{3.4}$$

where TI(1) − TI(2) is the difference between ground station counter and satellite counter, SCD − SCU is the Sagnac correction of the ground station, SPU − SPD is the difference between the uplink delay of the ground station and the downlink delay of the satellite, TX(1) − RX(1) is the difference between the transmission delay and reception delay of the ground station, and TX(2) − RX(2) is the difference between the transmission delay and reception delay of satellite.

The BDS relies on the radio-based two-way method to realize the time synchronization between satellite and ground. The ground station is equipped with the corresponding time synchronization device, which receives the reference time signal from the ground and the downlink time signal from the Beidou satellite. Based on this downlink signal, the measured value $\Delta_1$ can be obtained in accordance with the pseudo-code ranging principle. Then, the ground synchronization device sends time information to the satellite, where the measured value $\Delta_2$ can be obtained from the onboard synchronization device and then sent back to the ground station. By substituting $\Delta_2$, $\Delta_1$, ionospheric delay correction and Sagnac effect correction into Eq. (3.4), the clock difference between satellite and ground can be determined (Fig. 3.7).

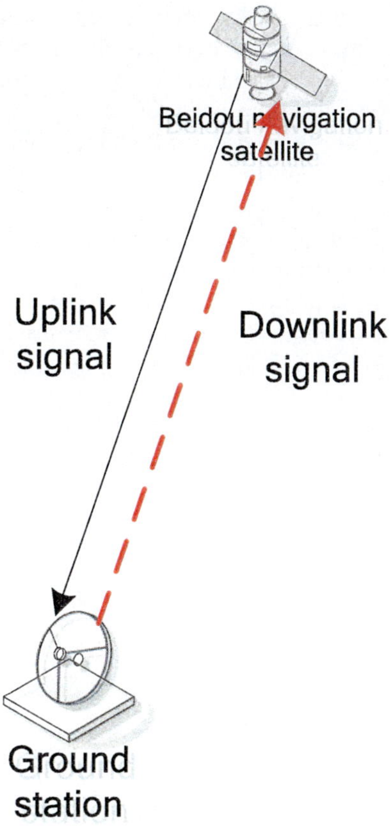


**Fig. 3.7** Schematic diagram of Beidou satellite-ground time synchronization technology

In this method, the uplink and downlink paths are approximately symmetrical so that the path delay can be mostly eliminated and the time synchronization accuracy can be greatly improved. The dual-frequency signal is used to effectively reduce ionospheric delay error, so the time synchronization accuracy of BDS can reach 0.6 ns.

### *3.2.3 Laser-Based Satellite-Ground Time Synchronization Technology*

The laser-based satellite-ground time synchronization technology has high-precision. The satellite-ground time synchronization accuracy of the US T2L2 project is 100 ps, while the accuracy of the time transfer link in the European ACES project is as high as 50 ps. China's BDS-2 satellite system has improved its measurement accuracy to 100 ps by using the spaceborne laser time-difference instrument to check the radio measurement results [43]. The main disadvantage of laser time synchronization technique is that it is highly susceptible to weather and cannot work in cloudy, rainy or foggy days. As shown in Fig. 3.8, the laser-based satellite-ground time synchronization technology relies on laser pulse signal and laser ranging principle to realize time synchronization.

The ground-based laser ranging station transmits short laser pulses to the satellite and records the difference $\Delta t_1$ between the time of laser pulse emission and the time of ground-based atomic clock measured by the time interval counter at the time of ground pulse emission. $\Delta t_1 = T_{JT} - T_G$, where $T_{JT}$ stands for the time of laser pulse emission and $T_G$ for the time of the ground-based atomic clock. After detecting a single photon, the satellite records the difference $\Delta t_2$ between laser pulse TOA and spaceborne atomic clock time measured by the time interval counter at the reception time. $\Delta t_2 = T_{JS} - T_s$, where $T_{JS}$ and $T_s$, respectively, stand for the time when the laser

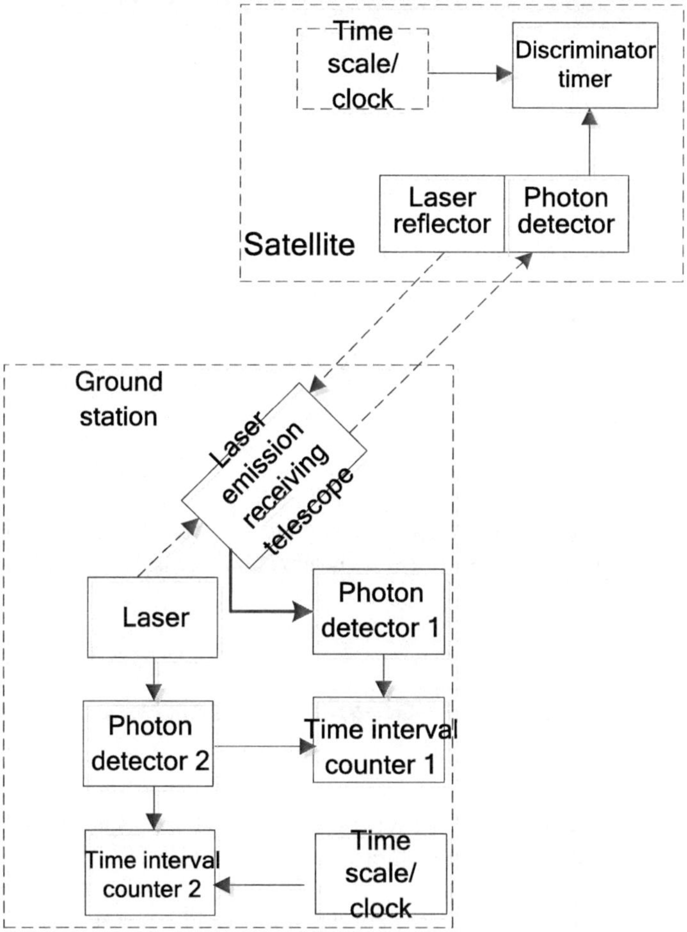


Fig. 3.8 Diagram of laser-based satellite-ground time synchronization technology

pulse reaches the satellite and the time of spaceborne atomic clock. At the same time, the onboard laser reflector returns the laser pulse to the ground station, and the time interval counter on the ground records the time difference $\Delta t_3$ between the received laser pulse and the ground-based atomic clock. $\Delta t_3 = T_{JR} - T_G$, where $T_{JR}$ is the time when the laser pulse reaches the ground. The difference between $\Delta t_3$ and $\Delta t_1$ is the delay of laser pulse transmission from the ground-to-satellite and then back to the ground. The clock difference between ground and satellite, $\Delta T_{S-G}$, is expressed as

$$\Delta T_{S-G} = T_s - T_G = \frac{\Delta t_3 - \Delta t_1}{2} - (\Delta t_2 - \Delta t_1) \tag{3.5}$$

To calculate accurately, it is necessary to add the corrections for relativistic effects, for atmospheric delays, and for the effects of the Earth's rotation. The difference $\Delta T_{S-G}$ between the ground clock $G$ and the satellite clock $S$ is

$$\Delta T_{S-G} = \frac{\Delta t_3 + \Delta t_1}{2} - \Delta t_2 + \frac{T_{sag}}{2} + T_{Rel} + T_{Atm} + T_{ICal} + T_{ECal} \tag{3.6}$$

where $T_{sag}$ is the correction of the Sagnac effect, i.e., the Sagnac difference between uplink and downlink; $T_{Rel}$ is the correction for the frequency drift caused by relativistic effects; $T_{Atm}$ is the correction for atmospheric delays; $T_{ICal}$ is the correction of the internal check; and $T_{ECal}$ is the correction of the external check.

### *3.2.4 Inverse Positioning Method*

In the inverse positioning method (see Fig. 3.9 for its principle), there are at least 4 ground time synchronization stations receiving signals from the satellite at the same time to obtain four pseudorange measurements. With the known positions and accurate time synchronization, the stations can obtain the data on the satellite orbit and clock offset in the positioning principle similar to GNSS. This is the main method used by GPS, where the time synchronization accuracy is within 10 ns. Its remarkable advantage is that it can achieve both accurate satellite positioning and satellite time synchronization.

The errors of inverse positioning mainly include ground station position error, equipment delay error, inter-station time synchronization error, ionospheric error, tropospheric error, and multipath error, among which the ionospheric error and equipment delay error are the main contributors. Single-frequency ionospheric error is significant, generally up to tens of nanoseconds, which cannot satisfy the requirement of satellite time synchronization. In this case, dual-frequency or tri-frequency ionospheric correction is used to reduce the error to less than 1 ns. So far, the ground inter-station time synchronization error has been reduced to 1 ns or less by using

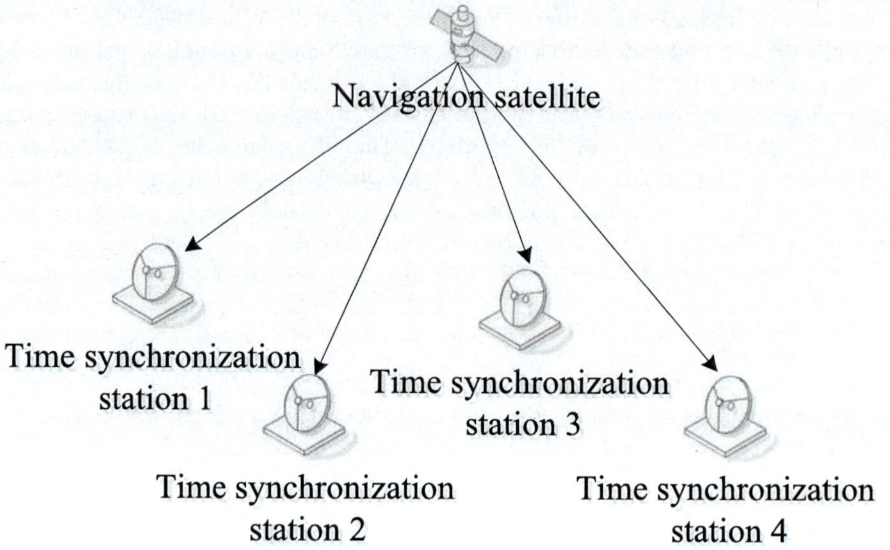


**Fig. 3.9** Schematic diagram of inverse positioning method

the Two Way Satellite Time and Frequency Transfer (TWSTFT) technique [32], and even to 100 ps or less by using optical-fiber time-frequency transfer link [31].

The satellite-ground time synchronization of GPS is achieved by inverse positioning. In order to implement navigation tracking and satellite-ground time synchronization, the GPS has established many monitoring stations in Ascension Island, Diego Garcia, Kwajalein, Hawaii, Colorado Springs, Cape Canaveral, and other areas. Those monitoring stations have precise position coordinates and maintain high-precision synchronization between each other.

This is a cost-effective method with strong operability, as it can determine the satellite orbit and measure the clock offset with simple equipment. However, it has two disadvantages, that is, the path delay cannot be canceled as the measurement link is a one-way link; and the time synchronization error between monitoring stations cannot be ignored. Therefore, its time synchronization accuracy is within 10 ns, generally 6 ns, with the maximum less than 18 ns.

### *3.2.5 Pseudorange and Radar Ranging Method*

In the pseudorange and radar ranging method, the satellite-ground time synchronization is completed by ground time synchronization station. The basic principle is that the time synchronization station measures the pseudorange according to the navigation signal transmitted by the satellite, and then measures the ground-to-satellite distance with the aid of responsive radar (Fig. 3.10). The difference between pseu-

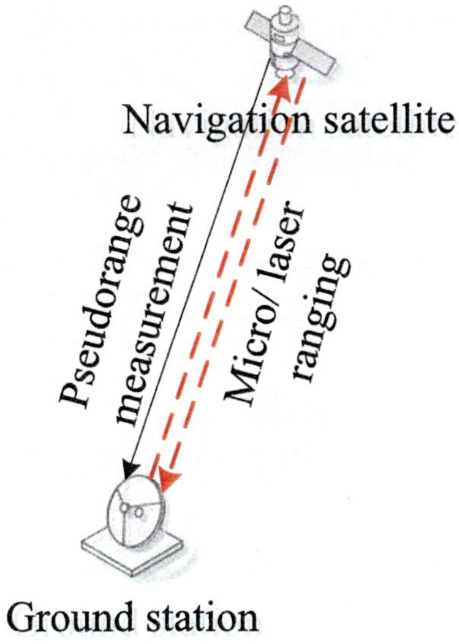


**Fig. 3.10** Schematic diagram of pseudorange and radar ranging method

dorange measurement and distance measurement is just a satellite-ground clock offset.

There are two methods for radar ranging. One is responsive radio radar ranging by using *C*-band or Ku-band. The GLONASS uses the *C*-band responsive radar ranging. In this method, the uplink and downlink signals have different frequencies, but the ionospheric propagation error correction is still incomplete. As a result, a 10 ns synchronization error is generated. The other method is laser radar ranging, a method used by the ground station in two-way laser ranging. In this method, the ground station transmits laser light for two-way distance measurement, while the satellite responds to the ranging signal by laser reflection. Since the laser signals in uplink and downlink paths are in the same frequency, the ionospheric propagation error can be neglected. In addition, the geometric center distance between satellite reflector and *L*-band antenna can be accurately measured and corrected. Therefore, the major ranging error depends on the pseudorange measurement by the ground station. The precision correction of this system error is up to 1–2 ns.

## 3.3 Uploaded Information and Uploading Strategy of BDS

Information uploading is the critical function of satellite navigation uplink. The realization of navigation and positioning mainly depends on the information uploaded by the uplink channel. The uploaded information generally concludes satellite ephemeris, satellite clock correction parameters, atmospheric delay correction parameters, satellite signal integrity information, etc. The uploaded information of

**Table 3.1** Classification of uploaded information in the BDS

| No. | Information type | Main function |
|---|---|---|
| 1 | Null information | Blank information |
| 2 | Basic navigation information | General information of BDS-3 |
| 3 | Satellite payload information | Commands and parameters of navigation satellite |
| 4 | Autonomous navigation information | Autonomous navigation messages, service instructions, and parameters |
| 5 | Information for smooth transition from BDS-2 to BDS-3 | BDS-2 navigation signal information |
| 6 | TT&C information | Commands and parameters related to TT&C |
| 7 | Test information | Reserved for testing and verification |

the BDS System (especially for BDS-3) is classified into seven categories, as shown in Table 3.1.

Each category is further divided into different types of parameters according to different uploading strategies. This section will present basic navigation information, mainly focusing on its collection method, types, and uploading strategy.

### *3.3.1 Collection of Basic Navigation Information*

In the process of BDS positioning, the accuracy is often limited by satellite orbit, satellite clock error, atmospheric troposphere, and ionosphere to only tens of meters for civil users. In order to improve the positioning accuracy, the ground monitoring station performs continuous observation and satellite data acquisition, and compares its own observations with those of a known precise reference station. Meanwhile, it collects the relevant meteorological data from environmental sensors, processes the measurement data, and transmits it to the ground control station. Then the ground control station processes and classifies the data collected by the monitoring station to obtain the information on the following parameters:

(1) Satellite ephemeris parameters.
(2) Satellite clock parameters and group delay parameter.
(3) Atmospheric delay parameters.
(4) Satellite signal integrity information.

The ground control station sends basic navigation information with various parameters to the ground upload station, where basic navigation parameters and control instructions are mapped to the uploaded signals in the format of the uplink message and then sent to the corresponding satellite. After receiving the uploaded information, the satellite broadcasts it in the format of navigation signal message to various user receivers and assists the user receivers in achieving high-precision positioning.

### 3.3.2 Types of Basic Navigation Information for Upload

The basic navigation information for upload sent by the BDS ground upload station is generally divided into nine types, as shown in Table 3.2.

Each type of information is further introduced as below

(1) Ephemeris parameters: a satellite ephemeris consists of 18 quasi-Kepler orbital parameters, plus the ephemeris age of data.
(2) Satellite clock parameters and group delay parameter:

a. Satellite clock correction parameters: mainly including the time parameters and correction parameters of the satellite clock.
b. Group delay parameter: the time delay between the navigation signal transmitted by the satellite (measured at the antenna phase center) and the satellite frequency source.

(3) Medium-precision almanac
(4) Simple almanac
(5) BGTO parameters: the time deviations between BDT and other GNSS systems including GPS, Galileo, GLONASS, etc.
(6) Ionospheric model parameters: 11 ionospheric delay correction parameters to be broadcast;
(7) Global basic integrity information, Class I (real time): Satellite health monitoring information;
(8) Global basic integrity information, Class II (fast): the monitoring information of spatial signal accuracy;
(9) Global basic integrity information, Class III (slow): the monitoring information of spatial signal accuracy and the spatial signal status information.

The format of the uploaded message is shown in Fig. 3.11.

The format of uploaded message is composed of two parts: constant information content and variable information content. The constant information content includes:

**Table 3.2** Types of basic navigation information

| No. | Information type | Receiving satellite |
|---|---|---|
| 1 | Ephemeris parameters | MEO/IGSO/GEO |
| 2 | Satellite clock parameters and group delay parameter | MEO/IGSO/GEO |
| 3 | Medium-precision almanac | MEO/IGSO/GEO |
| 4 | Simple almanac | MEO/IGSO/GEO |
| 5 | BGTO parameters | MEO/IGSO/GEO |
| 6 | Ionospheric model parameters | MEO/IGSO/GEO |
| 7 | Global basic integrity information, Class I (real time) | MEO/IGSO/GEO |
| 8 | Global basic integrity information, Class II (fast) | MEO/IGSO/GEO |
| 9 | Global basic integrity information, Class III (slow) | MEO/IGSO/GEO |

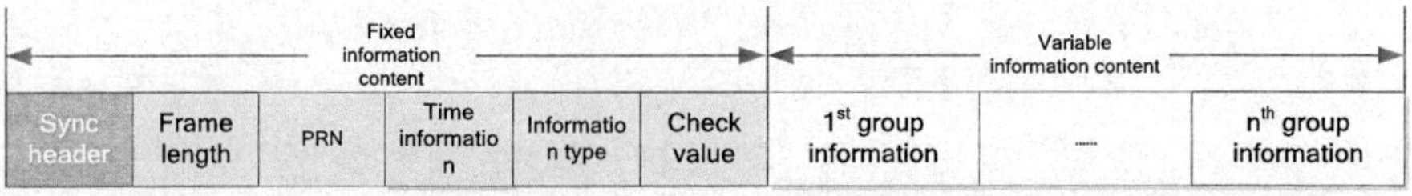


**Fig. 3.11** Format of uploaded message

sync header, frame length, PRN (target satellite number), time information, type of uploaded information, and check value. And the variable information content refers to the content of a specific type of uploaded information, with a length that may vary with information content.

1. Ephemeris Parameters

The BDS ephemeris consists of 18 quasi-Kepler orbital parameters plus the ephemeris age of data. The ephemeris parameters fall into the following three categories:

(1) Time information: week number, and the number of hours in the week.
(2) Valid reference information for ephemeris use: the ephemeris age of data, and the ephemeris reference time
(3) Orbit positioning parameters: the deviation of semi-major axis from the reference value, the change rate of semi-major axis, the difference between average satellite velocity and calculated velocity and its change rate, the mean anomaly of reference time, eccentricity, argument of perigee, the geographic longitude of ascending node at the reference time, the orbital inclination angle at the reference time, the geographical longitude change rate of ascending node, the change rate of orbital inclination angle, the sinusoidal correction amplitude of orbital inclination angle, the cosinusoidal correction amplitude of orbital inclination angle; the sinusoidal correction amplitude of orbital radius, the cosinusoidal correction amplitude of orbital radius, the sinusoidal correction amplitude of the argument of latitude, and the cosinusoidal correction amplitude of the argument of latitude.

2. Satellite clock parameters and group delay parameters

The satellite clock parameters of BDS, which are used for the correction of satellite time, mainly fall into the following three types:

(1) Time information: week number, and the number of hours in the week.
(2) Valid reference information for clock use: the clock age of data, and the clock reference time.
(3) Satellite time correction information: satellite time deviation parameters, satellite frequency offset parameters, and satellite frequency shift parameters.

The group delay parameter of BDS refers to the time delay between the navigation signal transmitted by the satellite (measured at the antenna phase center) and the spaceborne time-frequency reference source. It is mainly the group delay of the global system authorization signal or the group delay parameter of an open signal device.

3. Medium-precision almanac

The medium-precision almanac of BDS mainly covers the following five types of parameters:

(1) Satellite number.
(2) Almanac time parameters: almanac reference time and its week number.
(3) Satellite orbit parameters: eccentricity; the correction of reference orbit inclination angle at the reference time; the square root of semi-major axis; the geographic longitude of ascending node calculated at the reference time; the geographic longitude change rate of ascending node; the argument of perigee; and the mean anomaly at the reference time.
(4) Satellite clock parameters: satellite clock difference and clock velocity.
(5) Satellite health information.

4. Simple almanac

The simple almanac of BDS mainly covers the following four types of parameters:

(1) Satellite number
(2) Satellite orbit parameters: the correction for semi-major axis at the reference time; the longitude of ascending node at the reference time; and the argument of latitude at the reference time.
(3) Satellite health information.
(4) Almanac reference time parameters: almanac reference time and its week number.

5. BGTO parameters

The BDT-GNSS time offset parameters reflect the time differences between BDT and other GNSS systems including GPS, Galileo, and GLONASS. These parameters fall into the following three categories:

(1) GNSS system identification: the identification of GPS system, Galileo system, GLONASS system, and reserved systems.
(2) The week number of reference time and the time of the week at the reference time;
(3) The deviation coefficient, drift coefficient, and drift rate coefficient of the BDT time scale relative to the GNSS time reference.

6. Ionospheric model parameters

The broadcast model parameters of ionospheric delay correction in the BDS system are mainly classified into the following three types:

(1) Week number in the group number identification
(2) The group number of week in the group number identification
(3) Ionospheric model parameters.

7. Global basic integrity information

The global basic integrity information of BDS is divided into the following three types:

(1) Global basic integrity information, Class I: the satellite health statuses as monitored in real time, including "healthy", "unhealthy", "unmonitored", and "ineffective".
(2) Global basic integrity information, Class II: the monitoring information of spatial signal accuracy.
(3) Global basic integrity information Class III: the monitoring information of spatial signal accuracy and the spatial signal status information.

### 3.3.3 *Navigation Information Uploading and Activating Strategy*

The basic navigation information of BDS is collected by the ground monitoring station, processed by the ground control station, and then transmitted to the ground upload station, where the information is modulated into the uplink signal and sent out through the antenna. The spaceborne receiver acquires, tracks, and demodulates the uplink signal to get the navigation information, and then sends it in a package to the navigation task processing unit. The processing unit interprets the navigation information, corrects or discards wrong messages, and extracts and modulates the right messages into the navigation signal that will be broadcast to the user.

In the BDS system, basic navigation information is mainly uploaded with two optional strategies: timing upload and random upload. Timing upload is to regularly upload the information according to its specific timing requirement, for example, once every 1, 12 or 24 h. If a piece of uploaded information is checked as an error, it will be re-uploaded until it is correct. Random upload is to upload the information according to practical needs so that the upload time and frequency can be adjusted. If a piece of uploaded information is checked as an error, it will be re-uploaded until it is correct.

After receiving the uploaded information, the satellite will activate the new information in three modes. In the first mode (or normal mode), the information will be activated normally (at the beginning of the next hour) or immediately after being received. In the second mode (or real-time mode), the information will be activated at the beginning of the next data frame in the same information type. In the third mode, judgment will be made based on the information content and time, to activate the uploaded information immediately (within this hour) if commanded or activate it in the next hour if not commanded.

The time for enabling the injected universal navigation information of BDS falls into three categories: the first category, determine to use the normal enable mode (on the next hour after receiving this type of information) or real-time enable mode according to the downlink message broadcast enable mode command; the second category, enable at the beginning of next data frame after receiving information of the corresponding type; the third category, make a judgment based on the injected information content and time, and enable it immediately if it is commanded to enable within this hour and enable it in next hour if not.

## 3.4 BDS Satellite Uplink Reception Technology

In the BDS system, the satellite uplink receiver has 2 channels. The channel 1 is used for high-precision satellite-ground measurement and data upload, and the channel 2 for high-precision satellite-ground measurement and null-value monitoring. After receiving the uplink signal from the ground upload station via the antenna, the uplink receiver will process it through down-conversion, filtering and analog-to-digital conversion to accomplish the signal acquisition, tracking, demodulation, and frame synchronization. Meanwhile, the demodulated time data will be used with local time for precise satellite-ground measurement. In this section, technologies such as uplink signal receiving, pseudorange, and integral Doppler measurement, as well as null detection, will be addressed.

The two channels share the same physical channels such as RF front-end and analog-to-digital converter while using the same techniques—anti-interference, acquisition, tracking, demodulation, and frame synchronization—for baseband digital signal processing. The uplink signal can be expressed as

$$S(t) = D(t)P(t)\cos(2\pi f_0 t + \varphi) \tag{3.7}$$

where $D$ is the modulated data, $P$ is the spread spectrum code, $f_0$ is the center frequency of uplink signal, and $\varphi$ is the initial phase of the carrier. Fig. 3.12 shows the uplink receiver diagram based on a superheterodyne architecture. The null-value channel signal generated by the satellite itself is integrated with the uplink signal through the coupler and then sent to the receiving channel.

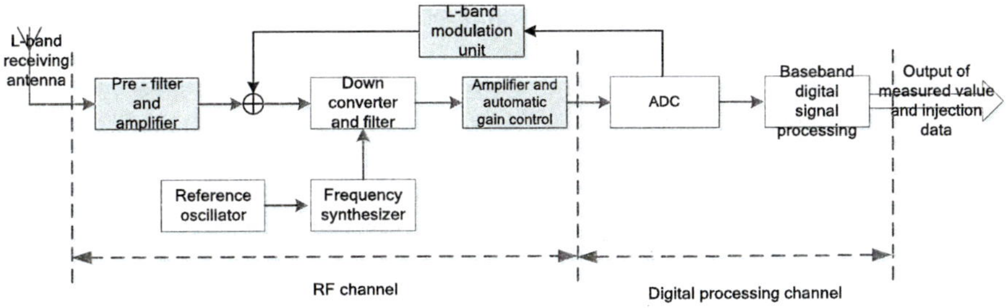


**Fig. 3.12** Diagram of uplink receiver

### 3.4.1 Uplink Signal Conditioning

After being received by the antenna, the uplink signal is required to be conditioned to obtain the digital signal that can be processed by the later baseband unit. The RF front-end is used for signal conditioning, which means, the pre-filter suppresses the interference out of band, and the LNA amplifies the uplink signal and then couples it with the null-value signal when the signal power is appropriate. Furthermore, the signal frequency is down-converted to intermediate frequency so that the uplink signal can be mixed with the local oscillator signal and then input to the ADC after the bandpass filtering and amplifying. Fig. 3.13 shows a typical RF front-end processing module based on the superheterodyne reception technique.

Three bandpass filters are installed in the RF front-end processing module.

**Bandpass filter 1**: To perform the bandpass filtering for the received signal to prevent the amplifier saturation caused by out of band interference.

**Bandpass filter 2**: To perform the bandpass filtering for the pre-amplified signal, to eliminate new frequency components and the image frequencies brought by amplifier nonlinearity, and to prevent the image interference from being introduced to the mixing process. Therefore, this filter is also called an image suppression filter.

**Bandpass filter 3**: To filter out the sum component of the mixed signal so as to make the received IF signal smoother.

The uplink signal quality can be measured by $C/N_0$ and signal distortion after conditioning. $C/N_0$ is the ratio of the power of the received signal to the system noise power, reflecting the sharpness of the received signal. If only the thermal noise in the circuit system caused by the thermal motion of the charged particles is considered, $N_0$ will be the average energy of the noise within the signal bandwidth, i.e., the noise's power spectral density. The higher the signal quality is, the better the signal acquisition, tracking, measurement, and demodulation will be performed. Generally, the noise figure of the entire uplink reception system is 2–3 dB, and the noise's power spectral density $N_0$ is within $-172$ dBm/Hz ~ $+171$ dBm/Hz at the 290 K ambient temperature.

In the receiving system, there are a large number of RF components such as filters and amplifiers. Owing to the band-limited characteristic of filters, high-frequency signal components are filtered out, and the "sharpness" of the correlated peaks is

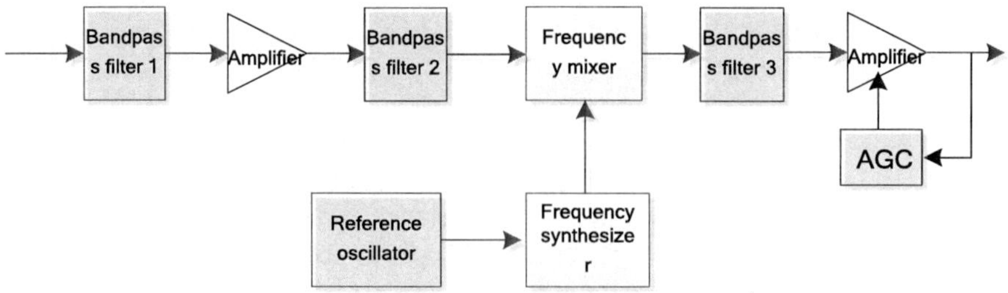


**Fig. 3.13** RF front-end processing module

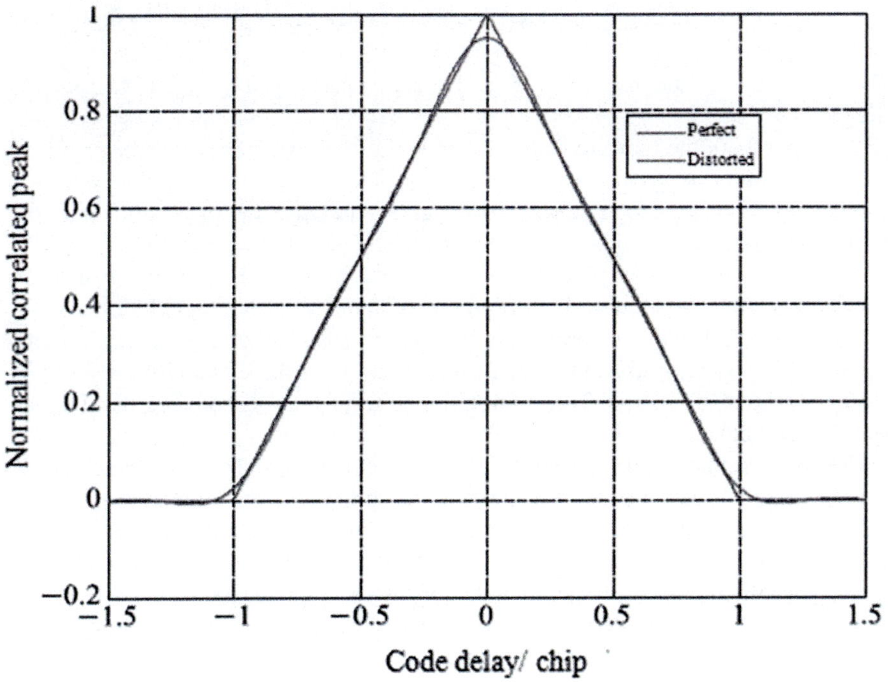


**Fig. 3.14** Correlated peak of distorted signal after reception

smoothed. The non-flat characteristic of filter bands and the nonlinear characteristic of amplifiers tend to cause the signal distortion, thus resulting in the asymmetry of the correlated peaks (Fig. 3.14).

Through the ADC device, the intermediate-frequency analog signal is converted into a digital signal, which, in turn, is turned into an orthogonal baseband signal after the down-conversion, digital filtering, and interference suppression. After being mixed, filtered, and amplified, the intermediate-frequency signal has high power and low center frequency enough for digital processing. Therefore, it can be generally converted into a digital signal by an analog-to-digital conversion (ADC) device for further processing. In the ADC, the sampling frequency $f_s$ and the number of quantization bits ($n$) are the key parameters to be selected.

The intermediate-frequency analog signal is a bandpass signal. When sampling the bandpass signal, the sampling rate shall be smaller than 2 times the highest signal frequency and greater than 2 times the signal bandwidth. The ADC sampling frequency $f_s$ in bandpass sampling is a very critical parameter. In order not to cause aliasing during sampling, the bandpass sampling frequency shall satisfy the following relationship:

$$\frac{2f_c + B_{fe}}{n+1} \leq f_s \leq \frac{2f_c - B_{fe}}{n}$$

$$f_s \geq 2B_{fe} \tag{3.8}$$

where $f_c$ is the center frequency of the intermediate-frequency signal, and $n$ is an integer.

While meeting the requirements for bandpass sampling frequency, the sampling frequency cannot be synchronized with the pseudo-code rate. A simple solution is to ensure that the sampling rate within the range of $(f_{code} \pm f_{d\_code})$ is not a multiple of the pseudo-code rate when the pseudo-code Doppler $f_{d\_code}$ is known.

Another parameter that plays an important role in ADC is the number of quantization bits ($n$). Quantization is to represent a sampled analog signal with a set of binary numbers. The quantization loss varies with the number of quantization bits. For the limited bandwidth signal, the quantitation loss is 3.5 dB for 1 bit, 1.2 dB for 2 bits, and 0.6 dB for 3 bits. The quantitation loss for four or more bits in ADC can basically be neglected.

However, it must be pointed out that the above quantitation losses are based on the assumption of no interference. If interference is considered, the number of ADC bits will have a huge impact on the quantitation performance. The multi-bit ADC design tends to have better anti-interference performance, as the noise of an interfering signal is usually tens of dB larger than that of a normal signal. By using the multi-bit ADC, the desired signals contained in the interfering sampling frequency can be recovered after the anti-interference suppression.

The intermediate-frequency signal after the ADC sampling can be expressed as

$$s_{IF}(t) = Ax(n - \tau)D(n - \tau)\cos(2\pi fn + \theta) \tag{3.9}$$

### 3.4.2 Uplink Baseband Processing

After the AD sampling, the intermediate-frequency digital signal will go through the anti-interference processing followed by signal acquisition, tracking, bit synchronization, frame synchronization, and other stages to realize the data demodulation and pseudorange measurement. The process of baseband processing includes acquisition and tracking.

In order to successfully track the uplink signal, the receiver must make sure that the carrier and pseudo-code phase of the initial duplicated signal match those of the received signal to be tracked. Otherwise, if the difference between the duplicated signal and the received signal in terms of carrier and pseudo-code phase exceeds the traction range of carrier loop and code loop, the tracking will fail. Therefore, the satellite receiver needs to initially estimate the carrier Doppler and pseudo-code phase to some extent before entering the tracking stage. This process is called signal acquisition.

The search step of the carrier frequency is $f_{bin}$, and each frequency band corresponds to a search value of carrier frequency. The search step of the code phase is called code bandwidth $t_{\text{bin}}$, and each code band corresponds to a search value of the code phase. Each intersection of frequency band and code band is called a search unit. If the signal search is successful, the estimated values of the corresponding carrier frequency and code phase will be not more than half of the actual frequency bandwidth and half of the actual code bandwidth, respectively.

The receiver tracks the signal mainly through the carrier tracking loop (carrier loop for short) and code tracking loop (code loop for short). The carrier loop usually takes two forms: phase-locked loop (PLL) and frequency-locked loop (FLL). The signal tracking stage starts with the captured estimates of carrier phase and pseudo-code phase. The tracking loop gradually refines these two signal parameters while outputting the pseudorange measurements and data bits that are contained in the demodulated message. The corresponding carrier and pseudo-code can be reproduced by the receiver. The carrier phase and pseudo-code phase of the reproduced signal is dynamically consistent with those of the input signal. The carrier of the input signal is multiplied by the duplicated carrier to be down-converted to baseband, while the pseudo-code of the input signal is multiplied by the duplicated pseudo-code to realize the pseudocode stripping and despreading. At this time, only the data bits are left in the received signal. In the tracking process, the receiver can estimate the Doppler shift and carrier phase of the uplink signal according to the recovered carrier phase, and obtain the pseudorange observations according to the recovered pseudo-code phase.

The typical principle of uplink signal tracking and processing is shown in Fig. 3.15.

1. Uplink signal acquisition algorithm

There are mainly three algorithms for signal search: linear search, parallel frequency search, and parallel code phase search. Next, these algorithms will be introduced one by one.

Linear search is the most basic acquisition algorithm that uses a linear correlator to search each cell in turn. The principle of this method is shown in Fig. 3.16. It can only search one cell at a time, so the acquisition time is long. Its advantage is that only a few digital correlators are needed, and the correlation results can be directly used for signal tracking. In the linear search, frequency and pseudo-code are searched by different strategies. After determining the carrier Doppler range of the input signal, the search for it can be started from the middle or both sides of the frequency range according to the search step $f_{\text{bin}}$, until the signal is detected or all the bands are searched.

Parallel frequency search is an algorithm that uses FFT to carry out a parallel search for the carrier frequency of the input signal. The specific process is to mix the input digital intermediate frequency with the locally duplicated nominal frequency carrier, and then correlate it with the duplicated pseudo-code to obtain the orthogonal coherent accumulation result $I + jQ$. FFT is used to calculate the intensity of the correlation results at each frequency component, and the signal amplitude at the frequency component is analyzed to find the corresponding residual carrier component. If the matching of pseudo-code phases fails, the amplitude after the FFT will

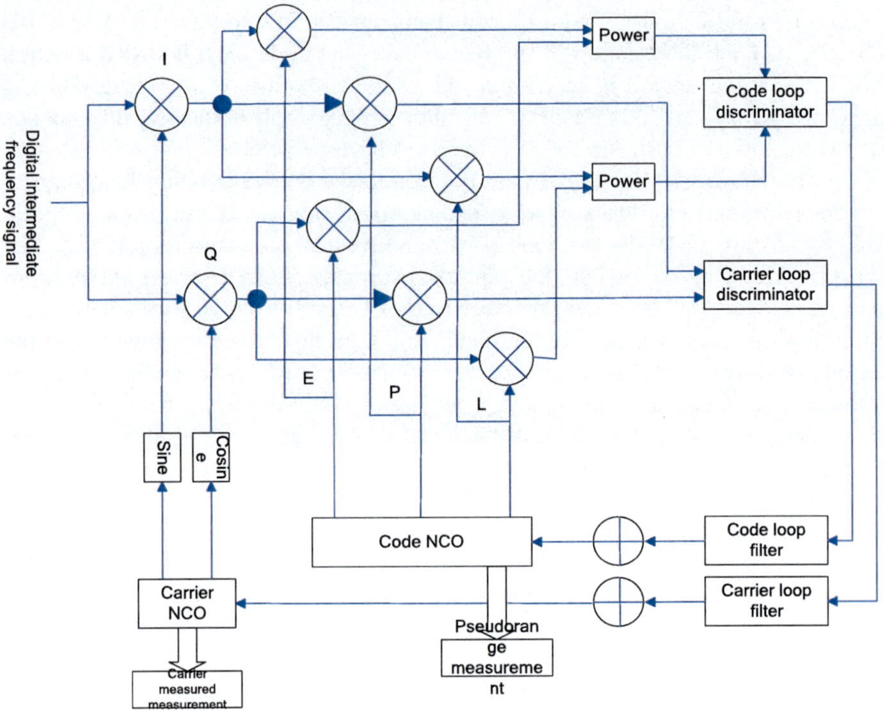


**Fig. 3.15** Schematic diagram of uplink signal tracking

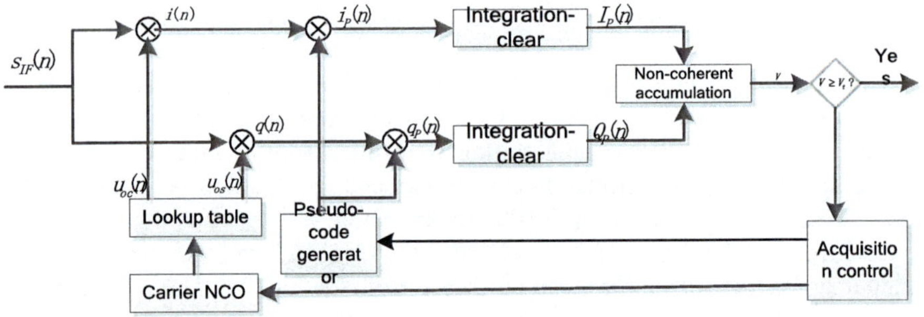


**Fig. 3.16** Block diagram of linear search in signal acquisition

be suppressed to below the threshold. Only when the code phases are consistent, the output amplitude after the FFT can have an obvious peak, i.e., the corresponding frequency error $f_e$. At this time, the locally duplicated pseudo-code phase is the estimated code phase of the received signal. The specific algorithm is shown in Fig. 3.17. The code phase estimation accuracy in parallel frequency search is the code phase search step $t_{\text{bin}}$, and the frequency resolution of FFT at the point $N$ is $1/NT_{\text{coh}}$.

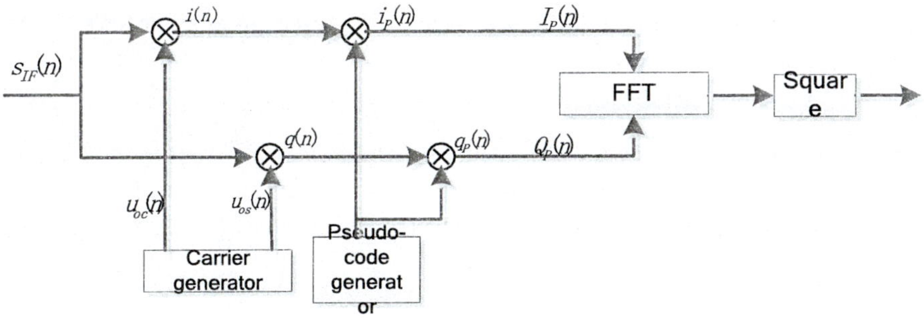


**Fig. 3.17** Acquisition algorithm flow of parallel frequency search

Parallel code phase search is an algorithm that uses FFT to carry out a parallel search for the pseudo-code phase of the input signal. The specific process is to mix the input digital intermediate frequency with the locally duplicated nominal frequency carrier and with the duplicated pseudo-code to obtain the orthogonal coherent integration result $i + jQ$; to perform the FFT operation on the orthogonal result, multiply the transferred result by the FFT conjugate value of the duplicated pseudo-code, and then perform the IFFT operation on the product to get the correlation result in the time domain; and finally to test these results to judge if the signal is present. The specific algorithm principle is shown in Fig. 3.18. This algorithm detects all the pseudo-code phases at a time by only adjusting the locally duplicated carrier frequency, so the capture speed can be greatly improved.

2. Design of uplink signal tracking

(1) PLL design

The carrier recovery of BPSK and QPSK spread spectrum signals requires the demodulation of the components $I$ and $Q$. When the pseudo-code and the modulated data are not considered, the intermediate-frequency signal obtained after the orthogonal

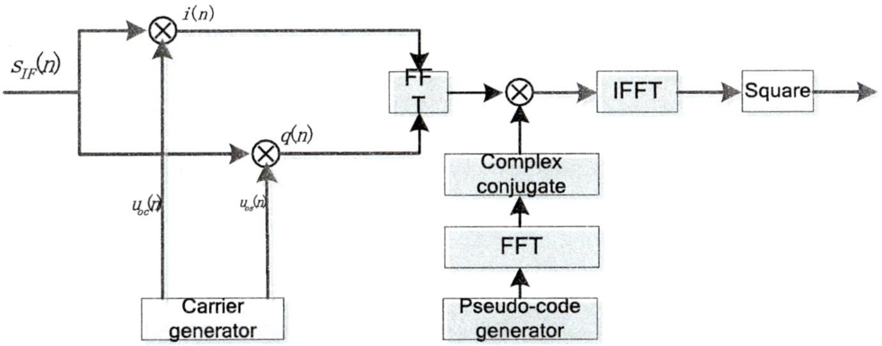


**Fig. 3.18** Acquisition algorithm of parallel code phase search

down-conversion can be expressed in the form of a complex number vector

$$r_{\mathrm{P}}(t) = I_{\mathrm{P}}(t) + jQ_{\mathrm{P}}(t) = A(t)\mathrm{e}^{j(\varpi_{\mathrm{e}}t+\theta)} \tag{3.10}$$

where $I_{\mathrm{P}}(t) = A\cos(\varpi_{\mathrm{e}}t + \theta)$ is called the in-phase component, and $Q_{\mathrm{P}}(t) = A\sin(\varpi_{\mathrm{e}}t + \theta)$ is called orthogonal component. The difference between orthogonal down-converted signal and locally recovered carrier phase, namely $\theta_{\mathrm{e}} = \varpi_{\mathrm{e}}t + \theta$, can be determined by calculating the angles of the in-phase component and orthogonal component.

$$\theta_{\mathrm{e}}(t) = \arctan\left(\frac{Q_{\mathrm{P}}(t)}{I_{\mathrm{P}}(t)}\right) \tag{3.11}$$

Fig. 3.19 shows a typical tracking loop for a digital carrier.

In the actual receiver design, the carrier noise is relatively low and the phase difference is susceptible to noise, so it is necessary to increase the signal-to-noise ratios $I_{\mathrm{P}}(t)$ and $Q_{\mathrm{P}}(t)$ by means of coherent integration. Coherent integration is to integrate the input signals $i_{\mathrm{P}}(t)$ and $q_{\mathrm{P}}(t)$ for a period of time, respectively, to obtain the integration results $I_{\mathrm{P}}(t)$ and $Q_{\mathrm{P}}(t)$. The time period for integration is called coherent integration time $T_{\mathrm{coh}}$. In digital processing, the discretized orthogonal signals can be expressed as

$$\begin{aligned} i_{\mathrm{P}}(n) &= aD(n)\cos(\varpi_{\mathrm{e}}(n)t(n) + \theta_{\mathrm{e}}) \\ q_{\mathrm{P}}(n) &= aD(n)\sin(\varpi_{\mathrm{e}}(n)t(n) + \theta_{\mathrm{e}}) \end{aligned} \tag{3.12}$$

(2) FLL Design

In the frequency-locked loop, the carrier stripping process is completed to reproduce the approximate frequency. Compared with the phase-locked loop, the frequency-locked loop requires that the frequency, rather than phase, of the duplicated carrier be the same as that of the recovered carrier. Therefore, the difference between the frequency-locked loop and phase-locked loop lies in the discriminator. The discriminator used in the FLL is called frequency discriminator.

The frequency discrimination result of the frequency-locked loop is filtered by the loop filter and then applied to the controlled input end of the numerically controlled oscillator of the carrier, so that the output frequency of the numerically controlled oscillator can be adjusted to the frequency of the received carrier. The frequency measurement error in the frequency-locked loop is mainly caused by thermal noise frequency jitter $\sigma_{n\mathrm{FLL}}$ and dynamic stress error $f_{\mathrm{e}}$.

(3) DLL design

The pseudo-code synchronization is implemented by a code tracking loop. The main function of this loop is to keep the duplicated pseudo-code and the received pseudo-code in phase, thereby obtaining the code phase of the received signal and the measured value of pseudorange. The code loop first duplicates a pseudo-code in

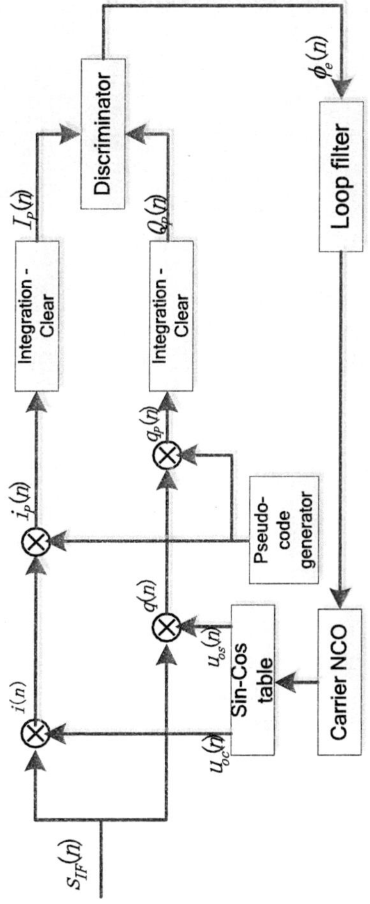


**Fig. 3.19** Typical structure of digital-carrier tracking loop

the same phase with that of the received signal and then correlates the received signal with the duplicated pseudo-code to strip the pseudo-code from the signal. The code loop is generally a delay-locked loop (DLL), as shown in Fig. 3.20.

The key to code loop operation is how to identify the phase difference between the received pseudo-code and the duplicated pseudo-code. This identification can be realized by copying three (or more) pseudo-codes in different phases, which are called early ($E$), prompt ($P$), and late ($L$) pseudo-codes, respectively. The phases of the three pseudo-codes are correlated with the received signal. The code loop compares multiple correlation results (usually for $E$ and $L$) to derive the position of the highest correlation peak (Fig. 3.21).

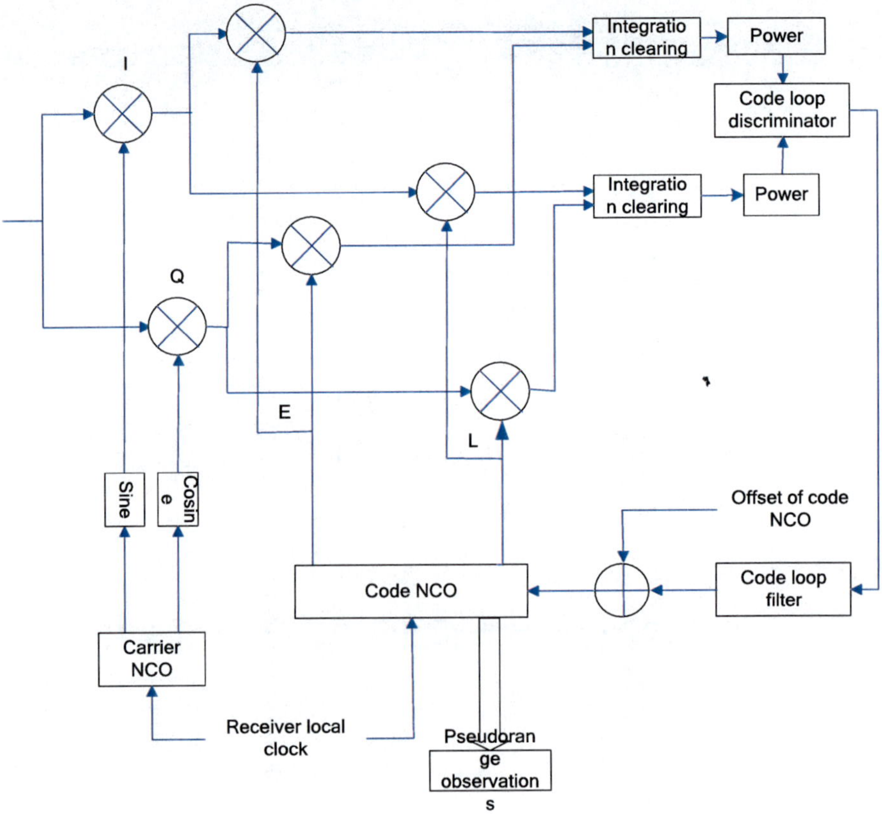


Fig. 3.20 Typical DLL loop

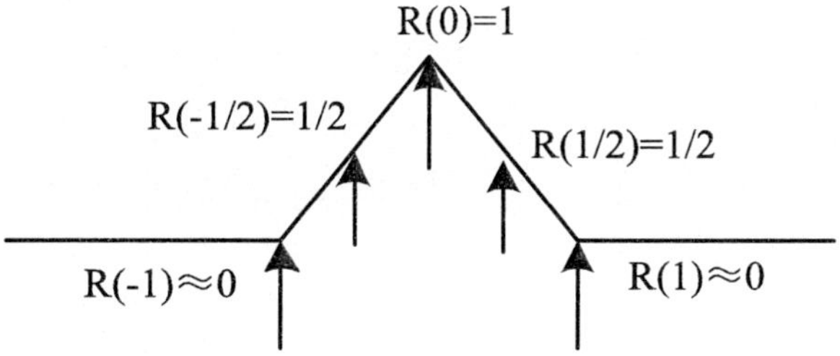


**Fig. 3.21** Correlation peaks of different phases of duplicated codes

### 3.4.3 *Upload Data Demodulation*

After the uplink signal is stably tracked, the information bits or symbol data must be demodulated with one of several optional methods. If the uplink signal is BPSK, the receiver can demodulate the information by demodulating the components $I$ and $Q$.

The demodulation performance and efficiency are evaluated by calculating the demodulation loss. The factors causing the demodulation loss generally include the quantization loss caused by the limited number of data bits when the receiver processes the digital signal, the possible loss of the coherent integration caused by the hopping of original navigation information bits, the correlation between noises, and various control error losses in the receiver. The sum of these losses is generally 1–2 dB. In order to improve the anti-interference ability and uplink efficiency, the system has been designed with a certain error correction capability to reduce the errors. Usually, the original navigation information is channel-coded in the signal system. Under this condition, the data bits directly demodulated by the tracking loop are not the original ones, so they must be decoded to obtain the original navigation information. The channel coding methods in common use include convolutional coding, Turbo coding, parity-check code, interleaving, cyclic redundancy check coding, and LDPC coding.

The demodulated information symbols are channel-decoded or de-encoded to obtain the navigation information bits, and then frame synchronization is performed to find the header of information frames. After that, the navigation information bits are decomposed and extracted according to the signal formatting and protocol. Meanwhile, judgment is made on whether the navigation information bits have been inverted by 180°.

### 3.4.4 *Pseudorange Measurement*

In radio navigation, pseudorange is defined as the difference between signal reception time and recovered-signal transmission time, multiplied by the speed of light. Since the receiver clock is not synchronized with the satellite clock, so this product is called pseudorange. If the pseudorange is $\rho(t)$, the transmission time of the recovered signal is $t^s$, the signal reception time is $t_u$, and the speed of light is $c$, then the pseudorange can be expressed as

$$\rho(t) = c(t_u - t^s) \tag{3.13}$$

If the receiver clock is synchronized with the satellite clock, the calculated range value will be called the true range. The spaceborne receiver usually uses the open-door/close-door ranging method for pseudorange measurement in one of the following two ways.

The first way is as follows: Select the pps of the receiving end as the opening moment for ranging and the pps of the transmitting end, which are later recovered by the receiving end, as the closing moment. Sample and save the recovered code NCO at the opening and closing moments separately. Subtract the NCO sampled at the opening moment from that sampled at the closing moment to obtain an original pseudorange measurement, i.e., the time difference between the opening moment and the closing moment. Then process the original pseudorange measurement to directly get the range measurement in second (s).

The second way is as follows: Select the pps of the transmitting end, which have been recovered by the receiving end, as the opening moment for ranging and the pps of the receiving end as the closing moment. Sample and save the recovered code NCO at the opening and closing moments separately. Subtract the NCO sampled at the opening moment from that sampled at the closing moment to obtain an original pseudorange measurement, i.e., the time difference between the opening moment and the closing moment. Then process the original pseudorange measurement to get the range measurement in second, which is then subtracted by 1 s to obtain the final pseudorange measurement.

The pseudorange measurement is mainly based on the code tracking loop, so its accuracy (random error) is related to the measurement accuracy of the code tracking loop. The measurement error of the code loop is mainly the code phase jitter error caused by thermal noise and dynamic stress. With the given code rate and input carrier-to-noise ratio, the satellite receiver reduces the mean square of thermal noise in the code loop by selecting the appropriate loop bandwidth, coherent integration time, and the adjacent correlator spacing, while reducing the error of dynamic stress by selecting reasonable loop parameters. The combination of the above two methods can reduce the random error in pseudorange measurement and improve the accuracy of pseudorange measurement.

The pseudorange measurement error in the spaceborne receiver system, namely system error, is mainly brought by different algorithms in the receiver and large changes in ambient temperature. The system error from different algorithms changes very little, so it can usually be calibrated and corrected easily. However, the system error from ambient temperature variation will change significantly with ambient temperature variation, thus playing a major role in the in-orbit operation of the onboard receiver and producing a great influence on pseudorange measurement. Therefore, this system error shall be corrected.

The method commonly used to correct the system error caused by ambient temperature variation is as follows. Equip the receiver with a temperature measuring

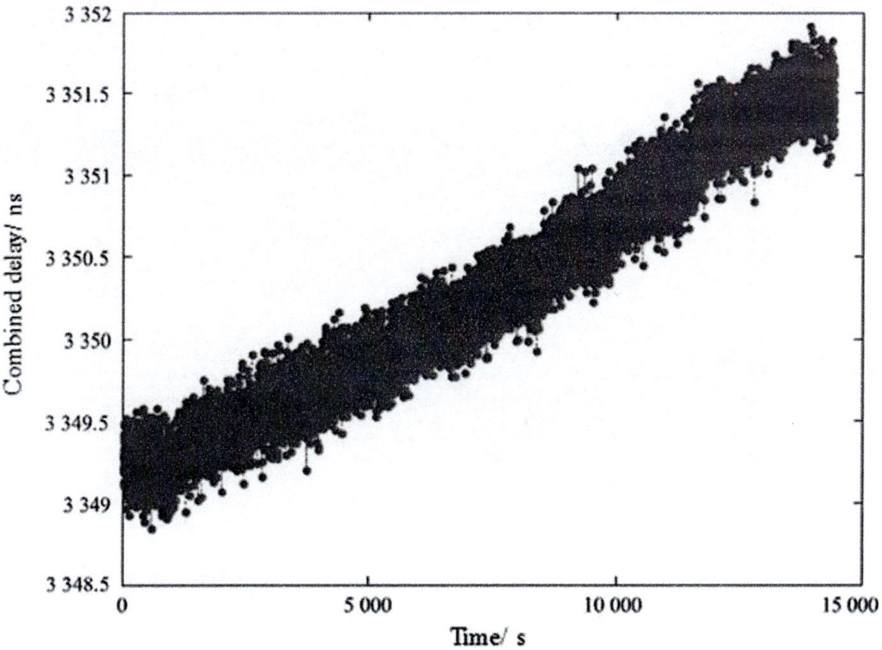


**Fig. 3.22** Combined delay measured before correction

device for temperature measurement, and measure the corresponding changes in time delay. Identify the relationship between the two measurements to establish a delay-temperature curve. In the in-orbit operation of the onboard receiver, correct the system error according to the measured temperature and the relationship curve.

Fig. 3.22 shows the change curve of measured combined delay (the measured satellite-ground pseudorange value minus the corresponding satellite-ground clock difference) in three temperature cycles (from −25 to +60 °C). It can be seen that, in the temperature cycle test, the peak of the combined delay variation is about 3 ns, which indicates that the system error is large. The curve of combined delay measured after the pseudorange measurement correction is shown in Fig. 3.23. It can be seen that, after the correction, the peak of combined delay variation is significantly reduced and the value of combined delay does not change substantially with temperature. This indicates that the system error has been successfully corrected.

### *3.4.5 Doppler Measurement*

Doppler shift, defined as the difference between signal reception frequency and signal transmission frequency, is caused by the relative motion between receiver and signal source. This phenomenon is called the Doppler Effect. If $f_d$ is Doppler shift,

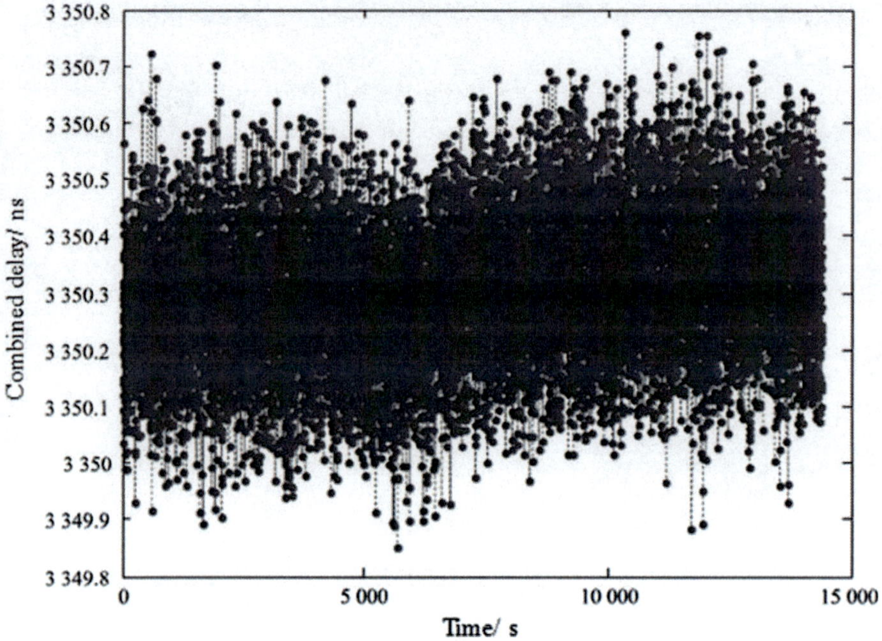


**Fig. 3.23** Combined delay measured after correction

$f_r$ is signal reception frequency, and $f$ is signal transmission frequency, then their relationship will be

$$f_{\mathrm{d}} = f_r - f \tag{3.14}$$

The integral Doppler $d\phi_k$ is the integral of Doppler shift $f_{\mathrm{d}}$ to time, as shown in Eq. (3.15)

$$d\phi_k = -\int_{t_0}^{t_k} f_{\mathrm{d}}(t)\mathrm{d}t \tag{3.15}$$

where $d\phi_k$ represents the integral Doppler measurement output by the receiver at the epoch $k$.

The actual integral Doppler measurements contain both random errors and system errors. The measurement of integral Doppler is mainly based on the phase-locked loop, so its accuracy (random error) is related to the measurement accuracy of the phase-locked loop that is mainly caused by phase jitter and dynamic stress. The sources of phase jitter are mainly thermal noise and the oscillation frequency jitter caused by mechanical vibration and Allan variance.

With the given RF carrier frequency, receiver Allan variance, and input carrier-to-noise ratio, the receiver can reduce the phase measurement error of phase-locked loop by selecting the appropriate loop bandwidth and coherent integration time, while reducing the error of dynamic stress by selecting reasonable loop parameters. The combination of the above two methods can reduce random error in the integral Doppler measurement and improve the measurement accuracy. Generally, there are two kinds of errors introduced to the integral Doppler measurement: the error in subsequent digital processing caused by the inexact frequency of receiver processing clock, and the truncation error caused by the rounding of frequency control word in digital processing. Next, the two errors will be described separately.

The error caused by the inexact frequency of receiver processing clock in the subsequent digital processing is related to the clock frequency accuracy. The worse the frequency accuracy, the greater the system error introduced. Because the processing clock is aging and the clock frequency accuracy may vary greatly with temperature, the system error is sure to exist and shall be corrected by the corresponding measures. Such error is called system error 1 for short. Currently, the method commonly used to reduce this error is to measure the real-time deviation of actual receiver clock frequency from nominal frequency and then correct the measured value in real time through the corresponding conversion.

The truncation error caused by the rounding of frequency control word in digital processing is related to the frequency resolution. Compared with system error 1, this kind of error is relatively small, but may have a cumulative effect in the long run. Therefore, it needs to be corrected in the integral Doppler measurement. This system error is simply referred to as system error 2. At present, the fixed error caused by the rounding of the fixed frequency control word is directly ignored. The system error 1 and the system error 2 are relatively independent but superimposed. The existence of the two system errors and their influence is manifested by the fact that the integral Doppler measurement does not match actual test result, and the range value fitted according to the integral Doppler measurement does not match the measured value.

Fig. 3.24 shows the integral Doppler measurements obtained without any correction under theoretical zero-Doppler test condition (the average of integral Doppler measurements shall be close to zero under this condition). The integral Doppler measurements obtained after correcting the system errors 1 and 2 are shown in Fig. 3.25.

### 3.4.6 Monitoring of Null-Value Self-closed Loop

The satellite receiver can measure the reception delay of the uplink signal by monitoring the null-value self-closed loop. The specific implementation steps are as follows: The digital processing terminal of the satellite receiver generates a null-value baseband signal, which is then up-converted to the RF frequency point of the uplink signal. The up-converted null-value signal is coupled with the uplink signal and sent to the RF front-end of the receiver for down-conversion processing. Then the analog intermediate-frequency signal is output by RF channel to digital processing

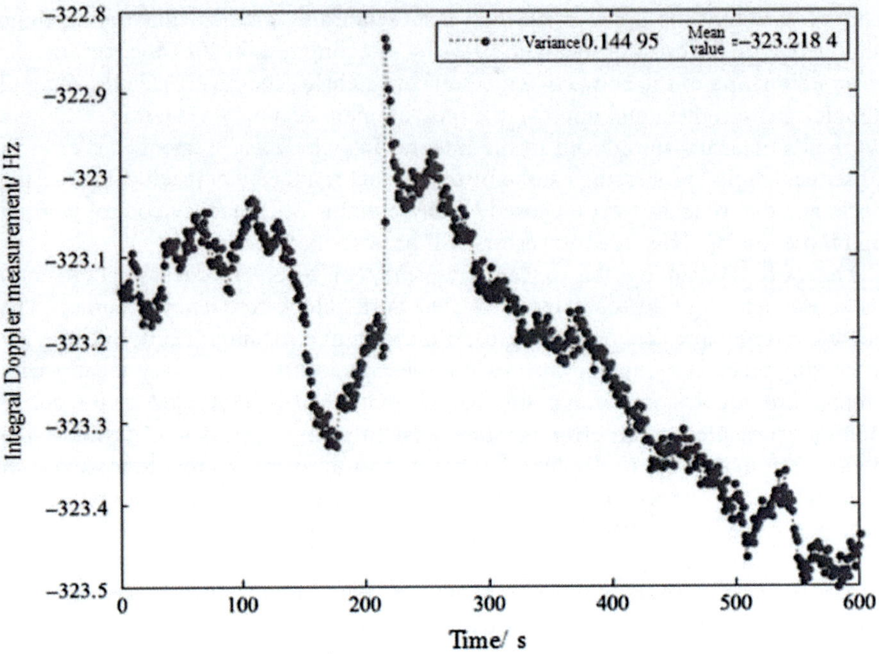


**Fig. 3.24** Integral doppler measurement under zero-doppler test condition, with system errors 1 and 2 not corrected

terminal for null-value signal acquisition and tracking. Finally, the reception delay is measured. The process flow is shown in Fig. 3.26.

In addition to measuring the reception delay, this monitoring process can also monitor the working state of the satellite receiver. If the tracking and reception of the null-value is in good condition, the RF channel and digital processing terminal in the receiver will function normally; otherwise, they will function abnormally in some loops. In such condition, it is necessary for the ground control station to identify and control the state of the satellite receiver.

## 3.5 Anti-interference Design of BDS Satellite Receiver

Since the uplink signal shares the same frequency band (*L*-band) with radar, aeronautical radio and tactical military communication link, other radio signals may fall into the *L*-band to cause unintentional interference with the uplink signal. The CDMA technique adopted by the uplink signal enables the spaceborne receiver to withstand the interfering signals that are several tens of dB stronger than the signal itself. However, the signal transmitted from the ground station to the receiver is so weak that even a small interference signal can have a huge impact on the received signal.

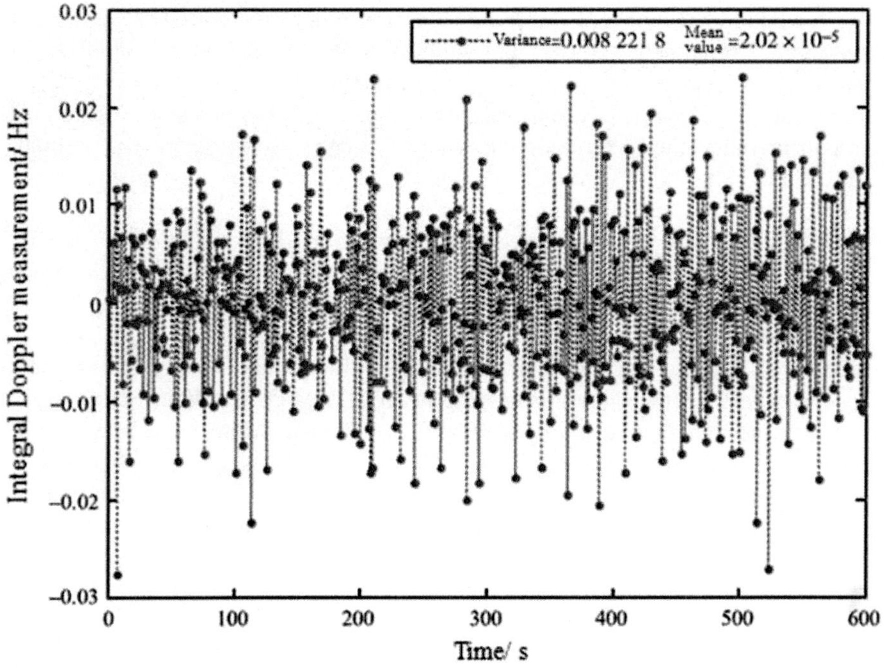


**Fig. 3.25** Integral doppler measurement under zero-doppler test condition, with system errors 1 and 2 corrected

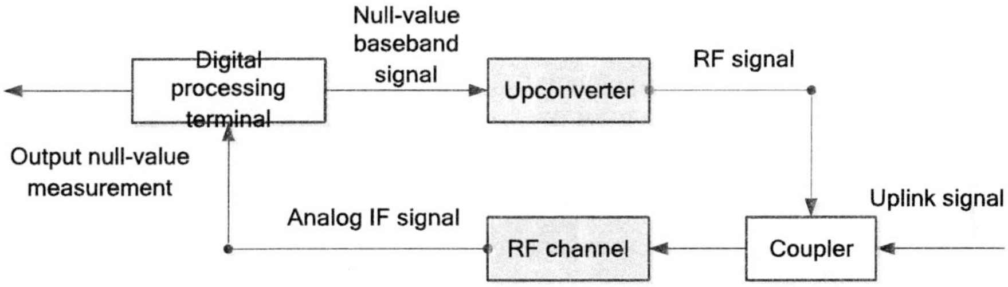


**Fig. 3.26** Monitoring flow of the null-value self-closed loop

This interference mainly affects the normal operation of the LNA, AGC, and AD chips at the RF front-end. It also affects the acquisition and tracking operations in baseband signal processing and ultimately the information transmission sensitivity and measurement accuracy. In serious case, it may even cause the malfunction of the uplink signal.

According to the signal type, interference can be classified into the following categories:

- Narrowband interference: the interference by a signal band narrower than that of the uplink signal. The typical band width of a narrowband interfering signal is 10–20% of that of the uplink signal. Narrowband interference is a relative concept. For the uplink signal with a bilateral bandwidth of 20 MHz, this interfering signal is considered a narrowband signal; whereas for the uplink signal with a bilateral bandwidth of 2 MHz, this interfering signal is considered a wideband signal. Common narrowband interference includes single-frequency interference (continuous waves in cosine form), sweep interference (continuous cosine waves with frequency fluctuations within a certain bandwidth), and narrowband band-limited interference (narrowband white noise interference, narrowband communication signals, etc.).
- Wideband interference: the interference by a signal band wider than or equal to that of the uplink signal, usually including UWB (Ultra-Wide Band) signal interference, matched spectral interference, wideband white noise interference etc.
- Pulse interference: the interference that has a short duration but huge energy, mainly from radar or radio. The common types of pulse interference are linear frequency modulation pulse interference and phase modulation pulse interference.

### *3.5.1 Analysis of Interference Effects on Uplink Signals*

The influence of interfering signals on the satellite receiver includes the influence on RF front-end, and the influence on signal acquisition and tracking. The main reason for the first influence is that the interference power is often much higher than normal signal power and thus can easily cause the saturation of RF front-end. The second influence is mainly the equivalent C/N0 loss caused by such interference.

1. Analysis of interference effects on RF front-end

The RF front-end of uplink signal receiver must be able to withstand the expected maximum RF interference level to ensure that no front-end gain is compressed by the expected interference, thus preventing the front-end gain from entering a saturated state. The RF interference mainly affects the Auto-Gain Control (AGC) and AD devices at the RF front-end. AGC is designed to ensure that the power of all the received signals (signal, noise, interference) remains substantially stable before the AD conversion, while the gain does not get saturated under high-power interference.

Continuous gain saturation may produce the following nonlinear effects on the input signal: distorting the input signal, generating new frequency components, reducing the equivalent carrier-to-noise ratio of the received signal, changing the *S*-Curve Bias (SCB) curve, and affecting the demodulation performance and measurement performance. Although the effects on data transmission and measurement performance can be evaluated through modeling and simulation, the AGC design must ensure that the gain will not enter the saturated state under the interference power, that is, the AGC must have a gain control range that is wide enough.

In the despreading process of spread spectrum signal, the interference can also be spread to reduce the amplitude of its power spectral density. Since the despreading is later than the ADC processing, strong continuous interference may increase the sensitivity loss of the ADC device operating at its maximum amplitude. For a 1-bit ADC with only Gaussian white noise, the loss caused by quantization noise is 3.50 dB. However, in the case of single-frequency interference, the loss to a 1-bit ADC is up to 6 dB. For the uniformly quantized ADC devices, only when they are higher-order (such as 12-bit or 10-bit) can the quantization loss caused by interference be suppressed. The ADC devices that are not uniformly quantized tend to be more advantageous than those uniformly quantized in dealing with strong interference.

2. Analysis of interference effects on signal acquisition and tracking

The effect of interference on acquisition and tracking depends on its effect on the SNR output by correlator. In the despreading process, the interference is also despreaded into noise to raise the signal noise level.

If the power of the received signal is denoted as $C/N_0$, the interference power as $J$, and the interference-to-signal ratio as $J/S$, the interference will cause equivalent $C/N_0$ lower. Narrowband interference,wideband interference,matched spectral interference have the different influence on the signal quality. It should be pointed out that equivalent $C/N_0$ is used to evaluate the impact of interference on signal acquisition and tracking from the perspective of SNR, but the signal acquisition is also affected by the cross-correlation suppression characteristic of pseudo-codes. Therefore, the effects of multiple access interference on signal synchronization are mainly the reduction of equivalent $C/N_0$ and the incorrect locking of code phase caused by cross-correlated side lobes.

### *3.5.2 Anti-interference Design of Spaceborne Receiver*

1. Anti-interference design of RF channel

The main functions of the RF channel are frequency conversion and power control. The AGC in the channel can enhance the anti-interference ability of the system to prevent the amplifier and AD from getting saturated, but the match between control response time and terminal shall be taken into consideration in the design.

Affected by various factors such as signal power, signal transmission path length, and external environment, the uplink RF receiving channel will enable the power of the input signal to change within a certain range. The input signal power often changes greatly by a margin up to several tens of dB, especially when it is interfered by electromagnetic signals in other frequency bands. In order to ensure that the digital processing terminal of the receiver works normally in spite of the large dynamic range of input signals, the receiver must have stable output power. Therefore, the uplink receiver must have the AGC function to increase the RF channel gain when receiving a weak signal and to decrease the gain when receiving a strong signal, thus keeping the output signal of the channel constant and avoiding abnormal operation with the presence of weak signal input or the signal distortion caused by receiver channel saturation with the presence of strong signal input.

In order to adapt the uplink receiver to a more complex electromagnetic interference environment, ensure normal signal reception in an electronic countermeasure environment and effectively suppress the interference, the design of the uplink receiver shall meet the following requirements:

(1) The receiver shall have the AGC function and the response time of AGC shall be reasonable;
(2) The uplink receiver shall have better linear characteristics to reduce the signal distortion under strong interference;
(3) The RF channel shall not cause the significant degradation of signal-to-noise ratio during the gain adjustment.

In order to meet the requirements of high gain and large dynamic range, two frequency converters ($LO_1$ and $LO_2$) and a two-stage AGC in serial connection can be used in the uplink receiver. To obtain better linear characteristics without significantly degrading the signal-to-noise ratio under strong interference, a typical RF channel with this function is designed for the uplink receiver. The use of an amplitude limiter circuit at the appropriate position of the receiver can also prevent the receiver saturation. The block diagram of such an RF channel is shown in Fig. 3.27.

In the typical AGC circuit, the response time can be changed by adjusting some parameters in the circuit. The AGC with fast response can suppress the pulse interference signals with different periods and duty cycles and improve the anti-pulse interference capability of the RF channel. However, the rapidly changing gain in the channel will render the operation of RF channel unstable, ultimately affecting the data processing function of the terminal. Therefore, the response time of the AGC circuit in the RF channel must be able to adapt to a large dynamic signal input level and suppress continuous wave (or pulse) interference with a long period and a large duty ratio.

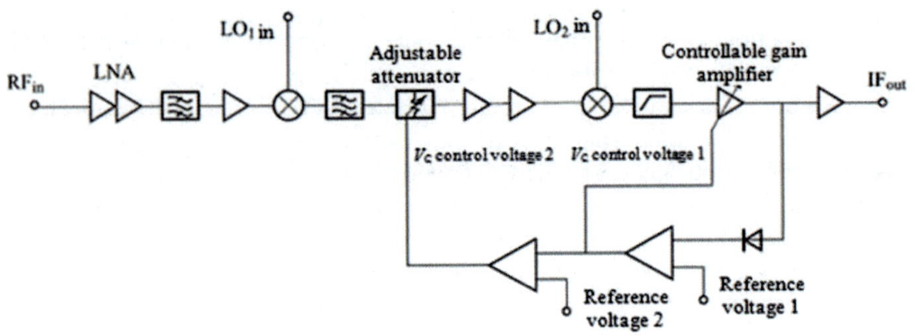


Fig. 3.27 Typical RF channel design

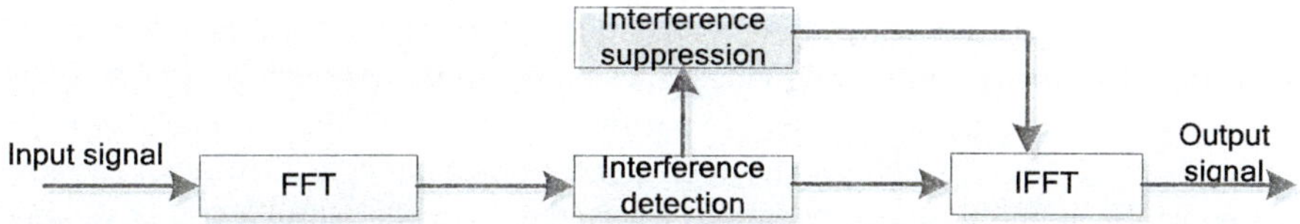


**Fig. 3.28** Schematic diagram of interference frequency domain detection and rejection technique

The AGC circuit in the channel cannot respond to a strong pulsed interference signal with a short period and duty cycle. This poses a threat to the AD sampling device in the signal processing unit. To solve this problem, an appropriate limiter is provided at the output end of the uplink receiver to confine the large output signal level to a certain range under narrow-pulse interference, thus ensuring that the output signal level can produce a safe range of input voltage for the AD sampling device in the processing terminal.

2. Narrowband interference suppression design

The narrowband interference suppression techniques for uplink signals mainly include interference frequency domain detection and rejection technique and linear adaptive interference suppression technique. Essentially, both techniques use the same notch filter to suppress the interference but in different ways.

(1) Interference frequency domain detection and rejection

Like white noise, the uplink signal exhibits a flat characteristic in the frequency domain, while the narrowband interference signal has concentrated energy and shows obvious spectral peaks in the spectrum. The basic principle of interference frequency domain detection and rejection is to separate narrowband interference signal from the uplink signal by using their significant difference in the frequency domain. The principle of this technique is shown in Fig. 3.28.

First, subject the input baseband signal (signal + noise + interference) to Fast Fourier transform (FFT) at $N$ points, where a time domain signal is converted into a frequency domain signal. Second, detect interference in the uplink signal in the frequency domain. Since both the uplink signal and the background white noise have flat spectra, while the narrowband interference signal is concentrated in a few high-amplitude spectral lines, a frequency domain spectrum threshold can be set in the interference detection module to distinguish the useful signal from the interference signal. When a signal line after the FFT processing is above this threshold, its corresponding signal shall be regarded as an interference signal; otherwise, the corresponding signal shall be judged as a useful navigation signal. Third, suppress the detected interference signal. Directly zero or attenuate the spectral lines to a level equivalent to background noise so as to suppress or weaken the frequency domain of narrowband interference. Finally, convert the frequency domain signal into a time domain signal through inverse fast Fourier transform (IFFT) after the interference suppression.

The interference frequency domain detection and rejection technique has many advantages. Compared with the complex filtering function designed for time domain processing, this technique only needs to perform the time-frequency domain conversion and set a threshold to achieve the interference suppression, in the form of zeroing or attenuation, in the frequency domain, where the algorithm complexity is significantly reduced. Besides, ideal filtering techniques (such as a rectangular filter) that cannot be realized in the time domain can be implemented in the frequency domain. In addition, the interference suppression in the frequency domain is able to quickly respond to fast-changing interference due to the elimination of convergence and is insensitive to interference models. As the interference frequency domain detection and rejection technique based on FFT has good performance in the suppression of multi-tone interference and narrowband interference, it has been widely used in engineering practice.

(2) Linear adaptive interference suppression technique

The linear adaptive interference suppression technique is mainly based on the fact that both the uplink signal and the Gaussian white noise are wideband signals with a weak correlation between adjacent data sample, so the spread spectrum signal component and Gaussian white noise component cannot be accurately derived from the correlation between adjacent data samples through an algorithm. Compared with the spread spectrum signal and Gaussian white noise, the narrowband interference is a narrowband signal with a strong correlation between adjacent data samples, so its component in the current data sample can be accurately estimated by using the neighboring data samples. The interference component in the received signal at the receiving end is first estimated and then canceled from the input signal, thus achieving the goal of suppressing the narrowband interference in the spread spectrum signal. The schematic diagram of this technique is shown in Fig. 3.29.

After passing through the adaptive filter, the input signal with a delay becomes an estimated interference signal. Then the estimated interference signal is subtracted from the input signal to obtain an error signal—the signal expected after anti-interference processing.

Adaptive filtering can be implemented in many ways, for example, by widely using the lattice filter and transversal filter in practical engineering. The transversal filter can be used as a predictive filter to mainly estimate the interference component in the current signal based on previous signal value, or as an interpolative filter to estimate the interference component in the current signal based on both previous and

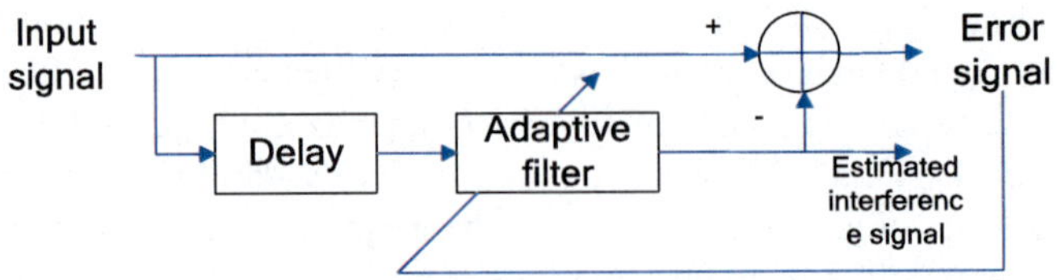


**Fig. 3.29** Schematic diagram of linear adaptive interference suppression technique

subsequent signal values. The structures of the two filters are shown in Figs. 3.30 and 3.31.

The main parameters of the adaptive filter are the number of filter taps and the update algorithm of the filter tap weight coefficient. The update algorithms of the filter tap weight coefficient in common use include Least Mean Square (LMS) algorithm and Recursive Least Squares (RLS) algorithm. The more widely used LMS algorithm is used here, as shown in Eq. (3.16). In this equation, $\mu$ is the convergence factor, $e(n)$ is the error signal, $x(n)$ is the input signal vector, and $w(n)$ is the tap weight coefficient vector.

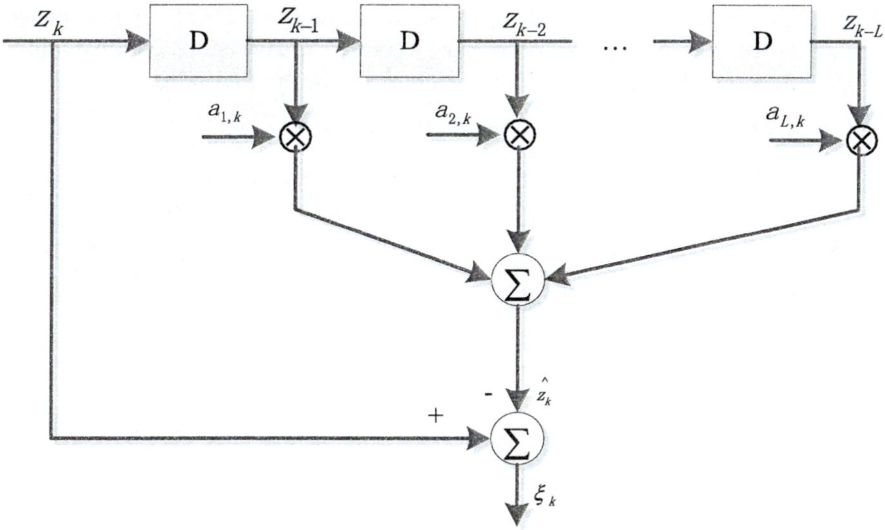


**Fig. 3.30** Structural diagram of linear adaptive predictive filter

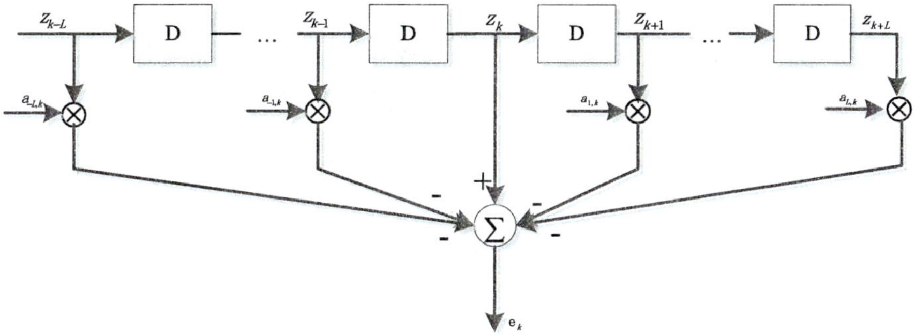


**Fig. 3.31** Structural diagram of linear adaptive interpolative filter

$$w(n+1) = w(n) + 2\mu e(n)x(n) \tag{3.16}$$

In Adaptive filtering, an optimization algorithm is used to automatically adjust the tap weight coefficient of the filter based on the time-varying characteristic of the signal. The Least Mean Square (LMS) is a simple and practicable algorithm that is widely used to make the weight coefficient converge to the minimum value of mean square error. However, in most of the adaptive algorithms, multiple iterations are required to reach the convergence point, and the convergence speed is slow. Therefore, the key to solve this problem is to increase the processing speed. Some people have adopted the self-orthogonal transform domain LMS algorithm, which can significantly reduce the number of iterations and greatly improve the ability of the adaptive filter to respond to fast-changing interference. But due to the rapid change of interference, it is still necessary to make a trade-off between the locking ability and the offset (i.e., the deviation of the weight from the optimal value).

3. Design of impulse interference suppression

Impulse interference is a kind of wideband interference. If not suppressed by applicable measures, it will have a serious destructive effect on the uplink measurement and communication. Compared with a useful signal, the energy of impulse interference is much greater than the signal strength over a discrete period of time, causing the AD device to get saturated for a long time and eventually destroying it (if no measures are taken at the front-end of the receiver). At the same time, the impulse interference makes it impossible for the receiver to extract the useful signal, thereby resulting in the failure of measurement and communication. Considering all these factors, it is necessary to add an anti-interference module before and after the AD device of the receiver to suppress the impulse interference before performing the signal reception and processing.

The suppression of impulse interference is the joint effort of AGC, limiter, and digital signal processing unit, the first two of which have been introduced in the previous sections. The technique of digital signal processing is mainly composed of efficient channel coding and stable loop tracking.

(1) Efficient channel coding: Since the bit error rate is significantly increased with the presence of impulse interference, efficient channel coding can improve the capability of random symbol error suppression by suppressing continuous symbol errors.
(2) Stable loop tracking: the equivalent $C/N_0$ under the impulse interference is very low. This greatly affects the loop tracking, degrades the tracking capability, and even causes the loss of lock. Therefore, the loop disturbance caused by this interference must be taken into consideration in the design of loop tracking. The impulse interference must be identified by certain detection means, and the loop tracking state shall be adaptively adjusted to ensure the stability of loop tracking.

# Chapter 4
# Satellite Navigation Broadcasting Link Technology

In the satellite navigation, the broadcasting link is a physical channel for navigation satellite to broadcast the navigation signals to the user so as to realize the positioning and timing.
Satellite navigation is to determine the user's position with the radio positioning method by receiving radio signal from a navigation satellite to measure the distance between the user and the navigation satellite in the principle of Time of Arrival (ToA) ranging. Except for Chinese BDS-1 that implements active positioning, all the other satellite navigation systems adopt the method of passive positioning, in which the user only passively receives the signal broadcasted by the satellite. In a satellite navigation system, the broadcasting link is the only interface between the satellite and the user. It determines the ranging accuracy and even the performance of the whole navigation system. Therefore, the broadcasting link technology is one of the key technologies in the design of a satellite navigation system. This technology contains a series of techniques related to the link from satellite to user. Among them, the techniques of signal structure design and signal generation and broadcasting are essential.

## 4.1 Overview of Satellite Navigation Broadcasting Link

In order to meet the needs of satellite-based PNT, the broadcasting link technology contains many techniques at four levels, as shown in Fig. 4.1. At the RF level are the RF characteristics of the navigation signal, including frequency band selection, frequency band design, signal bandwidth, and transmission power. The techniques at the baseband level include pseudocode design, modulation design, and constant envelope multiplexing design as well as compatibility and interoperability design. The techniques at the message level focus on the generation of navigation data

 
J. Xie et al., *Satellite Navigation Systems and Technologies*, Space Science and Technologies, https://doi.org/10.1007/978-981-15-4863-5_4

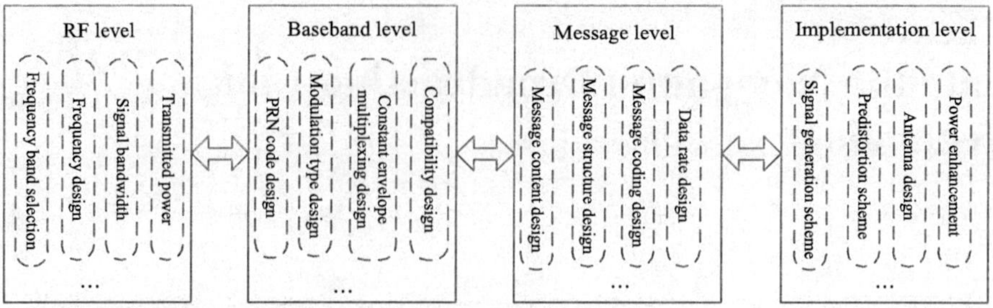


**Fig. 4.1** Broadcasting technology of navigation satellite system

messages, including message content design, message layout structure, message coding scheme, and message rate selection. All the techniques at these three levels belong to the broadcasting technology of navigation satellite system.

The main goal of the traditional communication system is to transmit data messages while considering whether the messages can be correctly demodulated. In the satellite navigation system, the ranging accuracy is a greater concern, and is mainly determined by modulation. This means that the baseband level is the core part of navigation signal design.

The techniques at the implementation level include the signal generation and broadcasting techniques. As the frequency bands for signal broadcasting and the service signal components increase, modern GNSS signals have become more and more complex, thus posing new challenges to the engineering implementation of spaceborne payload broadcasting and user receiver processing. In the design process, the signal implementability needs to be considered. Moreover, the payload baseband and RF channel are generally under unfavorable conditions, resulting in different degrees of signal distortion. To reduce the influences of these nonideal characteristics on the quality of navigation signal, the navigation signal should be designed with certain channel adaptability. Meanwhile, a predistortion technique should be adopted to improve the signal quality.

As the existing GNSS and RNSS systems in the world have to meet the requirements of compatibility and interoperability, the design of broadcasting link is restrained by frequency coordination, smooth transition, intellectual property, and engineering implementation, in addition to technical factors. With the increasing modernization of navigation signals, the Navigation Satellite Systems (NSSs) such as GPS, Galileo, GLNASS, and BDS have experienced or are experiencing the modernization of signal design. As far as the BDS is concerned, the system has developed a number of new technologies and methods in its evolution process from the determination of the BDS-3 signal baseline in 2009 to the adoption of the current signal system.

### 4.1.1 GPS Signal Structure

The spectrum of GPS navigation signals is shown in Fig. 4.2. The early GPS signals were broadcast in only two frequency bands, namely, the L1 frequency band (1575.42 MHz) for both C/A and P(Y) code signals, and the L2 frequency band (1227.6 MHz) for P(Y) code signal only. In the early GPS satellites such as Block I, Block II, Block IIA, and Block IIR, the BPSK (1) and BPSK (10) were used in the C/A and P(Y) code signals, respectively. With the implementation of the anti-spoofing (AS) policy, the P code was replaced by Y code. The Y code is obtained by encrypting the P code with the W code, whose code rate is lower than that of P code. Therefore, it is usually recorded as P(Y).

The modernization of GPS signals started from the modernization of GPS military signals. As a result, the modern military signals L1M and L2M were initially designed in the L1 and L2 frequency bands, to which the civil signals L1C and L2C were added later, respectively. In addition, a new L5 (1176.45 MHz) frequency was introduced to broadcast the civil signal L5C. Initially, the signals L1M, L2M, and L2C were broadcast by the satellite Block IIM, and the L5C signal by Block IIF. The L1C signal is broadcast by the next generation of GPS satellite "Block III".

The main parameters of the GPS signal system after the modernization are shown in Table 4.1 [13–15].

The design of GPS signals also reflects the US's resolution to strengthen the dominant position of GPS in the civil satellite navigation market. The newly added three civil signals have their own characteristics and can basically meet the performance requirements in all aspects. The emerging L2C is the first signal to meet the demand of the dual-frequency civil market. On the one hand, it employs a low code rate and time-division technique to meet the compatibility requirement; on the other hand, it caters to the low-end electronic consumer market by significantly reducing the cost and power consumption. Moreover, in this signal, the pilot components are

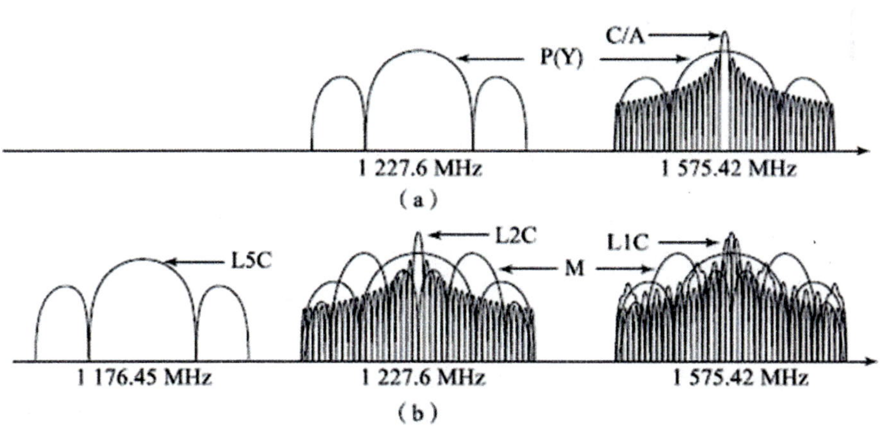


**Fig. 4.2** Spectrum of GPS navigation signal

**Table 4.1** Characteristics of GPS signal

| Signal | Center frequency/MHz | Modulation | Primary code family | Primary code length | Secondary code length | Code rate/(Mc s) | Message/symbol rate/(b $s^{-1}$/s $s^{-1}$) | FEC |
|---|---|---|---|---|---|---|---|---|
| L1C/A | 1575.42 | BPSK (1) | Gold | 1023 | N/A | 1.023 | 50/50 | No |
| L1P(Y) | | BPSK (10) | Overlay code | $6{,}187{,}104 \times 10^6$ | N/A | 10.23 | 50/50 | No |
| L1M | | $BOC_{sin}$ (10,5) | – | – | – | 5.115 | – | – |
| L1C | | TMBOC (6,1,1/11) | Weil | 10,230 | N/A | 1.023 | 50/100 | LDPC |
| | | | Weil | 10,230 | 1800 | 1.023 | N/A | No |
| L2P(Y) | 1227.6 | BPSK (10) | Overlay code | $6187104 \times 10^6$ | N/A | 10.23 | 50/50 | No |
| L2C | | TDDM + BPSK (1) | Truncated $_m$ | 10,230 | N/A | 1.023 (CM/CL) | 25/50 | Convolutional encoding |
| | | | Truncated $_m$ | 767,250 | N/A | | N/A | No |
| L2M | | $BOC_{sin}$ (10,5) | – | – | – | 5.115 | – | – |
| L5C | 1176.45 | QPSK (10) | Overlay code | 10,230 | 10 | 10.23 | 50/100 | Convolutional encoding |
| | | | Overlay code | 10,230 | 20 | 10.23 | N/A | No |

divided, convolutional coding is introduced, the code length is increased, and the reliability is enhanced. In the L5C signal, the modulation of a larger bandwidth BPSK (10) is introduced to improve the spectrum occupancy of GPS. The L5 signal in a protected frequency band targets the life-safety-related services (such as aerospace applications), while meeting the needs of tri-band civil use. TMBOC modulation is employed by L1C to cater to both mass and high-end applications by achieving the interoperability with Galileo and BDS on the one hand and maintaining full compatibility with L1C/A signals and ensuring the exclusiveness of L1C/A signal spectrum on the other hand.

### *4.1.2 Galileo Signal Structure*

The design of Galileo signals is relatively transparent, as the progress of each stage has been released in time. In March 2001, the European Commission established the Galileo Signal Design Task Force, which was responsible for the design of Galileo signals. In the same year, the team made preliminary planning for Galileo signals. In 2002, based on the signal planning in 2001, changes were made on Galileo signals, mainly on the E5 signal.

By 2006, the Galileo signals were basically determined. The Multiplexed Binary Offset Carrier (MBOC) was used by both E1 OS and L1C signals. In 2007, a joint statement was issued for confirmation. In 2006 and 2008, the drafts of Interface Control Document (ICD) for Galileo were issued separately. In February 2010, the Galileo OS SIS ICD (Version 1.0) was officially released [7].

The latest states of open signals in the Galileo system, which were defined in the Galileo OS SIS ICD (Version 1.3) released in 2016 [8], are shown in Table 4.2.

The modernization of Galileo signal design is basically synchronized with GPS modernization, and is unique in some aspects. For example, MBOC is used in interoperable signals. The constant envelope Alternate Binary Offset Carrier (AltBOC) modulation is used in E5 signals so that those signals can interoperate with GPS L5 and BDS B2 signals and provide better performances. Random code is used as a spreading code, and a criterion for spreading code optimization is established. A Minimum Shift Keying (MSK) signal and a Gaussian Minimum Shift Keying (GMSK) signal are introduced to the C frequency band to satisfy its strict compatibility requirement. Many studies of Galileo signal design can provide a reference for the design of BDS signals.

### *4.1.3 GLONASS Signal Structure*

The GLONASS and GPS were developed and put into use in the same decade. Early GLONASS military signals were broadcast in the L1 and L2 frequency bands, while civil signals only in the L1 band. Compared with GPS signals, the code rates of

**Table 4.2** Characteristics of Galileo signals

| Signal | Center frequency/MHz | Modulation type | Primary code family | Primary code length | Secondary code length | Code rate/(Mc $s^{-1}$) | Data rate/(s $s^{-1}$) | FEC |
|---|---|---|---|---|---|---|---|---|
| E5a_I | 1176.45 | AltBOC (15,10) 1191.795 MHz | Gold code | 10,230 | 20 | 10.23 | 50 | Convolutional encoding |
| E5a_Q | | | Gold code | 10,230 | 100 | 10.23 | N/A | N/A |
| E5b_I | 1207.14 | | Gold code | 10,230 | 4 | 10.23 | 250 | Convolutional encoding |
| E5b_Q | | | Gold code | 10,230 | 100 | 10.23 | N/A | N/A |
| E6 CS Data | 1278.75 | BPSK (5) | Memory code | 5115 | N/A | 5.115 | 1000 | Convolutional encoding |
| E6 CS Pilot | | BPSK (5) | Memory code | 5115 | 100 | 5.115 | N/A | N/A |
| E6 PRS | | $BOC_{cos}$ (10,5) | – | – | – | 5.115 | – | – |
| E1 OS Data | 1575.42 | CBOC (6,1,1/11) | Random code | 4092 | N/A | 1.023 | 250 | Convolutional encoding |
| E1 OS Pilot | | | Random code | 4092 | 25 | 1.023 | N/A | N/A |
| E1 PRS | | $BOC_{cos}$ (15,2.5) | – | – | – | 2.5575 | – | – |

GLONASS signals are low—the code rate of open service signal is 0.511 MHz and that of authorized service signal is 5.11 MHz. In the GLONASS system, the multiple access mode is FDMA. The center frequency of the L1 frequency band is 1602 MHz $+ n \times 0.5625$ MHz, and that of the L2 band is 1246 MHz $+ n \times 0.4375$ MHz, where $n$ is the channel number of the satellite. The satellite channels $n = -7, \ldots 0, \ldots,$ 13 were later adjusted to $n = -7, -6, \ldots 0, \ldots, 6$ under the guidance of the ITU in 2005, thereby reducing interference with other services such as radio astronomy. Russia has also being modernizing the GLONASS signals since 2002. In 2003, the GLONASS-M satellite was launched to begin broadcasting the open service signals in the L2 band. The signal characteristics of GLONASS FMDA signals are shown in Table 4.3.

One of the main reasons for the adoption of FDMA multiple-access method in the GLONASS system is to improve the anti-jamming capability of the system. Because the center frequency band of each GLONASS satellite is different, the receiver in this system is more complicated than that in GPS. This is not conducive to the marketing of GLONASS.

With the advancement of GPS and Galileo in signal compatibility and interoperability, it has become more difficult for FDMA signals to interoperate with them. Therefore, in the modernization of GLONASS, the addition of CDMA signals was considered. On February 26, 2011, Russia launched the first GLONASS-K satellite, which broadcasted the first CDMA signal L3OC at the L3 frequency. Other CDMA signals of GLONASS are still being designed. The modernized GLONASS CDMA signals that are described in open literature are shown in Table 4.4 [9–11].

### *4.1.4 BDS Signal Structure*

The regional system BDS-2 has been established and put into service from December 27, 2012. The development of global system BDS-3, whose signal design was started in 2009, is scheduled to be fully completed around 2020. The signals of BDS-2 are shown in Table 4.5. The BDS-2 signals are the "hybrids" of the original GPS signal and modern GNSS signal. They adopt the relatively mature QPSK modulation technique. Among them, the B1I and B2I use shortcodes (close to C/A code), and their pilot components are not divided. However, 3 frequency bands are available for BDS-2 signals. The navigation messages of these signals are encoded in two modes: BCH channel coding and NH secondary coding [1, 2]. The latter mode is the characteristic of modern GNSS signals. Obviously, the regional system BDS-2 is a transitional system to undertake the task of system verification and to accumulate techniques and experience for the construction of global system BDS-3. Therefore, the signals it adopts are relatively simple and mature.

China had established a Signal Design Taskforce to specially design the signals of the BDS-3 system while constructing the BDS-2 system. For this taskforce, the main

**Table 4.3** Characteristics of GLONASS FMDA signals [17]

| Signal | Center frequency/MHz | Modulation type | Primary code family | Primary code length | Secondary code length | Code rate/(Mc $s^{-1}$) | Message rate/(b $s^{-1}$) |
|---|---|---|---|---|---|---|---|
| L1OF | $1602 + n \times 0.5625$ | BPSK | m sequence | 511 | N/A | 0.511 | 50 |
| L1SF | $1602 + n \times 0.5625$ | BPSK | Long code | 33,554,432 | N/A | 5.11 | 50 |
| L2OF | $1246 + n \times 0.4375$ | BPSK | m sequence | 511 | N/A | 0.511 | 50 |
| L2SF | $1246 + n \times 0.4375$ | BPSK | Long code | 33,554,432 | N/A | 5.11 | 50 |

**Table 4.4** GLONASS CDMA signals

| Signal | Center frequency/MHz | Modulation type | Service type |
|---|---|---|---|
| L1OC | 1600.995 | BOC (1,1) | Open |
| | | BPSK (1) | |
| L1SC | 1600.995 | BOC (5,2.5) | Authorized |
| L1OCM | 1575.42 | BOC (1,1) | – |
| L2OC | 1248.06 | BOC (1,1) | Open |
| | | BPSK (1) | |
| L2SC | 1248.06 | BOC (5,2.5) | Authorized |
| L3OC | 1202.025 | BPSK (10) | Open |
| | | BPSK (10) | |
| L5OCM | 1176.45 | BPSK (10) | – |
| | | BPSK (10) | |

considerations are the requirements of various civil and military users, the compatibility with other GNSS systems and the smooth transition from BDS-2 signals to BDS-3 signals.

In order to achieve better compatibility and interoperability with other GNSS systems, the center frequency band of the B1 signal has been adjusted from 1561.098 to 1575.42 MHz, as experienced by GPS L1 and Galileo E1. The civil signal B1C uses QMBOC to modulate its two components. At the same time, the B1I signal of the BDS-2 system will still be broadcasted to achieve a smooth transition. And the signals B1I and B1C as well as the authorized B1A signal are multiplexed. In the B2 signal band, a variety of signals have been designed and verified. The B2a signal realizes the interoperability with the L5 signal defined by ICAO. The B2b signal has successfully broadcasted the unique message of the BDS-3 system. Through the modulation technique similar to AltBOC, the B2a and B2b are combined for broadcasting. In the B3 signal band, the B3A signal begins to be broadcasted, while the existing B3 signal in the BDS-2 system is still broadcasted.

At the beginning of September 2017, the China Satellite Navigation System Management Office released the "Interface Control Document for BDS Spatial Signals - Public Service Signals B1C and B2a (Beta)". This is the first announcement on B1C and B2a signals in the BDS-3 system. The B1C signal is modulated by QMBOC, while the B2a signal by QPSK (10). In December 2017, the official versions of "Interface Control Document for BDS Spatial Signals - Public Service Signal B2a (Version 1.0)" and "Interface Control Document for BDS Spatial Signals - Public Service Signal B1C (version 1.0)" were released [3, 4]. On February 9, 2018, the official version of "Interface Control Document for BDS Spatial Signals - Public Service Signal B3I (Version 1.0)" was released, marking the official confirmation of BDS-3 signals.

The characteristics of BDS-3 signals are shown in Table 4.6.

**Table 4.5** Characteristics of BDS-2 signals

| Signal | Center frequency/MHz | Modulation type | Primary code family | Primary code length | Secondary coding | Code rate/(Mc $s^{-1}$) | Message rate/(b $s^{-1}$) | FEC |
|---|---|---|---|---|---|---|---|---|
| B1I | 1561.098 | QPSK (2) | Gold code | 2046 | 20 bit NH code | 2.046 | 50 | BCH (15,11,1) |
| B1Q | | | – | – | – | 2.046 | – | – |
| B2I | 1207.140 | QPSK | Gold code | 2046 | 20 bit NH code | 2.046 | 50 | BCH (15,11,1) |
| B2Q | | | – | – | – | 10.23 | – | – |
| B3I | 1268.52 | QPSK (10) | Gold code | 10,230 | – | 10.23 | – | – |
| B3Q | | | – | – | – | 10.23 | – | – |

**Table 4.6** Characteristics of BDS-3 signals

| Configuration | Signal component | Center frequency/MHz | Modulation type | Primary code family | Primary code length | Secondary code length | Code rate/(Mc $s^{-1}$) | Message/symbol rate/(b $s^{-1}$/s $s^{-1}$) | FEC | Minimum received power/dBW |
|---|---|---|---|---|---|---|---|---|---|---|
| B1 | B1I | 1561.098 | BPSK(2) | Gold code | 2046 | 20 bit NH code | 2.046 | 50/50 | BCH | −163 |
| | B1C_data | 1575.42 | BOC(1,1) | Weil code | 10,230 | 1 | 1.023 | 50/100 | BCH + LDPC | −159 |
| | B1C_pilot | | QMBOC (6,1,4/33) | Weil code | 10230 | 1800 bit | 1.023 | No | No | |
| | B1A | | – | – | – | – | 2.046 | – | – | – |
| B2 | B2a_data | 1176.45 | QPSK (10) | Gold code | 10,230 | 5 bit | 10.23 | 100/200 | LDPC | −156 |
| | B2a_pilot | | | Gold code | 10,230 | 100 bit | 10.23 | No | No | |
| | B2b_data | 1207.14 | QPSK (10) | Gold code | 10,230 | – | 10.23 | – | – | – |
| | B2b_pilot | | | Gold code | 10,230 | – | 10.23 | No | No | |
| B3 | B3A | 1268.52 | – | – | – | – | 10.23 | – | – | – |
| | B3I | | QPSK (10) | Gold code | 10,230 | 20 bit NH code | 10.23 | 50/50 | BCH | 163 |
| | B3Q | | | – | – | – | 10.23 | – | – | – |

## 4.2 Design of Navigation Signal Structure

Generally, the design of the navigation signal structure includes the design of band, code, encoding, modulation, and other aspects. All these aspects should be traded off carefully. This section will address and discuss these elements.

### *4.2.1 Frequency Band*

The selection of carrier frequency band will influence the propagation properties (atmospheric loss, ionospheric delay and scintillation, tropospheric delay, etc.), Doppler shift, out-of-band rejection, and channel non-ideality, thus requiring a comprehensive assessment. It is mainly limited by the propagation properties of electromagnetic waves. Due to the existence of the atmosphere, electromagnetic waves will be absorbed and get lost when passing through the atmosphere. The range of optimal transmission wavelengths is 5 cm–10 m, and the corresponding frequency band is 30 MHz–6 GHz. The GPS navigation signals L1 (about 19 cm) and L2 (about 24 cm) are just located in this band.

The selection of the carrier frequency band is also limited by ITU frequency allocation and planning. Although multiple bands are available for satellite navigation, only the L1 band is exclusive for satellite navigation service. The other two bands L2 and L5 are also used by other services. In addition, the frequency band of GLONASS is shared by low-orbit satellite mobile communications and radio astronomy services. To reduce the impact on these services, the GLONASS frequency has been revised once. Considering all the performance aspects, the L frequency band is the best choice for satellite navigation service. As the result of "first declaration, first use", the GPS and GLONASS have actually occupied 80% of the "golden navigation frequency band".

The existing carrier frequencies of various GNSS systems are mainly concentrated in the L frequency band. With the increase of navigation signals, the L frequency band has already become overcrowded, leading to the increased difficulty in frequency coordination and system compatibility. In the future, the additional frequency bands optional for satellite navigation systems will also include the S frequency band (2483.5–2500 MHz) and C frequency band (5010–5030 MHz).

1. **L frequency band**

The navigation signals of GPS and GLONASS systems are all located in the L frequency band. The frequencies in the GPS L1 and L2 bands are 1575.42 MHz and 1227.6 MHz, respectively. The main reason that the L frequency band is selected as a navigation signal band is because in this band, the signal propagation performance is high and the RF devices are mature. The effects of atmospheric propagation on radio signals include: free space loss, ionospheric group delay and scintillation, group delay and shift in troposphere and stratosphere, atmospheric absorption loss, signal attenuation in rain cloud, and so on. Free space loss is the main form of propagation loss. The free space loss of omnidirectional antenna is calculated by Eq. (4.1):

$$L[\mathrm{dB}] = 32.4 + 20\log f + 20\log d \tag{4.1}$$

where $f$ is the frequency in MHz and $d$ is the propagation distance in km.

It can be seen that when the propagation paths are the same, the higher frequency will cause greater free space loss. The radius of the Earth is 6,378.137 km and the height of the satellite orbit is 21,528 km. Under this condition, the free space loss of S frequency band is about 4 dB higher than that of L1 band, and the free space loss of C frequency band is 10 dB higher than that of L1 band.

The atmospheric attenuation is related to carrier frequency, elevation angle, altitude, and humidity. Generally, the higher the frequency is, the greater the attenuation will be. To be specific, the maximum attenuation of C band caused by water vapor and oxygen is 0.5 dB, slightly higher than that of the L frequency band. In a tropical heavy rain, the attenuation of the C frequency band at an elevation angle of 90° is up to 5 dB, while that of the L frequency band is less than 0.2 dB. In dense fog and cloud, the attenuation of C frequency band in the worst case is 0.9 dB, while that of the L frequency band is less than 0.1 dB.

The effects associated with ionosphere, including ionospheric delay, ionospheric scintillation, ionospheric Doppler shift, and other effects, are generally inversely proportional to the carrier frequency $f$ or $f^2$. Among them, ionospheric group delay is one of the main error sources. The ionospheric delay can be calculated by [26]:

$$\Delta S = \frac{10.3}{f^2} \cdot \mathrm{TEC}\ (\mathrm{m}) \tag{4.2}$$

where $f$ is the frequency in Hz, and TEC is the total number of electrons in the path direction (unit: $10^{17}/\mathrm{m}^2$). Figure 4.3 shows the ionospheric delays in the L, S, and C frequency bands relative to the total number of electrons.

Taken together, the L frequency band is the best navigation band, as the frequencies in this band are neither too low, nor too high. As a result, it can maintain a relatively small ionospheric delay, and small attenuation in the atmosphere (rain Decay). The RF devices in this band are also relatively inexpensive and mature. Therefore, the L frequency band is the priority band for satellite navigation systems.

2. **S frequency band**

The S frequency band (2483.5–2500 MHz) is assigned to RDSS by the ITU. Since RNSS is a subset of RDSS, this band is available for satellite navigation services. According to the ITU Radio Regulations (2010 edition), S frequency band is used for not only RDSS service, but also for satellite mobile services, fixed services, mobile services, and radio positioning services. If different satellite radio navigation services are provided in this frequency band, their compatibility with each other needs to be considered. In addition, consideration should be given to the interference of the transmitted signals in the 2483.5–2500 MHz frequency band to the radio astronomy

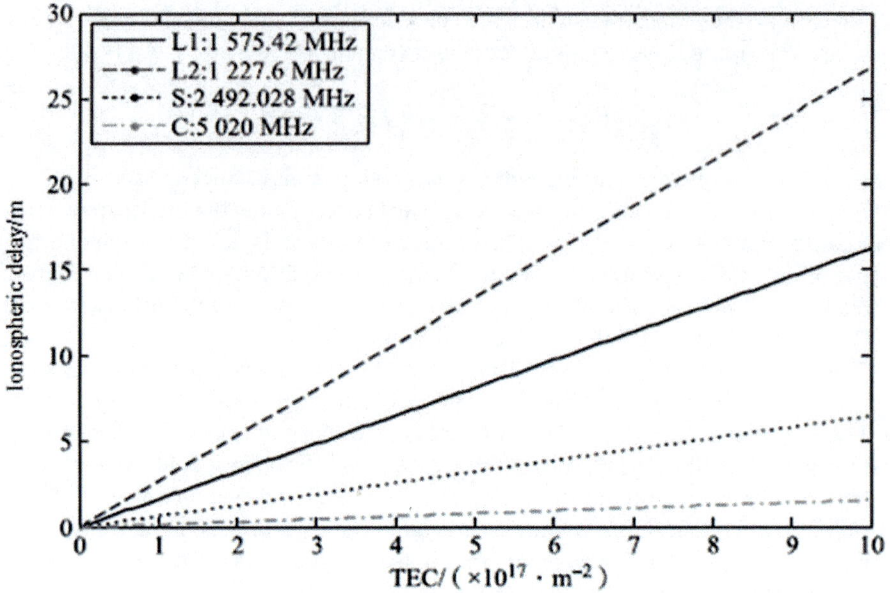


**Fig. 4.3** Ionospheric delay versus total number of electrons

service. In particular, the second harmonic of the transmitted signal may interfere with the radio astronomy service worldwide (in the 4990–5000 MHz frequency band).

The S frequency band has been applied in satellite navigation. The frequency of the BDS's GEO satellite carrier is 2491 MHz, and the frequency of the IRNSS's S band is 2492.028 MHz. Compared with the L frequency band, the biggest disadvantage of S frequency band is that its effective bandwidth is much narrower (only 16.5 MHz), its performance is less advantageous, and its free space propagation loss is greater. In addition, it faces the risk of interfering with the mobile service system and communication satellite system. However, the S frequency band also has some advantages. To be specific, this band has fewer navigation signals and simpler frequency coordination. In addition, it has a smaller ionospheric delay and higher accuracy than L frequency band in terms of dual-frequency correction.

3. **C frequency band**

In 2000, the C frequency band (5010–5030 MHz) was allocated to RNSS applications at the World Radiocommunication Conference. As shown in Fig. 4.4, radio astronomy services and microwave landing system services are in the vicinity of the C frequency band. They are extremely sensitive to interference.

In the microwave landing system, the power density radiated by satellite navigation signal to the Earth's surface in the 5030–5150 MHz frequency band shall not exceed −124 dB (W/m2) @ 150 kHz bandwidth. Meanwhile, the requirements for radio astronomy in this band are stricter. Therefore, the main limitation to the C frequency band for navigation use is the compatibility with neighboring services.

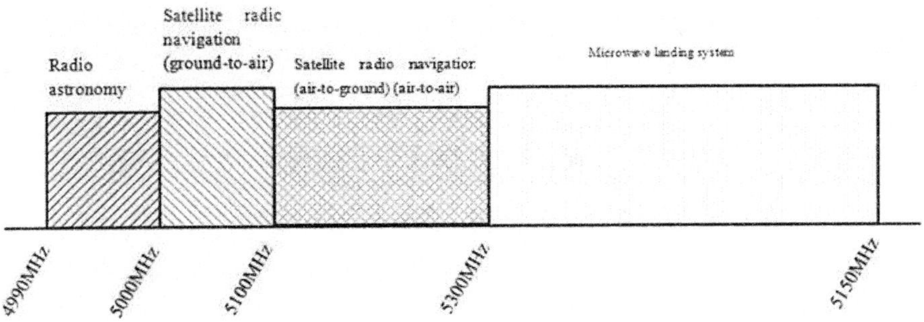


**Fig. 4.4** Services in C frequency band

The Working Group of Galileo Signal Design conducted extensive research on the C frequency band for satellite navigation and designed the navigation signal baseline in this band. Their analysis result considers the C frequency band as an ideal choice for the next generation of the satellite navigation system. The main advantages and disadvantages of the C frequency band for satellite navigation are as follows.

The advantages of the C frequency band includes: The ionosphere has a smaller impact on C frequency band. The ionospheric delay of the C frequency band is only about 1/10 of that of L frequency band. Dual-frequency correction using C frequency band and L frequency band has less noise. The carrier frequency of C frequency band is smaller—only 1/3 of that of L1 frequency band. The carrier smoothing performance of C frequency band is better in suppressing the code noise and multipath. The thermal noise produced by C frequency band has a smaller influence on carrier tracking, which is only 1/3 of that produced by L frequency band. The antenna size of C frequency band can be as small as 1/3 of that of L frequency band.

The disadvantages of C Frequency Band includes: The propagation loss of C frequency band is 10 times larger than that of L frequency band. The atmospheric loss (in clouds, rainfall, etc.) of C frequency band is larger. The penetration of C frequency band through a 12 cm concrete wall is poor, as the attenuation of 5 GHz signal is 9 dB more than that of 1.5 GHz signal. The carrier Doppler shift is larger and the required acquisition time is longer. Compared with L frequency band, the carrier tracking robustness of C frequency band is weaker, and the probability of its cycle slip is larger. The oscillator phase noise and dynamic stress are proportional to the carrier frequency of C frequency band, which is three times that of L frequency band.

### *4.2.2 PRN Code*

Random code sequence is an ideal sequence with an ideal correlation. However, it is a completely random nonperiodic sequence that cannot be reproduced, so it is impractical. In satellite navigation, a periodic repeatable sequence is adopted.

As its correlation characteristic is close to that of a random code sequence, it is called pseudo-random spreading code, or pseudocode for short. The pseudocode design determines the receiver's acquisition and tracking performance and thus is an important aspect of signal design demonstration.

1. **Indicators for PRN code evaluation**

To judge the quality of a set of binary pseudocode sequences, the most important indicators are their autocorrelation and cross-correlation properties. Typically, each code element of a binary pseudocode sequence is represented by the logic levels 1 and 0. The normalized correlation value of two sequences with equal length is defined as

$$R = \frac{N_A - N_D}{N_A + N_D} = \frac{N_A - N_D}{N} \tag{4.3}$$

where $N_A$ represents the number of the same code elements shared by the two sequences, $N_D$ represents the number of different code elements in the two sequences. In a navigation signal, the logic level of spreading code is usually mapped to the signal level. The common mapping relationship between them is shown in Table 4.7 [4]. In the following sections, the code elements will be expressed in the form of {1,−1}, unless otherwise stated.

For a periodic sequence $\boldsymbol{a}$ with the length of N, its normalized autocorrelation is defined as (without considering the Doppler shift) [24]:

$$\text{ACF}_{\text{even}}(\boldsymbol{a}; \tau) = \frac{1}{N}\sum_{i=0}^{N-1} a_i a_{i+\tau} \tag{4.4}$$

where $a_i$ is the ($i$ mod $N$) code element of symbol of the periodic sequence $a$, $a_i \in \{1, -1\}$. When $\tau \bmod N = 0$, the autocorrelation value is called main lobe value, otherwise, it is called sidelobe value. Similarly, the cross-correlation between two different sequences $a$ and $b$ is defined as

$$\text{CCF}_{\text{even}}(\boldsymbol{a}, \boldsymbol{b}, \tau) = \frac{1}{N}\sum_{i=0}^{N-1} a_i b_{i+\tau} \tag{4.5}$$

All the cross-correlation values are called sidelobe values. It can be seen that the Eqs. (4.4) and (4.5) define cyclic auto/cross-correlations, respectively, which can be calculated by FFT.

**Table 4.7** Mapping Relationship between Logic Levels and Signal Levels

| Logic level | Signal level |
|---|---|
| 1 | −1.0 |
| 0 | +1.0 |

For a good code family, the sidelobe value should be as small as possible. However, it is difficult to find a set of well-correlated codes. For a code family with the length $N$ and containing $M$ sequences, the maximum sidelobe value has a lower bound, called Welch bound [24]:

$$\theta_{\text{Welch}} = \sqrt{N}\sqrt{\frac{MN - N}{MN - 1}} \tag{4.6}$$

It can be seen after normalizing the Welch bound to $\theta_{\text{Welch}}^{\text{norm}} = \frac{1}{\sqrt{N}}\sqrt{\frac{MN-N}{MN-1}} \approx \frac{1}{\sqrt{N}}$ that, with the increase in code length and the reduction in the lower bound of the maximum cross-correlation sidelobe value, the correlation performance will become better.

Since the introduction of secondary code to modernized GNSS systems, half of the probability of symbol reversal has occurred in the received periodic primary code signal. To characterize the effect of symbol flipping, the concepts of odd correlation and even correlation are introduced here. The autocorrelation and cross-correlation defined by Eqs. (4.4) and (4.5) are even autocorrelation and even cross-correlation, respectively. The normalized odd autocorrelation and odd cross-correlation can be defined as [24]:

$$\begin{aligned}
\text{ACF}_{\text{odd}}(\boldsymbol{a}, \tau) &= \frac{1}{N}\sum_{i=0}^{N-1-\tau} a_i a_{i+\tau} - \sum_{i=\text{N}-\tau}^{N-1} a_i a_{i+\tau} \\
\text{CCF}_{\text{odd}}(\boldsymbol{a}, \boldsymbol{b}, \tau) &= \frac{1}{N}\sum_{i=0}^{N-1-\tau} a_i b_{i+\tau} - \sum_{i=\text{N}-\tau}^{N-1} a_i b_{i+\tau}
\end{aligned} \tag{4.7}$$

In general, the code family with even correlation is difficult to obtain. In the case of odd correlation, the Welch bound is still the lower bound of the maximum sidelobe value. As different satellites have different Doppler shifts, Eqs. (4.4)–(4.7) need to account for the influence of Doppler shift, as shown below [23]:

$$\begin{aligned}
\text{ACF}_{\text{even}}^{\text{Doppler}}(\boldsymbol{a}; \tau, \Delta f) &= \frac{1}{N}\sum_{i=0}^{N-1} a_i a_{i+\tau} \mathrm{e}^{j\frac{2\pi\Delta f i}{R_c}} \\
\text{CCF}_{\text{even}}^{\text{Doppler}}(\boldsymbol{a}, \boldsymbol{b}, \tau, \Delta f) &= \frac{1}{N}\sum_{i=0}^{N-1} a_i b_{i+\tau} \mathrm{e}^{j\frac{2\pi\Delta f i}{R_c}} \\
\text{ACF}_{\text{odd}}^{\text{Doppler}}(\boldsymbol{a}, \tau, \Delta f) &= \frac{1}{N}\sum_{i=0}^{N-1-\tau} a_i a_{i+\tau} \mathrm{e}^{j\frac{2\pi\Delta f i}{R_c}} - \sum_{i=N-\tau}^{N-1} a_i a_{i+\tau} \mathrm{e}^{j\frac{2\pi\Delta f i}{R_c}} \\
\text{CCF}_{\text{odd}}^{\text{Doppler}}(\boldsymbol{a}, \boldsymbol{b}, \tau, \Delta f) &= \frac{1}{N}\sum_{i=0}^{N-1-\tau} a_i b_{i+\tau} \mathrm{e}^{j\frac{2\pi\Delta f i}{R_c}} - \sum_{i=N-\tau}^{N-1} a_i b_{i+\tau} \mathrm{e}^{j\frac{2\pi\Delta f i}{R_c}}
\end{aligned} \tag{4.8}$$

where $R_c$ is the code rate, and $\Delta f$ is the carrier Doppler frequency difference. The even autocorrelation, even cross-correlation, odd autocorrelation, and odd cross-correlation defined by Eq. (4.8) are the basis for comparing the performance of two code families. The pseudocode can be evaluated in terms of time-domain characteristics and frequency-domain characteristics.

In the time domain, the correlation values under all possible code phase offsets and Doppler frequency offsets are calculated. The evaluation indicators include: maximum autocorrelation sidelobe, maximum cross-correlation sidelobe, autocorrelation cumulative-distribution probability, and cross-correlation cumulative-distribution probability. In the frequency domain, the spectrum of code sequence is obtained. The evaluation indicators include the maximum spectral amplitude and the cumulative probability distribution of amplitude.

2. **PRN code family**

The selection of primary code family determines the main performance of pseudocode and thus is the core content of pseudocode design. In the existing GNSS systems, the primary code family of a civil signal is mainly composed of an m sequence, a Gold code, a Weil code, and a random code (memory code).

(1) m sequence

The m sequence is the longest sequence, with autocorrelation similar to that of random binary codes. This sequence is generated by Linear Feedback Shift Register (LFSR), as shown in Fig. 4.5. In this figure, the blocks 1 to $r$, called the order of the m sequence, represent the $r$-stage shift registers. When the clock pulse arrives, the status of each register, either 0 or 1, will be shifted from left to right by 1 bit. Initially, the register status at each level is set as 1.

The m sequence can be converted into a polynomial $G = 1 + \sum_{i=1}^{r} C_i X^i$, where 1 means that the output of adder is fed back to the first stage register, and that the output of the $i$th stage register is just the input of the modulo 2 adder (otherwise it is 0) $C_i$ is 1.

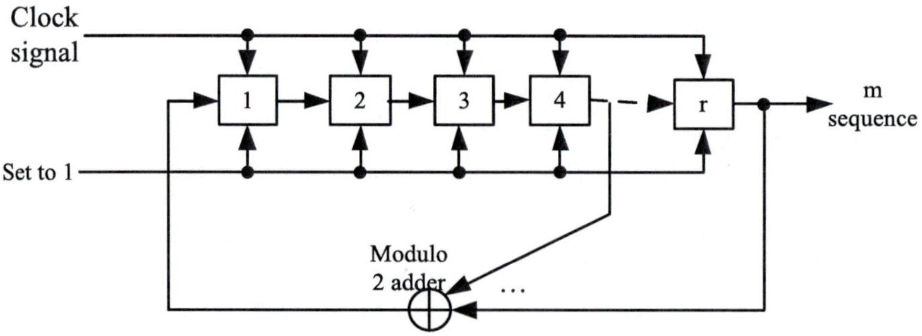


**Fig. 4.5** Block diagram of m sequence generator

The m sequence has the following properties. The length of the m sequence is $N = 2^r - 1$. The quantity of "1" in the m sequence is one bigger than the quantity of "0" to achieve a good balance. The translated two m sequences are added to obtain a new m sequence, which has the two-valued autocorrelation function, i.e.:

$$R(\tau) = \begin{cases} 1, & \tau \bmod N = 0 \\ -\frac{1}{N}, & \text{other} \end{cases} \tag{4.9}$$

Since the length of the m sequence is $N = 2^r - 1$, it often cannot satisfy the requirement of actual application. Therefore, the truncated code and the overlay code can be used to adjust the code length. The truncated code is shortened by adding a status detector. The overlay code is a short code with two or more lengths (mutual prime numbers) that is added by modulo 2 to obtain a long code sequence. If the truncation composite method is used, the original autocorrelation and cross-correlation properties will be destroyed.

The m sequence has been applied in GPS and GLONASS, with the shortened version used in GPS L2C and the original version used in GLONASS L1OF. In addition, a shortened m sequence will also be introduced to the secondary code of GPS L1C pilot component.

(2) Gold code

The m sequences have good autocorrelation, but nonideal cross-correlation values, and the number of m sequences is limited. To solve this problem, Gold code has been proposed. The Gold code is generated by adding two m sequences of the same order preferably to modulo 2, and changing the relative phase of the preferred pair of m sequences to obtain a new Gold sequence.

The number of Gold codes of the same order is much larger than that of m sequences. After optimization, the Gold code family can obtain better autocorrelation and cross-correlation. The two m sequences capable of generating a Gold code are called a preferred pair, such as two m sequences of order $r$, whose cross-correlation function has a bounded absolute value, as shown below:

$$|R(\tau)| = \begin{cases} 2^{\frac{r+1}{2}} + 1, & r \text{ is odd number} \\ 2^{\frac{r+2}{2}} + 1, & r \text{ is even number other than multiple of } 4 \end{cases} \tag{4.10}$$

Two m sequences of order r can produce $2^r - 1$ Gold codes plus two m sequences, i.e. $2^r + 1$ in total.

The Gold sequences in the same code family have a three-valued cross-correlation value. Due to excellent correlation, Gold codes (either shortened or unshortened) and overlay codes are widely used in GNSS systems. The early GPS C/A code, the primary code of Galileo's E5 signal and the code used by BDS-2 signal all belong to the Gold code family. In different satellites, the Gold code is different, and the relative phase of two m sequences is also different. The Gold codes, truncated codes, and overlay codes used by different GNSS systems are shown in Table 4.8.

**Table 4.8** Gold codes in current GNSS

| Signal | | Code family | Length of register sequence code | Register identification | Order | Generated polynomial |
|---|---|---|---|---|---|---|
| GPS L1C/A | | Gold | 1023 | G1 | 10 | $1 + X^3 + X^{10}$ |
| | | | | G2 | 10 | $1 + X^2 + X^3 + X^6 + X^8 + X^9 + X^{10}$ |
| GPS L1P | | Overlay code | 4092 | $X1$A | 12 | $1 + X^6 + X^8 + X^{11} + X^{12}$ |
| | | | 4095 | X1B | 12 | $1 + X^1 + X^2 + X^5 + X^8 + X^9 + X^{10} + X^{11} + X^{12}$ |
| | | | 4092 | X2A | 12 | $1 + X^1 + X^3 + X^4 + X^5 + X^7 + X^8 + X^9 + X^{10} + X^{11} + X^{12}$ |
| | | | 4095 | X2B | 12 | $1 + X^2 + X^3 + X^4 + X^8 + X^9 + X^{12}$ |
| GPS L5C | Data | Overlay code | 8190 | XA | 13 | $1 + X^9 + X^{10} + X^{12} + X^{13}$ |
| | | | 8191 | XBI | 13 | $1 + X^1 + X^3 + X^4 + X^6 + X^7 + X^8 + X^{12} + X^{13}$ |
| | Pilot frequency | Overlay code | 8190 | XA | 13 | $1 + X^9 + X^{10} + X^{12} + X^{13}$ |
| | | | 8191 | XBQ | 13 | $1 + X^1 + X^3 + X^4 + X^6 + X^7 + X^8 + X^{12} + X^{13}$ |
| E5a-I/E5a-Q | | Gold | 10,230 | R1 | 14 | $1 + X^1 + X^6 + X^8 + X^{14}$ |
| | | | | R2 | 14 | $1 + X^4 + X^5 + X^7 + X^8 + X^{12} + X^{14}$ |
| E5b-I | | Gold | 10,230 | R1 | 14 | $1 + X^4 + X^{11} + X^{13} + X^{14}$ |
| | | | | R2 | 14 | $1 + X^2 + X^5 + X^8 + X^9 + X^{12} + X^{14}$ |

(continued)

**Table 4.8** (continued)

| Signal | Code family | Length of register sequence code | Register identification | Order | Generated polynomial |
|---|---|---|---|---|---|
| E5b-Q | Gold | 10,230 | R1 | 14 | $1 + X^4 + X^{11} + X^{13} + X^{14}$ |
| | | | R2 | 14 | $1 + X^1 + X^5 + X^6 + X^9 + X^{10} + X^{14}$ |
| B1I/B2I | Gold | 2046 | G1 | 11 | $1 + X^1 + X^7 + X^8 + X^9 + X^{10} + X^{11}$ |
| | | | G2 | 11 | $1 + X^1 + X^2 + X^3 + X^4 + X^5 + X^8 + X^9 + X^{11}$ |

(3) Weil code

Weil code is a code family adopted by the GPS L1C. It is a relatively new code sequence with an excellent correlation. The BDS B1C signal also has a Weil code. The Weil code is generated based on the Legendre sequence.

The Legendre sequence with the length of $N$, where $N$ is a prime number, is defined as [25]

$$L_N(k) = \begin{cases} 1, & \text{where the integer } x \text{ exists and makes that } k \equiv x^2 \bmod N \\ -1, & \text{where the integer } x \text{ does not exists and makes that } k \equiv x^2 \bmod N \end{cases} \quad (4.11)$$

$L_N(k) \in \{1, -1\}$ is the value of the $k$th code element in the Legendre sequence, $k \in [1, N-1]$. $L_N(0)$ is defined as $L_N(0) = -1$. Obviously, the reciprocal transformation of all the code element values of $L_N(k)$ does not affect the correlation of Legendre sequence.

By multiplying the Legendre sequence by its own delay, a Weil sequence can be yielded, namely [26]

$$\mathrm{Weil}_N(k) = L_N(k) L_N(k + \tau) \quad (4.12)$$

Since the length of the Legendre sequence is a prime number, the length of the Weil sequence is also a prime number. The GPS L1C has a code length of 10,230 and requires 210 pairs of pilot codes/data codes. The prime number closest to 10,230 is 10223. The strategy adopted by L1C code design is to fill the Weil code (length: 10223) with 7 bits. There are 5111 Weil sequences within this length, 128 of which are 7-bit padding sequences. The insertion position has 10,223 options. The number of possible codes is very large. In fact, the 7 bits are quickly fixed, and the padding bits used by all the codes are 0 1 1 0 1 0 0.

The code of BDS B1C signal is obtained by truncating the Weil code with 10,243 code elements. The adopted interception method is cyclic interception, that is, the truncation sequence is

$$c(k;\tau;p) = \mathrm{Weil}_N((k+p-1) \bmod 10{,}243;\tau) \tag{4.13}$$

Equation (4.13) indicates that 10,230 code elements are intercepted from the $p$th Weil code, and that the value range of $p$ is 1–10,243.

The Weil code is generated based on the same Legendre sequence. Its generation requires only a small amount of storage resources, and can even be achieved by a shift register.

(4) Random code (memory code)

Random code, also known as memory code, is the pseudocode used by the Galileo E1 OS signal. A binary sequence with the length of $N$ has a total of $2^N$ options. The random code is directly selected from the $2^N$ options according to certain criteria.

Compared with the pseudorandom noise code based on fixed code family, random code has the following advantages. First, the code length is unrestricted and can be arbitrarily selected. Second, the number of candidate codes is huge. They can meet specific performance requirements as needed, such as balance, odd/even cross-correlation, and especially zero autocorrelation sidelobe (ASZ). The ASZ means that the relative delay at the position of $\pm 1$ chip is 0 [37].

However, as the code length increases, the number of candidate codes has increased exponentially. It is difficult to select the best code family from a large number of codes. For this reason, the Galileo signal design team finally adopted genetic algorithm to achieve the selection of random code and obtain a relatively good code family.

Since random code can only be found by the preferred algorithm rather than generated in real time, all the candidate codes need to be stored. Therefore, more memory resources are needed than those in the case of the Gold code.

3. **Code structure**

In modern GNSS signals, a tiered code structure is commonly used. This structure consists of a primary code and a secondary code. The block diagram of tiered code generation is as follows (Fig. 4.6) [7].

In this diagram, $f_c$ is the primary code rate, N is the primary code length, so the secondary code rate is $f_c/N$, and the width of a secondary code chip is equal to a primary code period. Compared with a single primary code, the tiered code destroys the periodic repetition of primary code and enhances the correlation performance. Its signal power spectral density is closer to the continuum spectrum so that the ability to resist narrowband interference is stronger and the bit synchronization process is faster. At the same time, this code structure does not significantly increase the

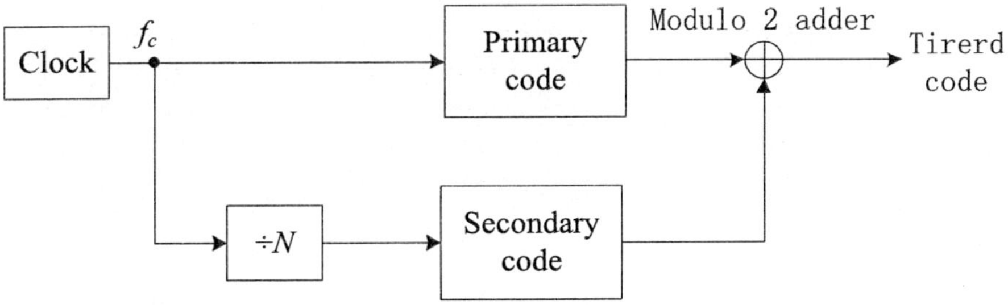


**Fig. 4.6** Block diagram of tiered code generation

capturing difficulty. In fact, the correlation performance of tiered code is between long code and shortcode, which is the result of a trade-off between the capturing complexity and the related performance.

The generation of a secondary code, such as the GPS L1C secondary code with a shortened m sequence and with the length of 1800, may be based on an existing code family or a preferred algorithm. In addition to tiered code structure, the overlay code structure that generates a longer code based on a shorter code is also used nowadays. Overlay code is a long code sequence composed of two or more short codes, and generated by adding a short code sequence modulo 2. For the $n$ shortcodes whose lengths are mutual prime numbers, their lengths are set as $L_1, L_2, \ldots, L_n$, and the generated overlay code length is $\prod_{i=1}^{n} L_i$.

4. **Code length**

The selection of code length is related to the code rate, which is generally a multiple of 1.023 Mc/s. In order to obtain the code period that is an integer multiple of 1 ms, the code length is usually taken as a multiple of 1023, such as 1023, 2046, 4092 … 10,230. Compared with the C/A code length of 1023, the code length in modernized GNSS systems has become longer in order to improve the autocorrelation and cross-correlation performance.

There is always controversy about the use of long primary codes and short primary codes. The main reason for using a long primary code is to improve the cross-correlation performance. For example, the code length of the GPS L1C signal is chosen to be 10,230, because there are both data components and pilot components in this length. In order to reduce the interference between them, the data code and pilot code broadcast by the same satellite should be nearly orthogonal to each other. To meet this requirement, increasing the code length is an inevitable choice.

### 4.2.3 Message Rate and Encoding

The message data in the navigation signal, called navigation message, is usually provided in the binary stream to the user. This section focuses on the message rate and message data coding.

1. **Message rate**

When designing a GNSS signal, a trade-off is required between the message rate and the power of data component (data pilot power allocation). The increase in bit rate can increase the Time To First Fix (TTFF), but reduces the stability or Bit Error Rate (BER) of tracking loop. On the other hand, the increase in the power of the data component can improve the data demodulation, but reduces the power of the pilot component and the available C/N0 region. Therefore, a compromise is needed between the message rate and signal power component [44].

2. **Encoding**

In order to detect the errors and reduce the data error rate, a check code and a Forward Error Correction (FEC) code are often used in an interleaving way to perform the error control coding of data bits. The main check codes used in navigation messages are parity check code and CRC-24Q check code. Usually, the CRC-24Q check code is used by more navigation messages, for example, by the CNAV navigation messages of GPS L2C and L5C, and by the CNAV messages of BDS system. Each of its 300-bit long message ends with a 24-bit CRC check code. The CRC-24Q can detect no more than 24 consecutive bit errors, including the 1–2 bit errors and any odd bit of error occurring anywhere in 300 bits. The generation process of L5C signal data component stream is shown in Fig. 4.7.

The FEC codes include BCH code, convolutional code, LDPC code, Turbo code, and other codes. Among them, LDPC is a special linear block code whose check matrix is a sparse matrix. This feature can reduce the decoding complexity and obtain the coding performance approximating to the Shannon limit.

The LDPC codes currently used or planned to be used in the navigation signals are mainly binary LDPC [such as LDPC (1200, 600) and LDPC (528,264) used by

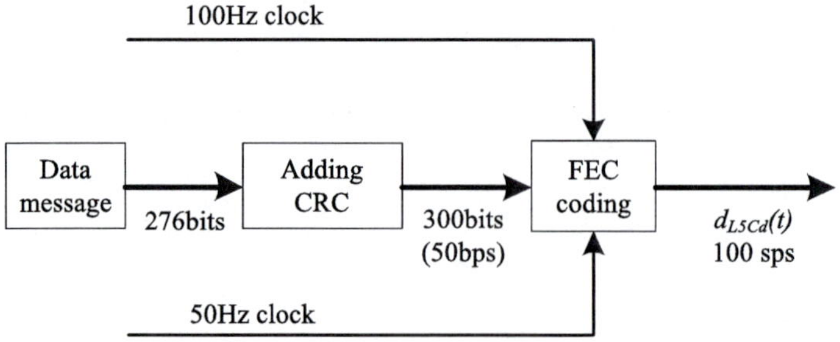


**Fig. 4.7** Generation process of L5C data component stream

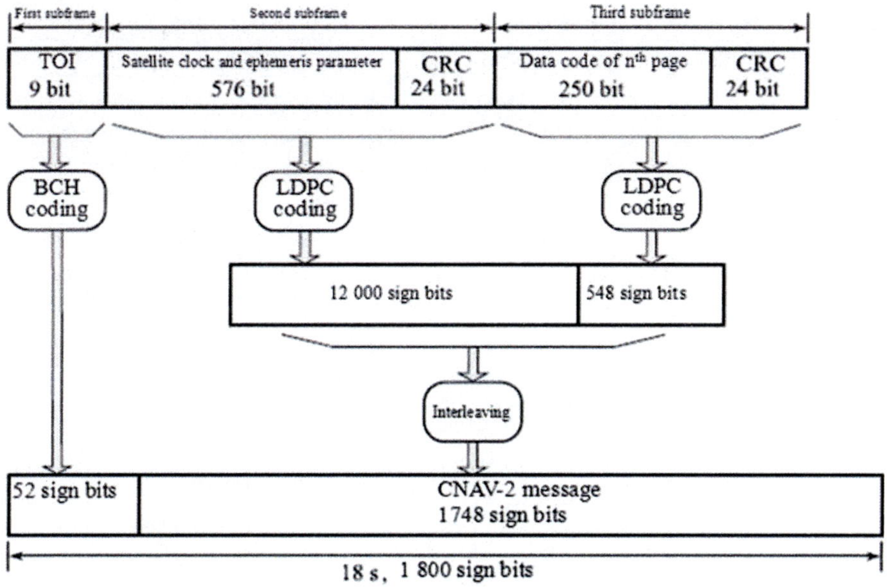


**Fig. 4.8** Generation process block diagram of L1C CNAV-2 signal data stream

GPS L1C CNAV-2] and 64-ary LDPC. The LDPC encoding is also used by BeiDou B1C. The generation process of GPS L1C CNAV-2 signal data stream is shown in Fig. 4.8.

The coding methods used for current GNSS signals are summarized in Table 4.9.

### *4.2.4 Modulation Waveform*

The modulation waveform determines the main performance indicators of navigation signals, including code tracking accuracy, anti-multipath performance, anti-interference, compatibility, and so on. In early GPS signals, the BPSK modulation method based on rectangular waveforms was used. BPSK is generally recorded as BPSK ($n$), where $n$ is the code rate of $n \times 1.023$ Mc/s.

In the process of GNSS signal modernization, new modulation signals, including BOC, LOC, BCS, MCS, PSWF, RC, and other signals, have been studied. Among them, BOC and its evolved version MBOC are adopted by GNSS. In addition, GMSK is also used for the Galileo signal baseline in the C frequency band.

1. **Modulation evaluation method and indicators**

For a direct sequence spread spectrum system, its baseband signal can be expressed as

**Table 4.9** Coding methods of GNSS signals

| System | Signal | Message type | Check | Coding |
|---|---|---|---|---|
| GPS | L1 C/A | NAV | Parity check hamming code (32,26) | – |
| | L5C | CNAV | CRC-24Q | Convolutional coding (600,300) |
| | L2C | CNAV | CRC-24Q | Convolutional coding (600,300) |
| | L1C | CNAV-2 | CRC-24Q | Subframe 1: BCH (51,8)<br>Subframe 2: LDPC (1200,600)<br>Subframe 3: LDPC (548,274) |
| GLONASS | C/A | | Hamming check | |
| Galileo | E1-B | I/NAV | CRC-24Q | Convolutional coding (1/2,7) |
| | E5a | F/NAV | CRC-24Q | Convolutional coding (1/2,7) |
| | E6-B | C/NAV | | Convolutional coding (1/2,7) |
| BDS | B1I | D1/D2 | | BCH(15,11,1) |
| | B2I | D1/D2 | | BCH(15,11,1) |
| | B1C | CNAV1 | CRC-24Q | Subframe 1: BCH (21,6), BCH (51,8)<br>Subframe 2: LDPC (1200,600), LDPC (528,264) |
| | B2a | CNAV2 | | LDPC(576,288) |

$$s(t) = \sum_k c_k p(t - kT_c) = \left(\sum_k c_k \delta(t - kT_c)\right) \otimes p(t) \tag{4.14}$$

where $c_k$ is a sequence of spreading codes, $p(t)$ is chip waveform, and $T_c$ is code period. For simplicity, the performance of a navigation signal is typically analyzed in the frequency domain. According to Eq. (4.14), the signal spectrum can be expressed as

$$S(f) = S_p(f) \cdot \sum_k c_k e^{-j2\pi f k T_c} \tag{4.15}$$

where $S_p(f)$ is the Fourier transform of $p(t)$. The power spectral density of the signal is

$$G(f) = \frac{E\left[|S(f)|^2\right]}{T} = \frac{\left|S_\mathrm{p}(f)\right|^2}{T_\mathrm{c}} \cdot \frac{1}{N} E\left[\left|\sum_{k=0}^{N-1} c_k \mathrm{e}^{-j2\pi f k T_c}\right|^2\right] = G_\mathrm{p}(f) \cdot A(f) \tag{4.16}$$

where $G_\mathrm{p}(f)$ is the power spectral density of the waveform $p(t)$ and determines the envelope of the signal power spectral density; $A(f)$ is the power spectral density fluctuation caused by spreading code (and data). As the spreading code is very long, it can be considered ideal when not strictly required. In this case, $G(f) \approx G_\mathrm{p}(f)$. In the general analysis, the power spectral density of waveform is replaced by that of navigation signal, and the evaluation indicators include code tracking accuracy, anti-multipath performance, signal compatibility, and anti-jamming performance.

(1) Code tracking error

Code tracking error is the code tracking accuracy in thermal noise environment. It can be evaluated according to the lower bound of coherent code tracking loop error (Carmer–Rao bound):

$$\sigma_\mathrm{CR}^2 = \frac{B_L}{C/N_0} \frac{1}{\left(2\pi \bar{\beta}_\mathrm{RMS}\right)^2} \tag{4.17}$$

where $\sigma_\mathrm{CR}^2$ is the code tracking error variance, $C/N_0$ is the carrier-to-noise ratio, $C$ is the carrier power, $B_\mathrm{L}$ is the loop single-side bandwidth, $\bar{\beta}_\mathrm{RMS}$ is the mean squared bandwidth of a signal, also known as Gabor bandwidth [20].

$$\bar{\beta}_\mathrm{RMS}^2 = \frac{\int_{-\beta_r/2}^{\beta_r/2} f^2 G(f)\mathrm{d}f}{\int_{-\beta_r/2}^{\beta_r/2} G(f)\mathrm{d}f} = \int_{-\beta_r/2}^{\beta_r/2} f^2 \bar{G}(f)\mathrm{d}f \tag{4.18}$$

where $\beta_\mathrm{r}$ is the bandwidth of RF front-end filter, $G(f)$ is the power spectral density of the signal with an infinite bandwidth, and $\bar{G}(f)$ is the power spectral density normalized within the bandwidth $\beta_\mathrm{r}$. It can be seen that with the increase in the proportion of high-frequency signal component and in RMS bandwidth, the lower bound of code tracking error will decrease.

(2) Anti-multipath performance

Multipath signals can cause the distortion of signal correlation function, destroy the symmetry of correlation function, and introduce the tracking error. The evaluation indicators of anti-multipath performance of a code loop generally include multipath error envelope and average multipath error. The multipath signal can be written as

$$s_\Sigma(t) = a_0 \mathrm{e}^{j\varphi_0} s(t - \tau_0) + \sum_{n=1}^{N} a_n \mathrm{e}^{j\varphi_n} s(t - \tau_n) \tag{4.19}$$

where $a_0$,$\varphi_0$ and $\tau_0$ are the amplitude, phase and propagation delay of the direct signal, respectively; $a_n$,$\varphi_n$ and $\tau_n$ are the amplitude, phase, and propagation delay of the nth multipath, respectively. The strongest multipath has the greatest impact. When evaluating the multipath performance, usually only one multipath is considered. Some scholars have analyzed the performance under two or three groups of multipath.

Under the conditions of only one multipath and coherent integration, when multipath and direct signals are in-phase or quadrature, the multipath error will be the maximum. The maximum multipath error under different multipath delay conditions is called multipath error envelope, and can be calculated by the following equation [40]:

$$\varepsilon_{\mathrm{r}} = \pm \frac{a_1/a_0 \int_{-B_r/2}^{B_r/2} G(f) \cdot \sin(\pi f d) \cdot \sin(2\pi f \hat{\tau}_1) \mathrm{d}f}{2\pi \int_{-B_r/2}^{B_r/2} f G(f) \cdot \sin(\pi f d) \cdot (1 \pm a_1/a_0 \cos(2\pi f \hat{\tau}_1)) \mathrm{d}f} \tag{4.20}$$

where $\hat{\tau}_1 = \tau_1 - \tau_0$ is the multipath delay relative to the direct signal, and $a_1/a_0$ is called multipath to direct ratio (MDR).

(3) Anti-interference performance

The sources of interference mainly include in-band interference, out-of-band interference, and environmental interference. Among them, in-band interference mainly refers to the energy portion of other microwave systems entering the bandwidth of the received signal. Out-of-band interference mainly refers to the strong energy of other signal systems close to the frequency band of the received signal, which can enter the baseband through the receiver front end to participate in the receiving process. Environmental interference mainly refers to the interference caused by space and geographical environments, such as multipath and block.

According to the characteristics of the frequency domain of interference signal, the radio frequency interference can be divided into continuous wave (CW) interference, narrowband interference, wideband interference, and modulation signal interference. Among them, CW interference can also be attributed to narrowband interference. Wideband interference signals can be regarded as the signals with a small signal-to-noise ratio.

Narrowband interference and matched spectral interference are the main types of interference. Anti-interference performance has been comprehensively analyzed and evaluated by using the quality factors of demodulation anti-narrowband jamming merit, demodulation anti-matched spectrum jamming merit, code tracking anti-narrowband jamming meri,t and code tracking anti-matched spectrum jamming merit. For example, the quality factors of demodulation anti-matched spectrum jamming merit is expressed as [40]

$$Q_{\mathrm{DEMAIMS}} = 10\log_{10}\left(\frac{1}{R_{\mathrm{d}} \int_{-\beta_r/2}^{\beta_r/2} (G_{\mathrm{s}}(f))^2 \mathrm{d}f}\right) (\mathrm{dB}) \tag{4.21}$$

where $G_s(f)$ is the normalized power spectral density over the infinite bandwidth of the signal.

(4) Signal compatibility

According to the ITU-R recommendation M.1831, the equivalent carrier-to-noise ratio is used to evaluate the compatibility between RF systems. The equivalent carrier-to-noise ratio is related to signal waveform, constellation configuration, and transmission power. In terms of compatibility, the spectral separation coefficient and the code tracking spectrum sensitivity coefficient are generally used as indicators for signal compatibility evaluation. The spectral separation coefficient is defined as [40]:

$$\chi_{\mathrm{J,S}} = \int_{-\beta_r/2}^{\beta_r/2} G_{\mathrm{l}}(f) G_{\mathrm{s}}(f) \mathrm{d}f \tag{4.22}$$

The code tracking spectrum sensitivity coefficient is defined as [38]:

$$\eta_{\mathrm{J,S}} = \frac{\int_{-\beta_r/2}^{\beta_r/2} f^2 G_{\mathrm{l}}(f) G_{\mathrm{s}}(f) \mathrm{d}f}{\int_{-\beta_r/2}^{\beta_r/2} f^2 G_{\mathrm{s}}(f) \mathrm{d}f} \tag{4.23}$$

where $G_s(f)$ and $G_l(f)$ are the normalized power spectral density of the desired signal and that of the interference signal, respectively; and $\beta_r$ is the received bandwidth of the desired signal.

2. **Modulation waveform design**

Modulation waveforms can be divided into three types. The first type is piecewise function modulation waveform with a limited number of level values, and is also the main modulation waveform used for GNSS signals. The waveforms in this type include BPSK, BOC, MBOC, etc. The second type is the continuous phase modulation waveform. The waveforms in this type include MSK and GMSK, of which GMSK has been used in Galileo C band signal. The waveforms in the third type include LOC, raised cosine, PSWF, etc. It is usually difficult for these modulation waveforms to achieve a constant envelope, so they are not used in current navigation signals.

(1) BPSK and BOC modulation

Binary Frequency Shift Keying (BPSK) is the first modulation waveform used in satellite navigation. As it is a mature technique with a simple generation and reception process, it is still widely used in navigation signals. The modulation waveform of BPSK is

$$p_{\mathrm{BPSK}}(t) = \begin{cases} 1, \ 0 \le t \le T_c \\ 0, \ \text{other} \end{cases} \tag{4.24}$$

where $T_c$ is the chip width. Assume that the spreading code has an ideal correlation feature. The power spectral density of the BPSK signal is

$$G_{BPSK}(f) = T_c \text{sinc}^2(\pi f T_c) \tag{4.25}$$

wheresin $c(x) = \sin(x)/x$. The corresponding correlation function is

$$R_{BPSK}(\tau) = \begin{cases} 1 - \frac{|\tau|}{T_c}, & |\tau| \le T_c \\ 0, & \text{others} \end{cases} \tag{4.26}$$

It is observed that BPSK has a triangular correlation function.

The Binary Offset Carrier (BOC) signal can be considered as a BPSK signal multiplied by a square wave subcarrier, i.e.,

$$s_{BOC}(t) = s_{BPSK}(t) \cdot \text{sign}(\sin(2\pi f_s t + \psi)) = \sum_{i=-\infty}^{\infty} c_i (-1)^{k \cdot i} p_{BOC}(t - i \cdot T_c) \tag{4.27}$$

According to the phase $\psi$ of the subcarrier, the BOC signal can be divided into a sine phase BOC and a cosine phase BOC. They are represented by $BOC_{sin}$ (*m*, *n*) and $BOC_{cos}$ (*m*, *n*), respectively, where *n* indicates that the subcarrier frequency *fs* is *n* times of 1.023 MHz, and *m* indicates that the code rate $R_c = 1/T_c$ is *m* times of 1.023 MHz. In addition, $k = 2m/n = 2f_s/T_c$, which is called the modulation index (or order) of BOC signal, is the number of half periods within one chip of a square wave subcarrier. For a sine BOC signal, its modulation waveform is

$$p_{\sin-BOC}(t) = \begin{cases} ign[\sin(2\pi f_s t)] = \sum_{i=0}^{k-1} (-1)^i p_{T_s}(t - iT_s), & 0 \le t \le T_c \\ 0, & \text{others} \end{cases} \tag{4.28}$$

where $T_s = 1/(2f_s)$ is the subcarrier's half period; $p_{T_s}$(t) is a rectangular function, which is 1 in case of $0 \le t < T_s$ or 0 in other cases. For a sinusoidal BOC signal, its autocorrelation function can be expressed as [42]

$$R_{\sin\text{-}BOC}(\tau) = \begin{cases} (-1)^l \left( \frac{(k-l)(2l+1)-l}{k} - \frac{2k-2l-1}{T_c} |\tau| \right), & lT_s \le |\tau| < (l+1)T_s \\ 0, & |\tau| \ge T_c \end{cases} \tag{4.29}$$

where *l* is an integer. The autocorrelation function of the cosine BOC signal of the order k is similar to that of sine BOC signal, i.e.,

$$R_{\text{cos-BOC}}(\tau) = \begin{cases} (-1)^{l}\left(\frac{2l-2k-1}{T_c}|\tau| + \frac{4kl+2k-4l^2}{2k}\right), & lT_s \le |\tau| < \frac{(2l+1)T_s}{2} \\ (-1)^{l+1}\left(\frac{2k-2l-3}{T_c}|\tau| + \frac{4l^2-4lk+8l-2k+4}{2k}\right), & \frac{(2l+1)T_s}{2} \le |\tau| < (l+1)T_s \\ 0, & |\tau| \ge \mathrm{T_c} \end{cases} \tag{4.30}$$

Figure 4.9 shows the autocorrelation functions of BPSK (1), $\mathrm{BOC_{sin}}(2,1)$, and $\mathrm{BOC_{cos}}(2,1)$. It can be seen that, unlike BPSK, the correlation function of the BOC signal has multiple side peaks. For the sinusoidal BOC signal of the order $k$, its autocorrelation function has a total of $2k - 1$ peaks. For the cosine BOC signal of the order k, its autocorrelation function has $2k + 1$ peaks.

(2) MCS

The modulation waveform of Multilevel Coded Symbol (MCS) is a step-shape waveform. When a chip waveform is equally divided into M segments, the MCS modulated waveform can be expressed as [26]

$$p_{\mathrm{MCS}}(t) = \sum_{i=0}^{M-1} s_i p_{T_c/M}\left(t - \frac{iT_c}{M}\right) \tag{4.31}$$

where $p_{T_c/M}(t)$ is 1 on $[0, T_c/M)$, and 0 in other places; $\{s_i, i = 0, 1, 2, \ldots, M-1\}$ is the value obtained at the $i$th segment. To ensure that the power is 1, the above equation can be divided by the factor $\sqrt{\sum_{i=0}^{M-1} s_i^2/M}$ for normalization. The MCS

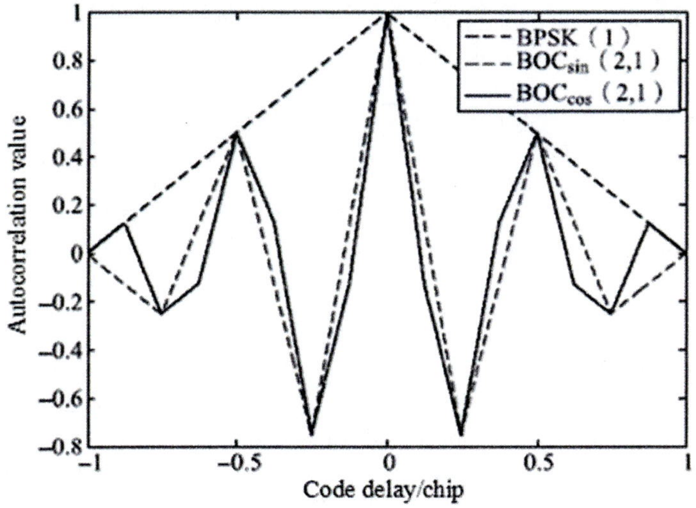


**Fig. 4.9** Autocorrelation functions of BPSK and BOC signals

can be written as MCS([$s$], $n$), where $n$ represents a code rate that is n times of 1.023 Mc/s, and the value of [$s$] represents $\left[s_0, s_1, \ldots, s_{M-1}\right]$.

(3) MBOC modulation

MBOC (Multiplexed Binary Offset Carrier) is a multiplexed binary offset carrier, and is also a variant of BOC modulation. It is a modulation type proposed by Galileo and GPS. In the MBOC, only the definition of the frequency domain is given, and the specific implementation is not limited. In terms of power spectral density, a MBOC is composed of a high-order BOC (m, n) and a low-order BOC ($n$, $n$) [19], i.e.,

$$G_{\mathrm{MBOC}(m,n,r)}(f) = (1-r)G_{\mathrm{BOC}(n,n)}(f) + rG_{\mathrm{BOC}(m,n)}(f) \tag{4.32}$$

The corresponding normalized autocorrelation function is

$$R_{\mathrm{MBOC}(m,n,r)}(\tau) = (1-r)R_{\mathrm{BOC}(n,n)}(\tau) + rR_{\mathrm{BOC}(m,n)}(\tau) \tag{4.33}$$

where $r$ is the power ratio of high-order BOC($m$, $n$), and $R_{\mathrm{BOC}(m,n)}(\tau)$ and $R_{\mathrm{BOC}(n,n)}(\tau)$ are the normalized autocorrelation functions of BOC($m$, $n$) and BOC($n$, $n$), respectively. The biggest advantage of MBOC modulation is that it provides both low-end and mid-to-high-end service capabilities. The larger the r is, the stronger the support for high-end performance will be provided. Both Galileo and GPS use MBOC (6,1,1/11) as well as $\mathrm{BOC_{sin}}(6,1)$ and $\mathrm{BOC_{sin}}(1,1)$ of sinusoidal phase.

The implementation forms of MBOC include CBOC (Multiplexed Binary Offset Carrier), TMBOC (Time-Multiplexed Binary Offset Carrier), and QMBOC (Quadrature Multiplexed Binary Offset Carrier). Among them, CBOC is used for Galileo E1OS, TMBOC for GPS L1C, and QMBOC for BDS B1C signal.

The CBOC is obtained by directly weighting and subtracting two BOC signals in the time domain. It can be expressed as

$$s_{\mathrm{CBOC}(m,n,r)}(t) = \sqrt{1-r}\,s_{\mathrm{BOC}(n,n)}(t) \pm \sqrt{r}\,s_{\mathrm{BOC}(m,n)}(t) \tag{4.34}$$

When added, it is called CBOC+; when subtracted, it is called CBOC−. It should be noted that the individual CBOC+ or CBOC− does not satisfy the power spectral density definition of MBOC. For this reason, the data component of Galileo E1OS signal uses CBOC+ (6,1,1/11), and the pilot component uses CBOC-(6,1,1/11) so that the power ratio of data component to pilot component is 1:1. The chip waveform for CBOC + or CBOC− with four levels is shown in Fig. 4.10 [7].

The TMBOC adopts the time-division method. The BOC (1,1) or BOC (6,1) chip has been used in different signals. The number of chips is determined by power ratio. The data component of this signal uses BOC(1,1), while its pilot frequency component uses TMBOC(6,1,4/33). The first and fifth chips in every 33 chips are illustrated. The 7th and 30th chips are BOC (6,1), while other chips are BOC (1,1). In this signal, the power ratio of data component and pilot component is 1:3. Generally speaking, this structure satisfies the spectral definition of MBOC (6,1,1/11). The

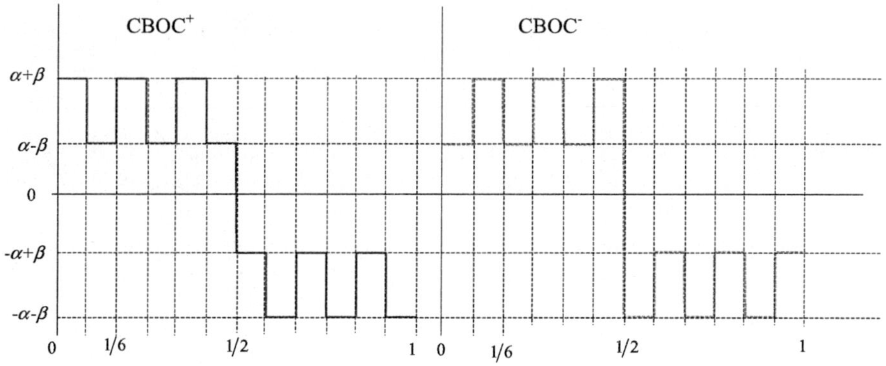


**Fig. 4.10** CBOC (6,1,1/11) chip waveform

positioning of BOC (6,1) chip in TMBOC is the result of optimization with the pseudocode design.

The QMBOC uses a phase division method to modulate the BOC($n$, $n$) and BOC($m$, $n$) components on two orthogonal phases of the carrier. The baseband signal of QMBOC(6,1, $r$) is expressed as

$$s_{\text{QMBOC}}(t) = \sqrt{1-r}\, s_{\text{BOC}(n,n)}(t) \pm j\sqrt{r}\, s_{\text{BOC}(m,n)}(t) \tag{4.35}$$

Similar to CBOC, the signal is called normal phase QMBOC or inverted QMBOC, and recorded as QMBOC+ or QMBOC−, respectively. Unlike CBOC, the autocorrelation function of the QMBOC signal is [45]

$$R_{\text{QMBOC}}(\tau) = E\left\{s_{\text{QMBOC}}(t)s^*_{\text{QMBOC}}(t)\right\} = (1-\gamma)R_{\text{BOC}(n,n)} + \gamma R_{\text{BOC}(m,n)} \tag{4.36}$$

regardless of whether QMBOC is positive or negative. Therefore, either QMBOC+ or QMBOC− can individually satisfy the definition of MBOC spectrum, so they can be used alone.

(4) MSK modulation

Minimum Shift Keying (MSK) is a modulation type in which the phase of time domain waveform changes continuously but with a constant envelope. In the satellite navigation system, the baseband expression of the MSK signal is

$$s_{\text{MSK-R}}(t) = A\sum_{i=-\infty}^{\infty} c_{\text{I}}(i)p_{\text{MSK}}(t-iT_{\text{c}}) + jA\sum_{i=-\infty}^{\infty} c_{\text{Q}}(i)p_{\text{MSK}}\left(t-iT_{\text{c}}-\frac{T_{\text{c}}}{2}\right) \tag{4.37}$$

where $c_i$ and $c_q$ are the code sequences of the I and Q branches, respectively, and $T_c$ is the chip width.

$$p_{MSK}(t) = \begin{cases} os\left(\frac{\pi t}{T_c}\right), & t \in [-T_c/2, T_c/2] \\ 0, & \text{other} \end{cases} \tag{4.38}$$

(5) GMSK modulation

GMSK is a special case of Continuous Frequency Shift Keying (CP-FSK). It uses Gaussian filter frequency pulses to smooth the transmission from one point to the next point on the signal constellation, while directly deriving and obtaining the Minimum Shift Keying (MSK) from the rectangle shaping of the frequency pulse. GMSK is also a continuous phase signal whose phase value is given by

$$\phi_{\mathrm{GMSK}}(t) = 2\pi h \sum_{i=-\infty}^{\infty} \alpha_i \int_{-\infty}^{t} g_{\mathrm{GMSK}}(t - iT_b)\mathrm{d}t \tag{4.39}$$

where $h$ is the modulation index, $h = 0.5$, $g_{\mathrm{GMSK}}(t)$ is the frequency pulse, $T_b$ represents the complex symbol width and is generated by multiplexing the signals in the I and Q branches. Therefore, the complex symbol width corresponding to the duration of the combination of one I chip and one Q chip is $T_b = 0.5T_c$, where $T_c$ is the chip width.

The frequency pulse waveform of GMSK is

$$g_{\mathrm{GMSK}}(t) = \frac{1}{2T_b}\left(Q\left[\frac{2\pi B(t - T_b)}{\sqrt{\ln 2}}\right] - Q\left[\frac{2\pi Bt}{\sqrt{\ln 2}}\right]\right) \tag{4.40}$$

where $Q(t) = \int_t^{\infty} \frac{1}{\sqrt{2\pi}} e^{-\frac{x^2}{2}} dx$.

$BT_b$ is the product of the $-3$ dB bandwidth and the symbol time. The higher its value, the clearer the eye diagram of the signal. But with the increase of OOB (Out of Band) emissions, the $BT_b$ will decrease. As a result, more energy will be concentrated in the middle of the frequency band, thus resulting in higher inter-code crosstalk overhead. A typical value for $BT_b$ in communication application is 0.3.

The GMSK-BOC signal similar to MSK can also be used. Figure 4.11 shows the power spectral densities of the GMSK-R(10) and GMSK-BOC(5,5) signals, respectively.

### 4.2.5 *Constant Envelope Multiplexing Method*

The power of the spaceborne broadcasting payload is limited. To maximize the efficiency, the High-Power Amplifier (HPA) typically operates in the saturated state. Due to its nonlinearity, the AM-AM and AM-PM distortion is caused. In order to reduce the influence of distortion, the signal is required to have a constant envelope.

Under normal circumstances, the constant envelope can be obtained only when the bandwidth is infinite. In fact, it is impossible to maintain a completely constant

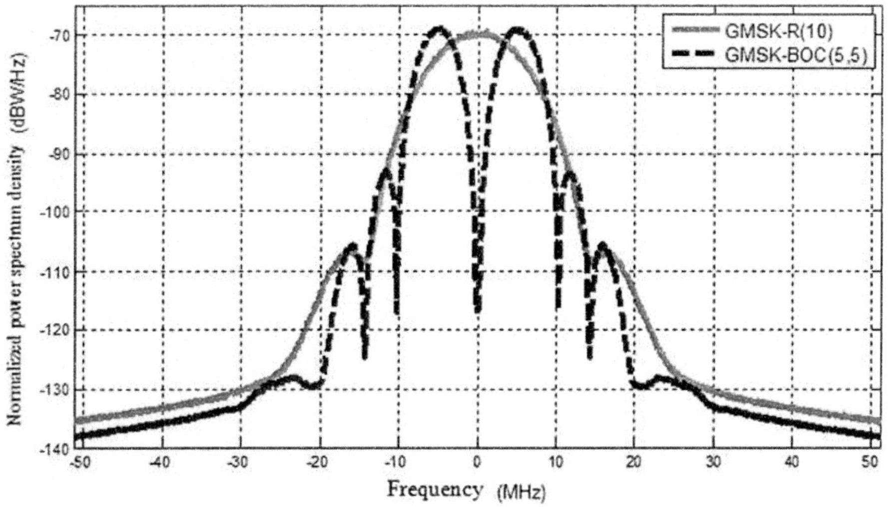


**Fig. 4.11** Power spectral densities of GMSK-R(10) and GMSK-BOC(5,5)

envelope due to the influence of the band limit. However, when the navigation signal has the constant envelope characteristic in the infinite bandwidth and the change of peak-to-average ratio is relatively small, the signal will be subjected to constant envelope multiplexing.

The methods of constant envelope multiplexing used in GNSS signals are mainly divided into single-frequency constant envelope multiplexing and dual-frequency constant envelope multiplexing. The dual-frequency constant envelope multiplexing involves multiple adjacent frequency bands. The individual components are multiplexed at baseband into a constant envelope signal to reduce the implementation complexity.

In the process of GPS signal modernization, a variety of single-frequency constant envelope multiplexing methods have been studied, including interplex, majority voting, mutual voting, Coherent Adaptive Subcarrier Modulation (CASM), POCET, and other methods. The Galileo system is a complex of constant envelope methods. Some people studied and proposed the use of AltBOC, a dual-frequency constant envelope multiplexing method, in the Galileo signal design. Chinese scholars also proposed the use of Dual-QPSK, TD-AltBOC, and ACE-BOC in the design of BDS-3 signal.

1. **Single-frequency constant envelope multiplexing**

Currently, single-frequency constant envelope multiplexing is mainly used for binary PRN code signals, for example, BPSK, BOC, and TMBOC. For multi-ary signals, scholars have also proposed the use of MPOCET. When using the method of single-frequency constant envelope multiplexing, an Intermodulation (IM) component, in addition to the desired signal, is introduced to achieve constant envelope multiplexing. The introduction of the IM component also consumes part of the power. To characterize the performance of a constant envelope multiplexed signal, this signal is usually evaluated in terms of multiplexing efficiency or multiplexing loss. The multiplexing efficiency is defined as the ratio of the power of all the useful signals to the total power [33], i.e,.

$$\eta = \frac{\sum_{n=1}^{N} P_n}{P} \tag{4.41}$$

where $P_n$ is the power of the $n$th signal component, and $P$ is the total power of the constant envelope signal. The corresponding multiplexing loss is defined as

$$\eta\_\text{loss[dB]} = -\lg \eta = -\lg\left(\frac{\sum\limits_{n=1}^{N} P_n}{P}\right) \tag{4.42}$$

It can be seen that, by increasing the multiplexing efficiency and the power ratio of the useful signals and decreasing the power ratio of the intermodulation component, the power loss caused by constant envelope multiplexing will become smaller.

(1) Interplex and CASM

Interplex is a phase modulation method whose general form is

$$s(t) = \sqrt{2P}\cos(2\pi f_c t + \theta(t)) \tag{4.43}$$

where $P$ is the total power, $f_c$ is the carrier frequency, and $\theta(t)$ is the phase modulation. For N unrelated binary signal components, $\theta(t)$ is defined as [33]

$$\theta(t) = \beta_1 \cdot s_1(t) + \sum_{n=2}^{N} \beta_n \cdot s_1(t)s_n(t) \tag{4.44}$$

where $s_n(t) = c_n(t)d_n(t) \cdot \text{sc}(t)$ is the $n$th signal component; $\text{sc}(t)$ represents a subcarrier; $d_n(t)$ represents a message symbol; $c_n(t)$ represents a spreading code waveform; $s_n(t) \in \{1, -1\}$; and $\beta_n$ is the modulation index that determines the power ratio between the signal components.

Coherent Adaptive Subcarrier Modulation (CASM) is a single-frequency constant envelope multiplexing method. Its characteristic is that all the signal components are separately modulated onto I and Q components, and that the power ratio between signal components can be adjusted according to the design. CASM is mathematically equivalent to Interplex. It differs from Interplex in implementation method. For N-way signal components, the CASM signal can be expressed as [21]:

$$s(t) = I_0(t)\cos(2\pi f_c t + \varphi_s(t)) - Q_0(t)\sin(2\pi f_c t + \varphi_s(t)) \tag{4.45}$$

where

$$I_0(t) = \sqrt{P_I}s_1(t),\ Q_0(t) = \sqrt{P_Q}s_2(t)$$

$$\varphi_s(t) = \sum_{k=3}^{N} m_k s_k(t) \cdot s_{m_k}(t)$$

$$m_k = \tan^{-1}\sqrt{\frac{P_{s_k}}{P_{s_{m_k}}}} \tag{4.46}$$

where $P_{\mathrm{I}}$ represents the signal power of the in-phase component, $P_{\mathrm{Q}}$ represents the power of the orthogonal component, so the total transmission power of a signal is $P_{\mathrm{I}} + P_{\mathrm{Q}}$; $s_{m_k}(t)$ indicates $s_1(t)$ or $s_2(t)$; $m_k$ is the modulation index that determines the power allocation between signals; $P_{s_k}$ represents the expected power of the $k$th signal; and $P_{s_{m_k}}$ represents the expected power of $s_1(t)$ or $s_2(t)$.

For $N$ signals, it can be known from Eq. (4.46) that a total of $N - 2$ modulation indices are used in the CASM multiplexing. When both the power distributed to each signal and the $P_{s_{m_k}}$ are known, the third equation based on Eq. (4.46) can be used to calculate the value of $m_k$.

Obviously, when the signal with the highest power is orthogonally multiplexed to all the other signals, that is, when the power of one signal is much higher than that of other signals in terms of power efficiency, the CASM multiplexing loss can be minimized so that the best effect of CASM multiplexing can be achieved.

(2) POCET method

In the Phase-Optimized Constant Envelope Transmission (POCET) method, the signal of constant envelope multiplexing is considered as phase modulation signal, and a signal vector S is constructed for $N$ binary PRN code signals $\{s_1(t), s_2(t), \ldots, s_N(t)\}$ and expressed as $\mathbf{s}(t) = [s_1(t), s_2(t), s_3(t), \cdots, s_N(t)]$. $\mathbf{s}(t)$ has only $2^N$ different values, each corresponding to one transmission phase $\theta_k, k = 1, 2, \ldots, 2^N$. In this design, the constraints on power and phase can be satisfied by optimizing these phase values. In the POCET method, the following assumptions are made. First, the $N$ binary PRN code signals are irrelevant to each other. Second, each PRN code is completely random, so the $2^N$ phase values have the same probability of occurrence. Then, the average correlation output of the constant envelope signal and the $n$th signal component can be obtained and given by [22]:

$$\mathrm{corr}_n(\boldsymbol{\theta}) = \mathrm{E}[s_{\mathrm{POCET}}(t) \cdot s_n(t)] = \frac{1}{T}\int s_{\mathrm{POCET}}(t) \cdot s_n(t)\mathrm{d}t = \frac{A}{2^N}\sum_{i=1}^{2^N} b_n(k)\mathrm{e}^{j\theta_k} \tag{4.47}$$

where $\theta$ is a set of $2^N$ phase values; $A$ is the envelope value of the multiplexed constant envelope signal, and is also the value of the $n$th signal component when

$b_n(k) = \pm 1$ is the $k$th value of $\mathbf{s}(t) = [s_1(t), s_2(t), s_3(t), \ldots, s_N(t)]$. In this case, the multiplexing efficiency is $\eta = \frac{\sum_{n=1}^{N} |\text{corr}_n|^2}{A^2}$.

The goal of POCET is to maximize the multiplexing efficiency while satisfying the power and phase constraints, that is, to minimize the envelope value A.

(3) DualQPSK

The DualQPSK can automatically multiplex four constant envelope binary signal components with equal power, whose phases are 0°, 90°, 45° and 135°, respectively. DualQPSK can be regarded as a special case of POCET. The coefficients of DualQPSK signal are not approximate numerical solutions. The DualQPSK baseband is given by [47]

$$\begin{aligned} s_{\text{DualQPSK}}(t) = & \frac{\sqrt{2+\sqrt{2}}}{4}\left(s_1(t) + e^{j\pi/2}s_2(t) + e^{j\pi/4}s_3(t) + e^{j3\pi/4}s_4(t)\right) \\ & + \frac{\sqrt{2-\sqrt{2}}}{4}s_3(t)s_4(t)\left(s_1(t) - e^{j\pi/2}s_2(t)\right) \\ & - \frac{\sqrt{2-\sqrt{2}}}{4}s_1(t)s_2(t)\left(e^{j\pi/4}s_3(t) - e^{j3\pi/4}s_4(t)\right) \end{aligned} \tag{4.48}$$

It can be seen that the multiplexing efficiency of DualQPSK is 85.36%.

2. **Dual-Frequency constant envelope multiplexing**

The dual-frequency constant envelope multiplexing is used to reduce the signal complexity. When two frequency bands are close to each other and filtered separately, the filter transition band is required to be steep so that a large group delay is caused. This performance can hardly meet the requirements. However, by using the dual-frequency constant envelope multiplexing, the multiplexed baseband is turned into a constant envelope signal, so the signal complexity is reduced.

Galileo E5 signal is the first signal using the dual-frequency constant envelope multiplexing. The Galileo E5 uses the AltBOC (15, 10) frequency bands including E5a and E5b, whose center frequencies are 1176.45 MHz and 1207.14 MHz, respectively. Either of E5a and E5b includes a data component and a pilot component, and uses the QPSK (10) modulation. The BDS B2 signal also faces the design requirements similar to Galileo E5. Therefore, Chinese scholars have proposed the use of such dual-frequency constant envelope methods as TD-AltBOC and ACE-BOC.

(1) AltBOC modulation

The baseband of the constant envelope AltBOC is given by [8]

$$\begin{aligned} s_{\text{AltBOC}}(t) = & \frac{1}{2\sqrt{2}}(s_1(t) + js_2(t))[\text{sc}_s(t) - j\text{sc}_s(t - T_s/4)] \\ & + \frac{1}{2\sqrt{2}}(s_3(t) + js_4(t))[\text{sc}_s(t) + j\text{sc}_s(t - T_s/4)] \end{aligned}$$

$$+\frac{1}{2\sqrt{2}}(s_2(t)s_3(t)s_4(t)+js_1(t)s_3(t)s_4(t))\left[\mathrm{sc_p}(t)-j\mathrm{sc_p}(t-T_s/4)\right]$$
$$+\frac{1}{2\sqrt{2}}(s_1(t)s_2(t)s_4(t)+js_1(t)s_2(t)s_3(t))[\mathrm{sc_P}(t)+j\mathrm{sc_P}(t-T_s/4)] \tag{4.49}$$

where $T_s$ is the period of the one-sided subcarrier function; $j=\sqrt{-1}$ is an imaginary unit; $s_1(t)$ and $s_2(t)$ represent the data and pilot components of the lower sideband, and $s_3(t)$ and $s_4(t)$ represent the data and pilot components of the upper sideband, respectively. $s_2(t)s_3(t)s_4(t)$, $s_1(t)s_3(t)s_4(t)$, $s_1(t)s_2(t)s_4(t)$ or $s_1(t)s_2(t)s_3(t)$ is a product signal or an intermodulation term. $\mathrm{sc_s}(t)$ and $\mathrm{sc_p}(t)$ represent the four-valued subcarriers of a single signal and of a product signal, respectively. One cycle of the subcarrier is shown in Fig. 4.12 [7].

Figure 4.13 shows the power spectral density of the periodic complex subcarrier of the useful signal component. It can be seen that in the periodic subcarrier, the power at the baseband (±*fs*) accounts for 94.96% of the complex subcarrier power. For dual-frequency constant envelope signals, their multiplexing efficiency has two definitions, depending on whether or not the subcarrier loss is considered. The first definition is to treat the power of complex subcarrier of the useful signal term as the power of the useful signal. In this case, the multiplexing efficiency of AltBOC is 85.36%. The second definition is to use only the power of the main lobe as the power of the useful signal, so the multiplexing efficiency of AltBOC is 85.36%*94.96% = 81.06%.

The AltBOC signal can be generated as expressed above. It is more convenient to generate this signal by referring to the phase lookup table. The two-dimensional lookup table of AltBOC is shown in Table 4.10 [7].

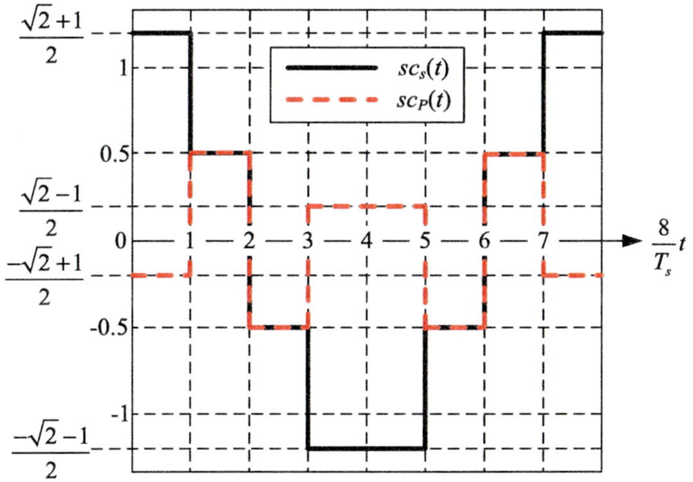


**Fig. 4.12** Periodic subcarrier waveform in AltBOC modulation

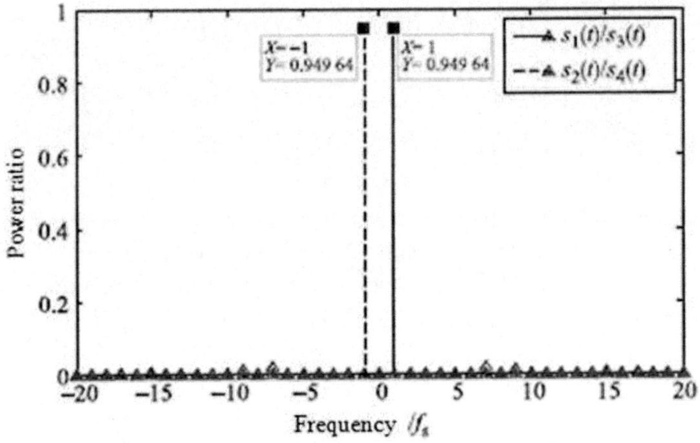


**Fig. 4.13** Power spectral density of complex subcarrier of useful AltBOC signal

(2) TD-AltBOC

In TD-AltBOC, the data/pilot components of the lower/upper sidebands are transmitted by time division. The mathematical expression of the TD-AltBOC (15, 10) baseband signal is [41]:

$$\begin{aligned} s_{\mathrm{TD-AltBOC}}(t) = {} & [S_{\mathrm{LD}}(t) + S_{\mathrm{LP}}(t)]\left[\mathrm{sc}(t) - j\mathrm{sc}\left(t - \frac{T_{\mathrm{s}}}{4}\right)\right] \\ & + [S_{\mathrm{UD}}(t) + S_{\mathrm{UP}}(t)]\left[\mathrm{sc}(t) + j\mathrm{sc}\left(t - \frac{T_{\mathrm{s}}}{4}\right)\right] \end{aligned} \tag{4.50}$$

When the power of $S_{\mathrm{TD-AltBOC}}(t)$ is 1, the four signal components are

$$s_{\mathrm{LD}}(t) = \sqrt{2}d_{\mathrm{LD}}(t)c_{LD}(t) = \sqrt{2}d_{\mathrm{LD}}(t)\sum_{l=\infty}^{\infty}\sum_{i=0}^{N_{\mathrm{LD}}-1} c_{\mathrm{ld}}(k)p_{T_{\mathrm{c}}}(t - (2N_{\mathrm{LD}}l + 2k + 1)T_{\mathrm{c}})$$

$$s_{\mathrm{LP}}(t) = \sqrt{2}c_{\mathrm{LD}}(t) = \sqrt{2}\sum_{l=\infty}^{\infty}\sum_{k=0}^{N_{\mathrm{LD}}-1} c_{\mathrm{LP}}(k)p_{T_{\mathrm{c}}}(t - (2N_{\mathrm{LD}}l + 2k)T_{\mathrm{c}})$$

$$s_{\mathrm{UD}}(t) = \sqrt{2}d_{\mathrm{UD}}(t)c_{UD}(t) = \sqrt{2}d_{\mathrm{UD}}(t)\sum_{l=\infty}^{\infty}\sum_{k=0}^{N_{\mathrm{LD}-1}} c_{\mathrm{Ud}}(k)p_{T_{\mathrm{c}}}(t - (2N_{\mathrm{LD}}l + 2k + 1)T_{\mathrm{c}})$$

$$s_{\mathrm{UP}}(t) = \sqrt{2}c_{\mathrm{UD}}(t) = \sqrt{2}\sum_{l=\infty}^{\infty}\sum_{k=0}^{N_{\mathrm{LD}}-1} c_{\mathrm{UP}}(k)p_{T_{\mathrm{c}}}(t - (2N_{\mathrm{LD}}l + 2k)T_{\mathrm{c}}) \tag{4.51}$$

where the rectangular function is defined as

**Table 4.10** Phase status lookup table of AltBOC signal

| | | | | | | | | | | | | | | | | | |
|---|---|---|---|---|---|---|---|---|---|---|---|---|---|---|---|---|---|
| $s_1(t)$ | | 1 | 1 | 1 | 1 | 1 | 1 | 1 | 1 | −1 | −1 | −1 | −1 | −1 | −1 | −1 | −1 |
| $s_3(t)$ | | 1 | 1 | 1 | 1 | −1 | −1 | −1 | −1 | 1 | 1 | 1 | 1 | −1 | −1 | −1 | −1 |
| $s_2(t)$ | | 1 | 1 | −1 | −1 | 1 | 1 | −1 | −1 | 1 | 1 | −1 | −1 | 1 | 1 | −1 | −1 |
| $s_4(t)$ | | 1 | −1 | 1 | −1 | 1 | −1 | 1 | −1 | 1 | −1 | 1 | −1 | 1 | −1 | 1 | −1 |
| $T' = \mathrm{mod}(t, T_s)$ | | $k^{S_{\mathrm{AltBOC}}(t) = e^{j\frac{k\pi}{4}}}$ | | | | | | | | | | | | | | | |
| $i_{Ts}$ | $t'$ | | | | | | | | | | | | | | | | |
| 0 | $[0,T_s/8)$ | 1 | 8 | 8 | 7 | 2 | 7 | 5 | 6 | 2 | 1 | 3 | 6 | 3 | 4 | 4 | 5 |
| 1 | $[T_s/8,2T_s/8)$ | 1 | 8 | 4 | 7 | 6 | 7 | 5 | 6 | 2 | 1 | 3 | 2 | 3 | 8 | 4 | 5 |
| 2 | $[2T_s/8,3T_s/8)$ | 5 | 8 | 4 | 3 | 6 | 7 | 5 | 6 | 2 | 1 | 3 | 2 | 7 | 8 | 4 | 1 |
| 3 | $[3T_s/8,4T_s/8)$ | 5 | 4 | 4 | 3 | 6 | 7 | 5 | 2 | 6 | 1 | 3 | 2 | 7 | 8 | 8 | 1 |
| 4 | $[4T_s/8,5T_s/8)$ | 5 | 4 | 4 | 3 | 6 | 3 | 1 | 2 | 6 | 5 | 7 | 2 | 7 | 8 | 8 | 1 |
| 5 | $[5T_s/8,6T_s/8)$ | 5 | 4 | 8 | 3 | 2 | 3 | 1 | 2 | 6 | 5 | 7 | 6 | 7 | 4 | 8 | 1 |
| 6 | $[6T_s/8,7T_s/8)$ | 1 | 4 | 8 | 7 | 2 | 3 | 1 | 2 | 6 | 5 | 7 | 6 | 3 | 4 | 8 | 5 |
| 7 | $[7T_s/8,8T_s/8)$ | 1 | 8 | 8 | 7 | 2 | 3 | 1 | 6 | 2 | 5 | 7 | 6 | 3 | 4 | 4 | 5 |

$$p_{\mathrm{T}}(t)=\begin{cases}1,\ 0\le t<T\\ 0,\ \text{other}\end{cases} \tag{4.52}$$

The subcarrier is

$$\mathrm{sc}(t)=\frac{1}{2\sqrt{2}}\mathrm{sign}(\cos(2\pi f_{\mathrm{s}}t)) \tag{4.53}$$

Given no intermodulation term, the TD-AltBOC satisfies the first definition. According to the first definition, the multiplexing efficiency of TD-AltBOC is 100%. However, according to the second definition, the power at the main lobe of its square wave subcarrier accounts for only 81.06%.

(3) ACE-BOC

The ACE-BOC (Asymmetric Constant Envelope - Binary Offset Carrier) can multiplex two QPSK signals in two adjacent frequency bands into one constant envelope signal. Moreover, any power ratio can be obtained from these signal components. $s_1$, $s_2$, $s_3$, and $s_4$ represent four binary PRN code signals whose power is $P_{s_1}$, $P_{s_2}$, $P_{s_3}$, and $P_{s_4}$, respectively. Considering the power allocation, the four signal components can be expressed as $s_{\mathrm{LI}}=\sqrt{P_{s_1}}s_1$,$s_{\mathrm{LQ}}=\sqrt{P_{s_2}}s_2$,$s_{\mathrm{UI}}=\sqrt{P_{s_3}}s_4$ and $s_{\mathrm{UQ}}=\sqrt{P_{s_4}}s_4$. The linear combination of the four signal components is [46]

$$\begin{aligned}s_{\text{linear}}(t)&=\left(s_{\mathrm{LI}}+js_{\mathrm{LQ}}\right)\mathrm{e}^{-j2\pi f_s t}+\left(s_{\mathrm{UI}}+js_{\mathrm{UQ}}\right)\mathrm{e}^{j2\pi f_s t}\\&=\left[\left(s_{\mathrm{UI}}+s_{\mathrm{LI}}\right)\cos(2\pi f_s t)-\left(s_{\mathrm{UQ}}-s_{\mathrm{LQ}}\right)\sin(2\pi f_s t)\right]\\&\quad+j\left[\left(s_{\mathrm{UQ}}+s_{\mathrm{LQ}}\right)\cos(2\pi f_s t)+\left(s_{\mathrm{UI}}-s_{\mathrm{LI}}\right)\sin(2\pi f_s t)\right]\end{aligned} \tag{4.54}$$

where $f_{\mathrm{s}}=1/T_{\mathrm{s}}$ is the frequency of the subcarrier. Equation (4.52) can be abbreviated as

$$s_{\text{linear}}(t)=\alpha_{\mathrm{I}}\sin(2\pi f_s t+\varphi_{\mathrm{I}})+j\alpha_{\mathrm{Q}}\sin\left(2\pi f_s t+\varphi_{\mathrm{Q}}\right) \tag{4.55}$$

where

$$\begin{aligned}\alpha_{\mathrm{I}}&=-\sqrt{\left(s_{\mathrm{UI}}+s_{\mathrm{LI}}\right)^2+\left(s_{\mathrm{UQ}}-s_{\mathrm{LQ}}\right)^2};\\ \alpha_{\mathrm{Q}}&=\sqrt{\left(s_{\mathrm{UI}}-s_{\mathrm{LI}}\right)^2+\left(s_{\mathrm{UQ}}+s_{\mathrm{LQ}}\right)^2};\\ \varphi_{\mathrm{I}}&=-\mathrm{atan2}\left(s_{\mathrm{UI}}+s_{\mathrm{LI}},s_{\mathrm{UQ}}-s_{\mathrm{LQ}}\right);\\ \varphi_{\mathrm{Q}}&=\mathrm{atan2}\left(s_{\mathrm{UQ}}+s_{\mathrm{LQ}},s_{\mathrm{UI}}-s_{\mathrm{LI}}\right)\end{aligned}$$

where arctan2 $(\cdot)$ is a four-quadrant arctangent function.

It can be seen that the envelope of the direct linear combination is not constant. In order to obtain a constant envelope, we take the sign of the linear combination and normalize the power to calculate the baseband of ACE-BOC modulation as below:

$$
\begin{aligned}
s_{\text{ACE-BCC}}(t) &= \frac{\sqrt{2}}{2}\text{sign}[S_{\text{linear}}(t)] \\
&= \frac{\sqrt{2}}{2}\left\{\alpha_{\text{I}}\text{sign}[\sin(2\pi f_{\text{s}}t + \varphi_{\text{I}})] + j\alpha_{\text{Q}}\text{sign}\left[\sin\left(2\pi f_{\text{s}}t + \varphi_{\text{Q}}\right)\right]\right\}
\end{aligned}
\tag{4.56}
$$

where sign(·) is the symbol function. It is easy to prove that the envelope of $s_{\text{ACE-BOC}}(t)$ in Eq. (4.56) is constant. Since the power of any one of the signal components can be set as 0 without affecting the constant envelope property, no more than four signal components of ACE-BOC can achieve constant envelope multiplexing. Given a specific power ratio, ACE-BOC can be generated by referring to a two-dimensional phase lookup table like AltBOC. According to the first definition, the multiplexing efficiency of ACE-BOC is 87.20%. For any power ratio of the main lobe, the multiplexing efficiency of ACE-BOC is 81.06%.

(4) ES ACE-BOC

The ES ACE-BOC with Equal Length Subcarrier Segments can also achieve the constant envelope multiplexing of dual-frequency four signal components, but by solving the subcarrier coefficients through a numerically optimized process. In the ES ACE-BOC, the dual-frequency constant envelope signal is firstly assumed to be [46]

$$
\begin{aligned}
s_{\text{ES-ACE-BOC}}(t) = {} & s_2(t)[u_{\text{s}2}(t) + ju_{\text{s}2}(t - T_{\text{s}}/4)] + s_1(t)[u_{\text{s}1}(t) - ju_{\text{s}1}(t - T_{\text{s}}/4)] \\
& + js_4(t)[u_{\text{s}4}(t) + ju_{\text{s}4}(t - T_{\text{s}}/4)] + js_3(t)[u_{\text{s}3}(t) - ju_{\text{s}3}(t - T_{\text{s}}/4)] \\
& + s_1(t)s_3(t)s_4(t)\left[u_{\text{p}2}(t) + ju_{p2}(t - T_{\text{s}}/4)\right] \\
& + s_2(t)s_3(t)s_4(t)\left[u_{\text{p}1}(t) - ju_{\text{p}1}(t - T_{\text{s}}/4)\right] \\
& + js_1(t)s_2(t)s_3(t)\left[u_{\text{p}4}(t) + ju_{\text{p}4}(t - T_{\text{s}}/4)\right] \\
& + js_1(t)s_2(t)s_4(t)\left[u_{\text{p}3}(t) - ju_{\text{p}3}(t - T_{\text{s}}/4)\right]
\end{aligned}
\tag{4.57}
$$

where $u_i(t)$ is a periodic subcarrier, $i \in \{s_1, s_2, s_3, s_4, p_1, p_2, p_3, p_4\}$, which determines the relationship between the power ratio and the phase of each signal component. Each period of the subcarrier is divided into M segments that can be expressed as

$$
u_i(t) = \sum_{k=0}^{M-1} x_i^{(i)}\text{rect}\left(t - \frac{kT_{\text{s}}}{M}\right) \tag{4.58}
$$

where rect($t$) is a rectangular function that is 1 in case of $[0, T_{\text{s}}/M]$ and is 0 in other cases; $M$ is a multiple of 4, typically chosen to be 8 for ease of implementation. In order to obtain these subcarrier coefficients, the highest multiplexing efficiency (power at the main lobe) needs to satisfy the constraints on constant envelope, power ratio, and phase.

(5) Generalized AltBOC

Generalized AltBOC is the phase rotation of generalized DualQPSK, and is also a method of four-signal dual-frequency constant envelope multiplexing. Among the four signal components, the data component and pilot component of the lower sideband are orthogonal to each other with equal power, so are the data component and pilot component of the upper sideband. The total power ratio between the upper and lower sidebands is adjustable, as shown below [47]:

$$\begin{aligned} s_{\text{Generalized\_AltBOC}}(t) = {} & (s_1(t) + js_2(t))e^{-j\pi/8k_s(t)} \\ & + p(s_3(t) + js_4(t))e^{j\pi/8k_s(t)} \\ & + b(s_2(t)s_3(t)s_4(t) + js_1(t)s_3(t)s_4(t))e^{j\pi/8k_p(t)} \\ & + a(s_1(t)s_2(t)s_4(t) + js_1(t)s_2(t)s_3(t))e^{-j\pi/8k_p(t)} \end{aligned} \tag{4.59}$$

## 4.2.6 Navigation Signal Compatibility Design

The definition of compatibility between GNSS systems has been widely studied and discussed. At the Fourth ICG Meeting held in St. Petersburg, Russia, in September 2009, compatibility was defined as "global or regional navigation systems and their augmentation systems, which can be used alone or together without introducing unacceptable interference or other damage to a standalone system or service." The narrow compatibility mainly refers to the compatibility between GNSS systems. The generalized compatibility includes, in addition to the compatibility between GNSS systems, also the compatibility with other systems such as satellite communication system and terrestrial communication system and even with radio astronomy services.

1. **Compatibility assessment method**

The compatibility assessments include GNSS signaling system compatibility assessment and RF compatibility assessment. The main indicators used in signal system compatibility assessment are Spectral Separation Coefficient (SSC) and Code Tracking Spectrum Sensitivity Coefficient (CTSSC). The SSC is defined as [40]

$$\kappa_{s,i} = \int_{-\beta_r/2}^{\beta_r/2} G_s(f)G_i(f)\mathrm{d}f \tag{4.60}$$

The CTSSC is defined as [38]

$$\chi_{s,j} = \frac{\int_{-\beta_r/2}^{\beta_r/2} G_s(f)G_j(f)\sin^2(\pi f d)\mathrm{d}f}{\int_{-\beta_r/2}^{\beta_r/2} G_s(f)\sin^2(\pi f d)\mathrm{d}f} \tag{4.61}$$

where $G_s(f)$ is the normalized power spectral density of the desired signal within its transmission bandwidth, $G_i(f)$ is the normalized power spectral density of the interference signal within its transmission bandwidth, and $\beta_r$ is the received bandwidth. The actual signal power spectral density is mainly determined by message, spreading code and modulation waveform. However, for convenience, the power spectral density of the modulation waveform is usually used instead of that of the actual signal. This may cause a relatively large error for a signal with short code length.

The RF compatibility is evaluated by using equivalent carrier-to-noise ratio degradation. According to the ITU-R Recommendation M.1831, the carrier-to-noise ratio degradation with the presence of another system is calculated to be [15]

$$\Delta_{\text{dB}} = \left(\frac{C}{N_0 + P_0}\right)_{\text{dBHz}} - \left(\frac{C}{N_0 + P_0 + I_0}\right)_{\text{dBHz}} = \left(\frac{N_0 + P_0 + I_0}{N_0 + P_0}\right)_{\text{dB}} = \left(1 + \frac{I_0}{N_0 + P_0}\right)_{\text{dB}} \tag{4.62}$$

where $N_0$ is the thermal noise spectral density, $P_0$ is the interference caused by the navigation signal system, and $I_0$ is the interference caused by another system. To get accurate RF compatibility assessment results, we need to consider the satellite constellation, antenna pattern, and even transmission power. In the M.1831, the aggregate gain $G_{agg}$ is given. In the worst case scenario, the aggregate gain is defined as

$$G^{\text{agg}} = \frac{\max\limits_{t}\left[\max\limits_{i}\left(p_i^r(t)\right)\right]}{P_{\max}^R} \tag{4.63}$$

where $P_{\max}^R$ refers to the maximum received power of a single satellite at the input end of its receiver antenna; and the numerator is the sum of the maximum received power of all the satellites at their antenna input ends at all times (excluding that of the desired signal). By using the aggregate gain, the interference is calculated to be

$$I_0\ [\text{dB W/Hz}] = G_{\text{agg}}\ [\text{dB}] + P_{\max}\ [\text{dB W}] + \text{SSC}\ [\text{dB/Hz}] + L_x\ [\text{dB}] \tag{4.64}$$

where $P_{\max}$ is the maximum power of interference signal received by a single satellite; $L_x$ is the processing loss; and SSC is the spectral separation coefficient. When considering the impact of other systems on the navigation signal system, $P_0$ can also be calculated in a similar way.

2. **GNSS compatibility**

GNSS compatibility is primarily concerned with the compatibility with the existing satellite navigation systems. According to the criteria set by the US L5 Electromagnetic Compatibility Research Group, L5 is considered compatible with other systems when its signal-to-noise ratio drops by no more than 5.8 dB with the presence of other systems. The signals of current GNSS systems are mainly concentrated in the L frequency band. The actual compatibility assessment needs to consider the signal system parameters, constellation configuration, antenna characteristics, ground received power, and other factors. The carrier-to-noise ratio degradation values at different positions on the ground will also be different. When studying the worst case, the evaluation indices can be simplified by considering only SSC, terrestrial received power, and aggregate gain (the approximate number of satellites).

The SSCs of typical signal components are shown in Table 4.11. Considering that the bandwidths of the desired signal and the interference signal are both 40.92 MHz, the influence of other GNSS systems can be analyzed. When calculating the intra-system interference, the received power of the desired signal is taken as the minimum value (such as −157 dB W). When calculating the inter-system interference, the received power of the interference signal is taken as the maximum value (such as 154 dB W). If 10 satellites can be seen at the same time within the system and the other systems can see the satellite $N_1$ at the same time, then the aggregate gain of intra-system interference will be about 10 dB and the total gain of inter-system interference will be about $10 \times \lg(N_t)$ dB. For the MBOC (6,1,1/11) signal with the thermal noise power spectral density of −201.5 dB W/Hz and the processing loss of 0, the carrier-to-noise ratio will decrease with the number of satellites, as shown in Fig. 4.14. In other words, as the number of interfering satellites increases, the resulting carrier-to-noise ratio will decrease.

3. **Compatibility with radio astronomy**

The Radio Astronomy (RA) band (4990–5000 MHz) is particularly sensitive to interference. The navigation signal in this band must meet its in-band interference requirements. Taking the C frequency band as an example, the ITU has specified the maximum value of Power Flux Density (PFD) of C-band signal as [15]

$$\mathrm{PFDmax} = -171 - X\left[\mathrm{dB\ W/m^2}\right] \tag{4.65}$$

where

$$X = 32 - 25 \log 10(\psi/2) \tag{4.66}$$

**Table 4.11** Spectral separation coefficient dB

| Desired signal | Interference signal | | | | |
|---|---|---|---|---|---|
| | BPSK(1) | BPSK(1O) | MBOC(6, 1, 1/11) | $\mathrm{BOC_{sin}}(10, 5)$ | BOCcos(15, 2, 5) |
| BPSK(1) | −61.82 | −70.00 | −68.16 | −87.08 | −96.42 |
| BPSK(IO) | −70.00 | −71.42 | −70.53 | −79.89 | −83.76 |
| MBOC(6, 1,1 /11) | −68.16 | − 70.53 | −65.46 | 81.98 | −91.34 |
| $\mathrm{BOC_{sin}}(10, 5)$ | −87.08 | −79.89 | −81.98 | −71.58 | −86.48 |
| $\mathrm{BOC_{cos}}(15, 2.5)$ | −96.42 | −83.76 | 91.34 | −86.48 | −68.45 |

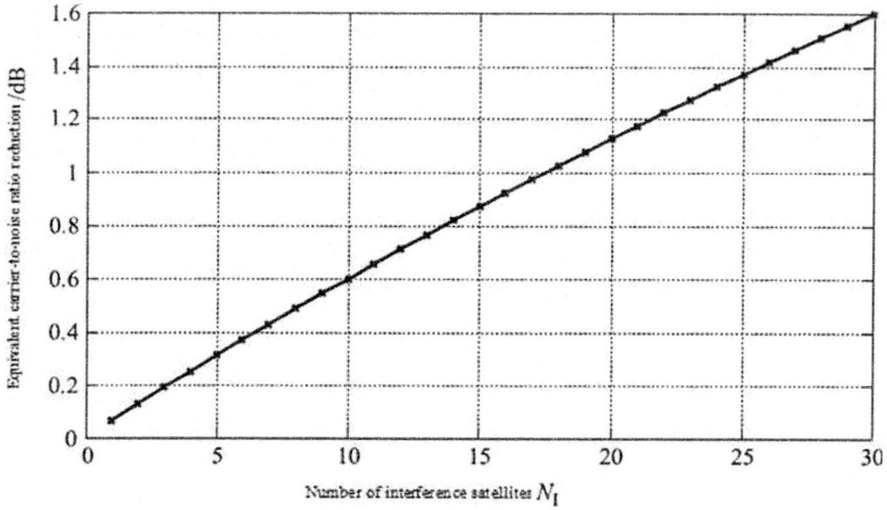


**Fig. 4.14** Equivalent carrier-to-noise ratio as a function of the number of interfering satellites

where

$$\psi = \arccos(1 - 0.02/N_{sat}) \tag{4.67}$$

where $N_{sat}$ is the number of the satellites that are simultaneously within the beam of the radio telescope. Taking a satellite as an example, the maximum value of its signal PFD will be $PFD_{max} = -227.981\,8$ dB W/m$^2$. To meet the attenuation requirements, the power of a signal in the frequency band of 4990–5000 MHz (10 MHz frequency bandwidth) needs to be strictly designed.

4. **Compatibility with non-GNSS systems**

The S frequency band has been applied in satellite navigation, for example, in China's BDS-3 test satellite and India's IRNSS. An important user of the S frequency band is Globalstar Communications Satellite System. The Globalstar provides voice and data services by using the 2483.5–2500 MHz frequency band as a satellite downlink to communicate with the user terminal. Within each beam, this 16.5 MHz frequency bandwidth is divided into 13 Frequency-Division Multiple Access (FDMA) channels, each having a bandwidth of 1.23 MHz and a center frequency of $f_c = 2491.77 \pm k{*}1.23$ MHZ $(k = 0, 1, \ldots, 6)$.

For the $k$th FDMA channel, the baseband power spectral density can be obtained using a raised cosine filter:

$$G^k_{\sin C}(f) = \begin{cases} 1, & f - kB| \le \frac{B}{2}(1-\rho), \quad k = -6, \ldots, 6 \\ 0, & |f - kB| \le \frac{B}{2}(1-\rho), \quad k = -6, \ldots, 6 \\ g(f), & |f - kB| \ge \frac{B}{2}(1+\rho), k = -6, \ldots, 6 \end{cases} \tag{4.68}$$

**Table 4.12** WiMAX parameters

| Item | WiMAX base station | WiMAX mobile station |
|---|---|---|
| Channel bandwidth/MHz | 5 | – |
| Average power/dBW | 36 | 20 |
| Antenna gain/dB | 18 | 0–6 |
| Antenna height/m | 15–30 | 1.5 |
| Line loss/dB | 2 | 0 |

where $g(f) = 1/2 - 1/2\cos\left(\frac{\pi}{2\rho f_c}(f - \rho B - kB)\right)$, in which $f_c$ is the cutoff frequency of the filter, $f_c = B/2$; $B$ is the bandwidth of a single FDMA channel, $B = 1.23$ MHz; $\rho$ is the roll-off factor, $\rho = 0.2$.

The total PSD of the Globalstar signal can be expressed as the sum of the PSDs of the 13 FDMA channels:

$$G_{\mathrm{GLOB}}(f) = \sum_{k=-6}^{6} G_{\mathrm{SRC}}^{k}(f) \tag{4.69}$$

In addition, the 2483.5–2500 MHz frequency band is also planned to be distributed to the Worldwide Interoperability for Microwave Access System (WIMAX system). The parameters of the WiMAX base station and mobile station specified in ITU-R M.2116 are shown in Table 4.12.

## 4.3 Design of Navigation Signal Messages

Navigation messages are important data used for positioning and navigation and broadcasted by the satellite to the user in the binary stream. The most important and basic components of a navigation message are system time, satellite ephemeris, satellite clock correction (i.e., space clock error), etc.

In addition, the navigation messages also contain other data, such as the system/satellite status data to improve the performance and reliability of the system service; the almanac to assist the user receiver in capturing the satellite signals; the earth orientation parameters to facilitate the user to realize the conversion between geocentric solid coordinate system and geocentric inertial coordinate system; and the time deviation information to achieve the compatibility and interoperability between different GNSS systems.

The design of navigation message is directly related to the accuracy and the efficiency of system positioning calculation, and thus has a great impact on the performance of the entire positioning system. According to the timeliness, the data contained in the navigation message can be divided into real-time data and non-real-time Data. In general, the user wants to receive, as soon as possible, the positioning data such as satellite ephemeris and satellite clock and the integrity data (such as signal integrity, message integrity, system working status, etc.). However, this is not the case for other data. The design of the navigation message should be able to meet the broadcasting requirements of different time-sensitive data.

The navigation message also provides the system time and error detection and correction code for the user to realize the data synchronization and find and correct the errors in the transmitted data after capturing the navigation signal.

The navigation message is generally transmitted in fixed-length data segments (the basic message unit, often referred to as frame). The frame transmission time corresponds explicitly to the system time acquired by the user in the process of message reception. The relationship between the contents of navigation message and the system time may be fixed or not. If not, the identification information of the message contents will also be given in each frame.

In summary, the navigation message design should consider the timeliness, amount and mutual relationship of various data to be sent, the scalability and flexibility of the message structure, and the utilization rate of communication channel resources. The navigation message structure has the following characteristics:

(1) Generally, the navigation message structure should meet the data timeliness requirements by following different update periods.
(2) The contents of the navigation message include synchronization code, navigation information, time information as well as error detection and correction code.
(3) If the relationship between frame contents and system time is no fixed, the identification information of the contents should be included in each frame.

### *4.3.1 Development History of GNSS Navigation Message*

In October 1997, the first GPS ICD, called IS-GPS-200, was released to define NAV, the earliest commercial message of GPS L1 signal. In November 2003, a new GPS ICD, called IS-GPS-705, was released to define the CNAV (Civil Navigation) message of the L5 signal that had a data block structure. In March 2006, IS-GPS-800 was released to define the CNAV-2 message of the L1C signal. The modification of navigation message is usually parameter modification, while the structure modification can be achieved only when a new signal is designed.

In 1998, GLONASS publicly released an ICD to define the navigation message of L1/L2 frequency. The signal structures and messages at both frequencies are the same, while supporting the GLONASS dual-frequency user receiver. The navigation message of GLONASS, in a fixed frame format, is basically similar to the NAV navigation message of GPS.

In 2002, the Galileo system of the European Union released an ICD to define three frequency bands for navigation signal broadcasting, and different frequencies for different service and navigation messages. Galileo navigation messages include C/NAV (E6) commercial service navigation message, F/NAV (E5a) basic navigation message and I/NAV (E1-B and E5b) integrity navigation message. Compared with NAV navigation messages, Galileo navigation messages can realize the efficient broadcasting of almanac and integrity alarm parameters through inter-satellite and inter-frequency cross-broadcasting while inheriting the fixed frame structure of NAV messages. In February 2010, the Galileo ICD1.0 was officially released.

In 2012 and 2013, the BeiDou Satellite Navigation System (BDS) released the ICD 1.0 and 2.0 for public service signals to define the navigation messages of B1I and B2I signals, which were similar to the GPS NAV fixed frame navigation message.

In the BDS global system under construction, a flexible message structure will be adopted and different messages will be designed for different services and demands. In December 2017, the public service signals B1C, B2a (v1.0) were released, In February 2018, the B3I signal was made public according to the system construction plan. The signal specifications and navigation messages of the BDS system have been elaborated on in this book.

With the development of satellite navigation systems, the design of GNSS navigation messages is also constantly improving. The structure of early navigation messages is based on a fixed frame format, which is simple and easy to use. However, the fixed frame has the disadvantage such as poor data timeliness, low data transmission rate, and limited flexibility and scalability. In the GNSS modernization, the following significant improvements have been made in the message structure:

(1) The GPS CNAV adopts the data block format. 14 kinds of data blocks, each with 300 bit, have been built for different types of contents. Moreover, new data block types can be defined to extend the system functions.
(2) The GPS CNAV-2 adopts the optimized hybrid frame structure based on a fixed frame (composed of three subframes with different lengths) and data blocks. This combination has the advantages of both fixed frame and data block, in addition to high flexibility and scalability
(3) The Galileo F/NAV adopts the "free format". On the basis of a three-level format (mainframe, subframe, and page), different types of information can be flexibly combined and broadcasted by defining different page types.
(4) The signal message design of the BDS global system has also been improved.

### 4.3.2 Design of Fixed frame Navigation Message

In the fixed frame structure, the subframe (pages, strings), frame, and superframe are strictly corresponding to the relevant system time. This is convenient for users to receive and use the data. The GPS L1 (also Called NAV message), GLONASS, and BDS B1I, B2I have all used the fixed frame structure. This section will take GPS NAV and BDS B1I/B2I as examples to explain the fixed frame message.

### 1. GPS NAV message

The GPS L1C/A signals carry the NAV messages. NAV is the earliest navigation message, as shown in Fig. 4.15. It includes frames, subframes, words, and pages. Each frame is 1500 bits long, and is transmitted at a rate of 50 bps for a period of 30 s. One frame includes 5 subframes. Each subframe is 300 bits long, and contains 10 words, each of which is 30 bits long. The 5 subframes are sequentially transmitted from the subframe 1. Either of the subframes 4 and 5 have 25 pages. One page is sent in each frame period. The page 1 of subframe 4 or 5 is transmitted in the first frame period, and the page 2 transmitted in the second frame period, and so on. Thus, the contents of the first 3 subframes are repeated once every 30 s. When the first 3 subframes are repeatedly transmitted for 25 times, the transmission of all the pages in subframes 4 and 5 will be completed once. Therefore, the transmission of the complete GPS NAV navigation message needs 750 ($30 \times 25$) seconds, and the superframe period of GPS NAV is 750 s.

The contents of each GPS NAV subframe is related to real-time performance. The subframes 1, 2, and 3 are used to transmit basic navigation information with high real-time requirement, including satellite time, satellite clock correction parameters, satellite ephemeris parameters, satellite accuracy, and satellite health information (integrity information). The subframes 4 and 5 are used to transmit supplementary positioning and navigation information with a low real-time requirement, including the almanacs and health information of 32 stars, and ionospheric model parameters, UTC data, and other information.

The overall structure of the GPS NAV message is shown in Fig. 4.16. The first two words of each subframe are Telemetry Word (TLW) and Handover Word (HOW). The 3rd–10th words of each subframe are message data.

### 2. BDS message structure

According to the information rate, the messages of BDS B1I and B2I signals fall into two types: D1 (50 bps) and D2 (500 bp). The D1 is composed of a superframe,

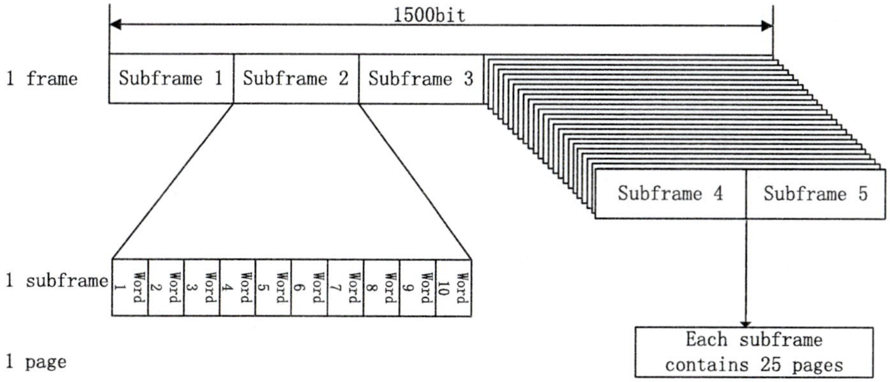


**Fig. 4.15** Structure of GPS NAV navigation message frame

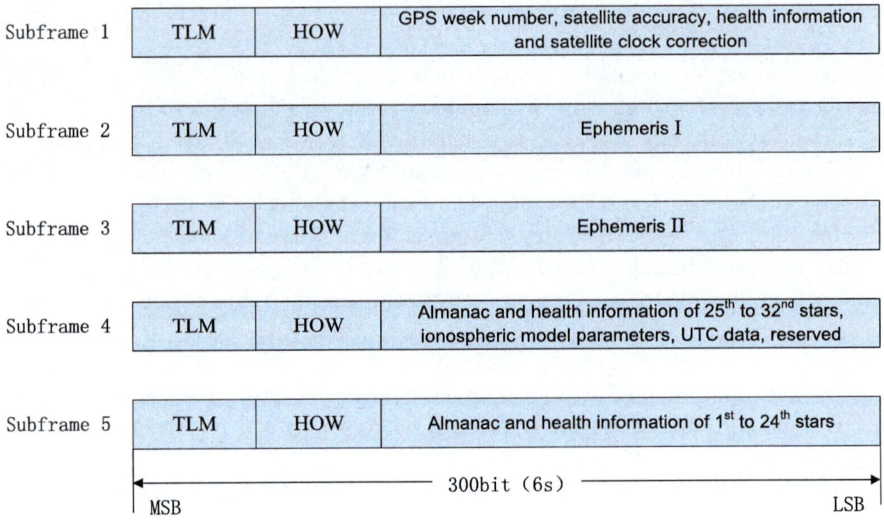


**Fig. 4.16** Format and content of GPS NAV subframe

multiple frames, and multiple subframes. The D1 superframe consists of 24 frames, totaling 36000 bits, and lasting for 12 min. Each D1 frame consists of 5 subframes, totaling 1500 bits and lasting for 30 s. Each D1 subframe consists of 10 words, totaling 300 bits and lasting for 6 s. And each word has 30 bits and lasts for 0.6 s.

The five subframes are sequentially transmitted from the subframe 1 to the subframe 5. Among them, either of the subframes 4 and 5 has 24 pages. One page is transmitted in each frame period. Page 1 of subframe 4 or 5 js transmitted in the first frame period, and page 2 transmitted in the second frame period, and so on. Thus, the contents of the first 3 subframes are repeated once every 30 s. When the first 3 subframes are repeatedly transmitted for 24 times, the transmission of all the pages of subframes 4 and 5 will be completed once. Therefore, the transmission of the complete BDS D1 navigation message needs 720 (30 × 24) s, and the superframe period of BDS D1 is 720 s. The structure of the D1 message frame is shown in Fig. 4.17.

The D2 is also composed of a super frame, multiple frames and multiple subframes. However, the D2 superframe consists of 120 mainframes, totaling 180,000 bits and lasting for 6 min. Each D2 frame consists of 5 subframes, totaling 1500 bits and lasting for 3 s. Each D2 subframe consists of 10 words, totaling 300 bits and lasting for 0.6 s. And the D2 word has 30 bits and lasts for 0.06 s.

The five subframes are sequentially transmitted from the subframe 1 to the subframe 5, and one page is transmitted in each frame period. The subframe 1 has 10 pages, the subframes 2–4 each have 6 pages, and the subframe 5 has 120 pages. Page 1 of each subframe is transmitted in the first frame period, and page 2 transmitted in the second frame period, and so on. 10 frame periods are needed for the transmission of the 10 pages of subframe 1, 6 periods for the 6 pages of subframes

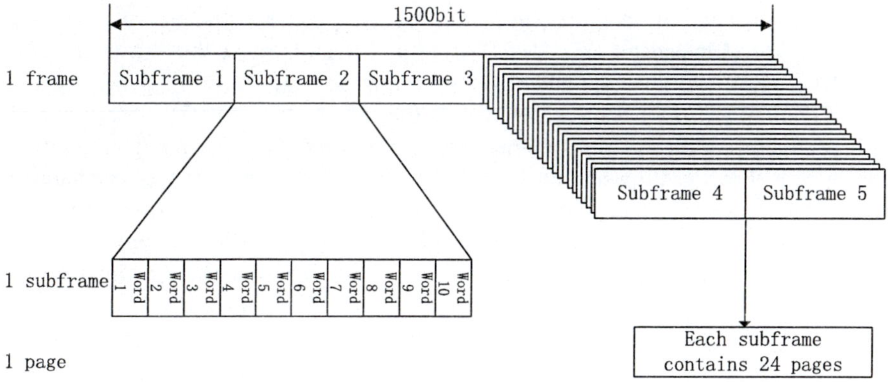


**Fig. 4.17** BDS d1 frame structure

2–4, and 120 periods for the 120 pages of subframe 5. Therefore, 360 (3 × 120) s are needed to complete the transmission of BDS D2 message, and the superframe period of BDS D2 is 6 min. The structure of the D2 message frame is shown in Fig. 4.18.

### 3. **BDS message content**

The contents of each subframe in BDS D1 and D2 are real-time data. The basic navigation information of the broadcasting satellite, including Second of Week (SOW), Week Number (WN), User Range Accuracy Index (URAI), autonomous Satellite Health Flag (SatH1), ionospheric delay model correction parameters (αn, βn), ephemeris and age of data, satellite clock correction parameters and age of data, and satellite equipment group delay differential, as well as the information on BDS integrity and differential correction is transmitted in subframes with relatively high repeat frequency. The almanacs and time parameters of other systems are transmitted in subframes with low repeat frequency.

The contents of the D1 message include the basic navigation information of the broadcasting satellite, the almanac of all satellites, and the time parameters from other systems. The D1 subframes 1–3 broadcast the basic navigation information, while the pages 1–24 of subframe 4 and the pages 1–10 of subframe 5 transmit almanac information and time parameters. The pages 11–24 of subframe 5 are currently reserved.

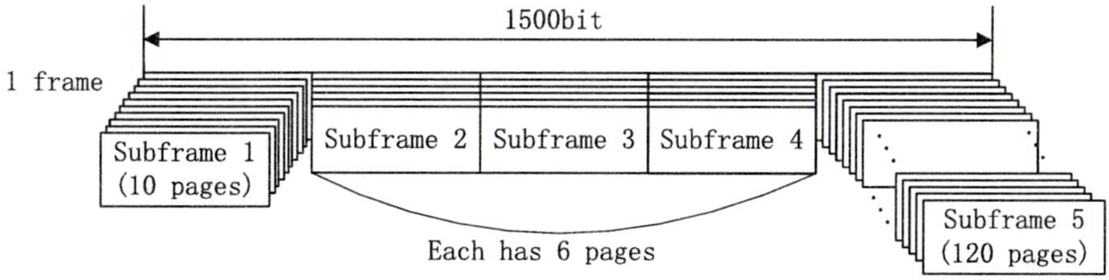


**Fig. 4.18** BDS D2 frame structure

In the future, they will be expanded to suit the BDS construction—from currently supporting the almanacs of only 30 BDS satellites to supporting the almanacs of up to 63 BDS satellites. The contents of D1 subframes are shown in Fig. 4.19.

The contents of the D2 message include the information on BDS integrity and differential correction and ionospheric grid, in addition to the basic navigation information of the broadcasting satellite, the almanac of all satellites, and the time parameters of other systems. The subframe 1 broadcasts the basic navigation information of the broadcasting satellite, the subframes 2–4 broadcast the information on BDS integrity and differential correction, and the subframe 5 broadcasts the almanacs of all satellites, the ionospheric grid, and the time parameters of other systems. The contents of the D2 subframes are shown in Fig. 4.20.

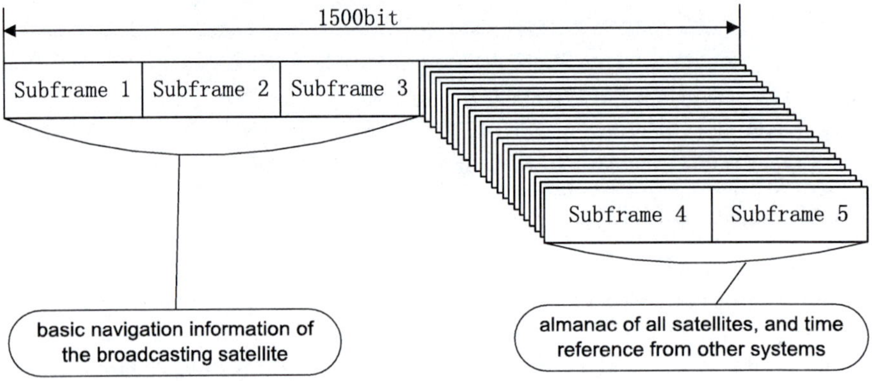


**Fig. 4.19** BDS D1 message contents

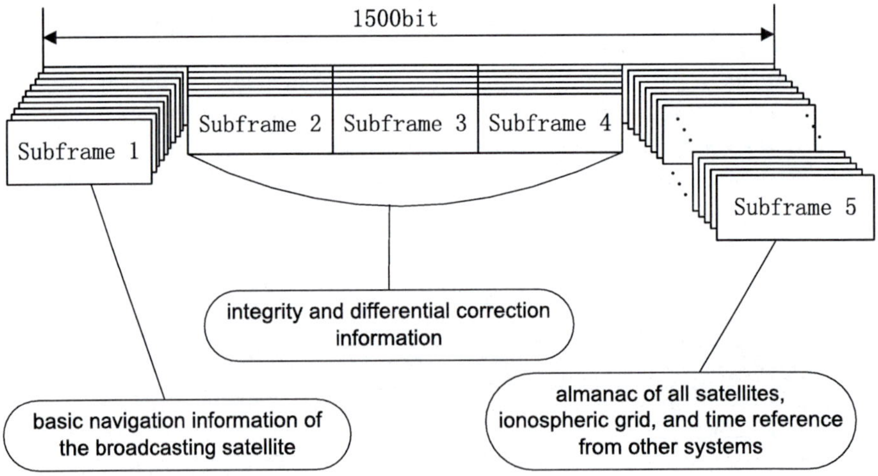


**Fig. 4.20** BDS D2 message contents

(1) Basic navigation information of the broadcasting satellite

The basic navigation information of the broadcasting satellite in the BDS D1/D2 messages includes preamble, SOW, satellite clock correction parameters, etc.

(1) Preamble (Pre)

The first 11 bits of each subframe are the preamble "11100010010" for frame synchronization.

(2) Subframe identification

A total of 3 bits are used for the identification of the 5 subframes.

(3) BDS time (BDT)

It consists of a 13-bit Week Number (WN) and a 20-bit Second Of Week (SOW). WN is the integral week count of BDT in the range of 0–8191, starting from 00:00:00 on January 1, 2006. SOW is the count of seconds in the week of BDT, and is counted from zero at BDT 00:00:00 on Sunday.

(4) User Range Accuracy Index (URAI)

User Range Accuracy (URA), denoted by User Range Accuracy Index (URAI), is used to describe the accuracy of satellite space signals in meter. The URAI is a 4-bit number in the range of 0–15.

The calculation formula for the user to derive URA from URAI is as follows:

If $0 \leq N < 6$, then $X = 2^{N/2+1}$;
If $6 \leq N < 15$, then $X = 2^{N-2}$;
If $N = 15$, it means that the satellite is in orbit maneuver status or there is no accuracy forecast.

(5) Autonomous Satellite Health flag (SatH1)

The SatH1 has 1 bit: 0 or 1. "0" indicates that the broadcasting satellite is normal, while "1" indicates not.

(6) Ionospheric Delay Model Parameters ($\alpha 0$, $\alpha 1$, $\alpha 2$, $\alpha 3$, $\beta 0$, $\beta 1$, $\beta 2$, $\beta 3$)

The 8 parameters have 64 bits in total. The user can calculate the vertical ionospheric delay correction of the B1I signal by using the 8 parameters and the Klobuchar model. For the B2I signal, the ionospheric delay needs to be multiplied by a frequency-dependent factor $k(f) = \frac{f_1^2}{f_2^2}$, where $f_1$ is the nominal carrier frequency of B1I signal and $f_2$ is the nominal carrier frequency of B2I signal.

For the B1I/B2I dual-frequency user, the group delay caused by ionospheric effect can be corrected by the following B1I/B2I dual-frequency pseudorange correction formula:

$$\mathrm{PR} = \frac{\mathrm{PR}_{\mathrm{B2I}} - k(f) \cdot \mathrm{PR}_{\mathrm{B1I}}}{1 - k(f)} - \frac{C \cdot (T_{\mathrm{GD2}} - k(f) \cdot T_{\mathrm{GD1}})}{1 - k(f)} \tag{4.70}$$

where PR is the pseudorange after the correction of ionospheric effects, $\mathrm{PR}_{\mathrm{B1I}}$ is the B1I signal pseudorange (corrected by satellite clock correction rather than $T_{\mathrm{GD1}}$), $\mathrm{PR}_{\mathrm{B2I}}$ is the B2I signal pseudorange (corrected by satellite clock correction rather than $T_{\mathrm{GD2}}$), $T_{\mathrm{GD1}}$ is the equipment group delay differential of B1I, $T_{\mathrm{GD2}}$ is the equipment group delay differential of B2I, $C$ is the speed of light, i.e., 2.99792458 × 108 m/s.

(7) Equipment Group Delay Differentials ($T_{\mathrm{GD1}}$, $T_{\mathrm{GD2}}$)

The $T_{\mathrm{GD1}}$ and $T_{\mathrm{GD2}}$ have 10 bits, respectively.

(8) Age of Data, Clock (AODC)

AODC is a 5-bit number and is the extrapolated interval of clock correction parameters. It is the time difference between the reference epoch of clock correction parameters and the last observation epoch of those parameters. AODC is updated at the start of each hour in BDT.

(9) Clock correction parameters ($t_{oc}$, $a_0$, $a_1$ and $a_2$)

The four clock correction parameters have 74 bits in total. They are defined in Table 4.13.

The BDT at the time of signal transmission can be calculated by the following equation:

$$t = t_{\mathrm{sv}} - \Delta t_{\mathrm{sv}} \tag{4.71}$$

where $t$ is the BDT (in seconds) at the time of signal transmission, $t_{sv}$ is the effective satellite ranging code phase time (in seconds) at the time of signal transmission, $\Delta t_{sv}$ is the offset of satellite ranging code phase time (in seconds) and is given by

$$\Delta t_{\mathrm{sv}} = a_0 + a_1(t - t_{\mathrm{oc}}) + a_2(t - t_{\mathrm{oc}})^2 + \Delta t_{\mathrm{r}} \tag{4.72}$$

**Table 4.13** Clock correction parameters

| Parameter | Definition | Number of bits | Scale factor (LSB) | Effective range | Units |
|---|---|---|---|---|---|
| $t_{oc}$ | Reference time of clock | 17 | $2^3$ | 604,792 | s |
| $a_2$ | Frequency drift | 11* | $2^{-66}$ | – | $s/s^2$ |
| $a_1$ | Frequency offset | 22* | $2^{-50}$ | – | s/s |
| $a_0$ | Time offset | 24* | $2^{-33}$ | – | s |

[a]Two's complement representation with the sign bit (+ or –) occupying the MSB

where $a_0$, $a_1$, and $a_2$ are satellite clock offset, satellite clock frequency offset and satellite clock frequency drift, respectively; $\Delta t_r$ is the correction term to relativistic effect (in seconds); and $t_{oc}$ is the reference time of clock parameters (in seconds) in the effective range of 0–60,4792.

$\Delta t_r$ can be calculated by the following equation. It changes with the sinusoidal variation of the eccentric anomaly of satellite orbit near-point angle $E_k$.

$$\Delta t_r = F \cdot e \cdot \sqrt{A} \cdot \sin E_k \tag{4.73}$$

where *e* is the orbit eccentricity; *A* is the semimajor axis of satellite orbit; *e*, *A* and $E_k$ are given in the ephemeris of the broadcasting satellite; *F* is given by

$$F = -2\mu^{1/2}/C^2 \tag{4.74}$$

where $\mu = 3.986004418 \times 10^{14}\ \text{m}^3/\text{s}^2$ is the Earth's universal gravitational constant, and $C = 2.99792458 \times 10^8$ m/s is the light speed.

The users of single-frequency B1I and B2I should make a further correction as follows:

$$(\Delta t_{sv})_{B1I} = \Delta t_{sv} - T_{GD1} \tag{4.75}$$

$$(\Delta t_{sv})_{B2I} = \Delta t_{sv} - T_{GD2} \tag{4.76}$$

(10) Ephemeris data age (AODE)

AODE is a 5-bit number and is the extrapolated interval of ephemeris parameters. It is the time difference between the reference time of ephemeris parameters and the last observation time of those parameters. AODE is updated at the start of each hour in BDT.

11) Ephemeris parameters

The ephemeris parameters, including 15 orbital parameters and 1 ephemeris reference time, describe the satellite orbits obtained at a certain curve fit interval. The update period of ephemeris parameters is usually 1 h, as defined in Table 4.14.

The user receiver can calculate the coordinates of the satellite in the CGCS2000 coordinate system based on the received ephemeris parameters.

(2) Almanac information

The almanac includes the approximate position of the BDS satellites. The user receiver uses the almanac and local position to predict the satellite's operational status, thus shortening the satellite acquisition time. The specific definitions are shown in Table 4.15.

When the user receives the almanac, the reference time corresponding to the group of almanac parameters can be derived from the almanac reference time $t_{oa}$ and the almanac reference week number $WN_a$.

**Table 4.14** Ephemeris parameter definitions

| Parameter | Definition |
|---|---|
| $t_{oe}$ | Ephemeris reference time |
| $\sqrt{A}$ | Square root of semimajor axis |
| $e$ | Eccentricity |
| $\omega$ | Argument of perigee |
| $\Delta n$ | Mean motion difference from computed value |
| $M_0$ | Mean anomaly at reference time |
| $\Omega_0$ | Longitude of ascending node of orbital plane computed according to reference time |
| $\dot{\Omega}$ | Rate of right ascension |
| $i_0$ | Inclination angle at reference time |
| $\dot{i}$ | Rate of inclination angle |
| $C_{uc}$ | Amplitude of cosine harmonic correction term to the argument of latitude |
| $C_{us}$ | Amplitude of sine harmonic correction term to the argument of latitude |
| $C_{rc}$ | Amplitude of cosine harmonic correction term to the orbit radius |
| $C_{rs}$ | Amplitude of sine harmonic correction term to the orbit radius |
| $C_{ic}$ | Amplitude of cosine harmonic correction term to the angle of inclination |
| $C_{is}$ | Amplitude of sine harmonic correction term to the angle of inclination |

**Table 4.15** Definitions of almanac parameters

| Parameter | Meaning |
|---|---|
| $t_{oa}$ | Almanac reference time |
| $\sqrt{A}$ | Square root of the long semiaxis |
| $e$ | Eccentricity |
| $\omega$ | Argument of perigee |
| $M_0$ | Mean anomaly of reference time |
| $\Omega_0$ | Reference ascending node of orbital plane computed according to reference time |
| $\dot{\Omega}$ | Rate of right ascension |
| $\delta_i$ | Correction of reference orbital inclination at reference time |
| $a_0$ | Satellite clock bias |
| $a_1$ | Satellite clock rate |

The almanac week number WNa, an 8-bit number, is the BDT integer week count (Modulo 256) with the effective range of 0–255.

(3) Time parameters relative to UTC and other GNSS systems

These parameters reflect the time relationships between BDT and UTC or other GNSS systems (GPS, Galileo, and GLONASS). The specific definitions are shown in Table 4.16 [1].

**Table 4.16** Definitions of time parameters relative to UTC and other systems [1]

| Parameter | Definition |
|---|---|
| $A_{0UTC}$ | BDT clock bias relative to UTC |
| $A_{1UTC}$ | BDT clock rate relative to UTC |
| $\Delta t_{LS}$ | Delta time due to leap seconds before the new leap second becomes effective |
| $WN_{LSF}$ | Week number of the new leap second |
| $DN$ | Day number of week of the new leap second |
| $\Delta t_{LSF}$ | Delta time due to leap seconds after the new leap second becomes effective |
| $A_{0GPS}$ | BDT clock bias relative to GPS time |
| $A_{1GPS}$ | BDT clock rate relative to GPS time |
| $A_{0Gal}$ | BDT clock bias relative to Galileo time |
| $A_{1Gal}$ | BDT clock rate relative to Galileo time |
| $A_{0GLO}$ | BDT clock bias relative to GLONASS time |
| $A_{1GLO}$ | BDT clock rate relative to GLONASS time |

After receiving the BDT and time parameters, the user can calculate the UTC time, GPS time, GLONASS time, and Galileo time by using the conversion formula. The broadcast time parameters relative to other GNSS systems indicate that the BDS has considered interoperability with other GNSS systems.

(4) Integrity and differential correction information

The information on integrity and differential correction, which serves for regional users at present, is only broadcast on D2 messages, This information includes the BDS satellite identification of integrity and differential correction information (BDID), BDS regional user range accuracy index (RURAI), BDS differential correction, and differential integrity information ($\Delta$t, UDREI).

At present, the BDS D2 message contains the integrity and differential information of 18 satellites, which may be expanded as needed.

(5) Ionospheric grid information (Ion)

The ionospheric grid information is only broadcast on the D2 message to mainly serve regional users. Each Ion includes the vertical delay at grid point ($d\tau$) and its error index (GIVEI). The ionospheric grid covers 70°–145° east longitude and 7.5°–55° north latitude. It is divided into 320 grid points of 5° $\times$ 2.5°.

$d\tau_i$ is the vertical ionospheric delay of B1I signal at the *i*th grid point, expressed in meters. The user needs to interpolate the ionospheric correction at the grid point and obtain the ionospheric correction at the pierce point of the observed satellite in order to correct the observed pseudorange. The Grid Ionosphere Vertical Error (GIVE), represented by Grid Ionosphere Vertical Error Index (GIVEI), is used to describe the accuracy of delay correction at the ionospheric grid point. According to the $d\tau_i$ value and GIVEI, the user can select the effective data of the grid points adjacent to or nearby the observed intersection point to design the model and obtain the delay correction at the ionospheric pierce point by interpolation.

(6) Message broadcasting strategy

The subframes are broadcasted from subframe 1 to subframe 5 cyclically, and rebroadcasted from subframe 1 at the beginning of the next week regardless of which subframe is being broadcasted before the end of the current week.

For a subframe with pages, the pages are cyclically broadcasted in the order of 1 to *N* (*N* is the total number of pages, e.g., *N* is 24 for D1 subframe 4), and rebroadcasted from page 1 at the beginning of the next week, regardless of which page is being broadcasted before the end of the current week.

### 4.3.3 Design of Data Block Navigation Message

The fixed frame message can be expanded by using the reserved pages in the subframe, at the price of the loss of channel resources. The message with the data block structure has been presented in the design of GPS CNAV message. The BDS B-CNAV2 also uses the data block structure to provide the user with an expandable and flexible navigation message. In this section, GPS CNAV and BDS B-CNAV2 will be taken as examples to explain the data block message.

1. **GPS CNAV message**

CNAV navigation message is broadcasted in GPS L2C and L5 signals. The CNAV adopts the data block format based on message type, instead of traditional fixed format based on frame, subframe, and page. In GPS CNAV, 14 kinds of data blocks of 300 bits length are available for different types of content. The structure of the CNAV message is shown in Fig. 4.21.

The CNAV message starts with a 38-bit data head, which contains the following information:

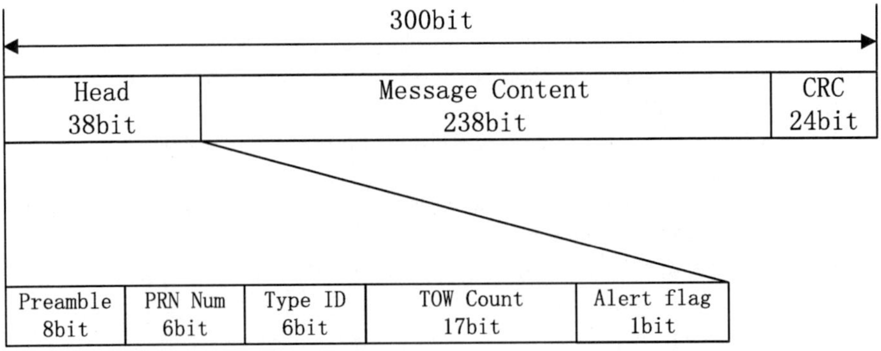


**Fig. 4.21** GPS CNAV message structure

(1) Preamble: "10001011" with 8 bits, the same as NAV subframe preamble.
(2) PRN number: 6 bits, the PRN number corresponding to the satellite.
(3) Message type ID: 6 bits. Up to 64 types of message can be defined, each corresponding to a different type of data block content. The user identifies the message content and format by message type ID.
(4) Time Of Week (TOW) count: 17 bits, similar to that of NAV message. The CNAV also describes the system time with WEEK NUMBER (WN) and the Time Of Week (TOW). TOW (unit: 6 s) represents the number of seconds in the GPS week corresponding to WN.
(5) Alarm flag: 1 bit, similar to the alarm identifier in the NAV message prefix (HOW). When the alarm flag is raised, the user is reminded to use the SV signal at the user's own risk. By quickly informing the user of the condition of satellite signal, alarm flag is of great significance for those applications related to life safety.

2. **Contents of GPS CNAV message**

By using the 6-bit message type ID, the CNAV can define up to 64 message types. So far, 15 message types have been defined, as shown in Table 4.17. The remaining message types are temporarily reserved. The CNAV message is convenient to extend by defining new message types.

**Table 4.17** Definitions of CNAV message types

| Message type ID | Message type | Message content |
|---|---|---|
| 0 | Message type 0 | 1, 0 alternated |
| 10 | Message type 10 | Ephemeris I |
| 11 | Message type 11 | Ephemeris II |
| 30 | Message type 30 | Clock difference parameters, and ionospheric and group delays |
| 31 | Message type 31 | Clock difference parameters, and simple almanac |
| 32 | Message type 32 | Clock difference parameters, and EOP parameters |
| 33 | Message type 33 | Clock difference parameters, and UTC parameters |
| 34 | Message type 34 | Clock difference parameters, and satellite clock and ephemeris correction |
| 35 | Message type 35 | Clock difference parameters, and time differences from GPS and other GNSS systems (GGTO) |
| 36 | Message type 36 | Clock difference parameters, and text message |
| 37 | Message type 37 | Clock difference parameters, and middle almanac |
| 12 | Message type 12 | Simple almanac |
| 13 | Message type 13 | Clock Difference Correction (CDC) |
| 14 | Message type 14 | Ephemeris Differential Correction (EDC) |
| 15 | Message type 15 | Text message |

(1) Message type 0

The message content of message type 0 has no practical meaning. The 238-bit message content in the data block is composed of alternated "1" and "0". When the satellite has an error in the generation or transmission of an effective message, the message of type 0 will be sent in the corresponding time slot. So the message type 0 can be considered as a type of invalid message.

(2) Message types 10, 11

The message types 10 and 11 jointly broadcast the information on satellite ephemeris and health. They are equivalent to the data in NAV subframes 2 and 3. Either of the message types 10 and 11 includes an ephemeris reference time $t_{oe}$. Only when the two message types share the same ephemeris reference time can they be combined into a complete set of ephemeris parameters. When using the ephemeris, it is necessary for the user to confirm the ephemeris reference time $t_{oe}$ and avoid the use of unmatching ephemeris parameters.

In addition to ephemeris parameters, the message types 10 and 11 also contain the following information.

(1) WN: The Week Number (WN) is increased from 10 bits (originally held in NAV) to 13 bits, so the week number of CNAV message is the actual GPS week modulo 8192. It will be cleared once every 157 years and then restarted.
(2) Health symbols L1, L2, and L5: The health status of the three frequency bands L1, L2, and L5 is identified by a 1-bit number.
(3) User range accuracy: Similar to the NAV using URA, the CNAV uses the user range accuracy index $URA_{ED}I$ to characterize the user range accuracy. The $URA_{ED}I$ has 5 bits, ranging from −16 to 15.

(3) Message types 30–37

Each of the message types 30–37 consists of 238 bits. The first 89 bits are satellite clock correction parameters (clock differential parameters), while the latter 149 bits are different contents in different message types.

Similar to NAV, the clock correction parameters of CNAV also include four clock correction parameters ($t_{oc}$,$a_{f_2}$,$a_{f_1}$ and $a_{f_0}$), satellite clock difference $a_{f_0}$, frequency offset $a_{f_1}$, and frequency shift $a_{f_2}$. The more the bit positions are, the more accurately the parameters can be expressed.

In addition to providing the clock correction parameters, the message types 30–37 also provide other data and information, as detailed below.

(1) Message type 30: Group delay and ionospheric correction parameters. The group delay parameters include a group delay correction value TGD and four inter-frequency and intra-frequency correction terms ISC. The GPS CNAV group delay correction is based on the L1 P(Y) signal, so the $ISC_{L1C/A}$, $ISC_{L2C}$, $ISC_{L5I5}$, and $ISC_{L5Q5}$ indicate, respectively, the delay deviations of L1C/A, L2C, L5I5, and L5Q5 signals from L1 P(Y) signal. The ionospheric model of CNAV is

exactly the same as that of NAV message, namel,y an 8-parameter Klobuchar model.

(2) Message type 31: Parameters of reduced almanac. Compared with NAV almanac, the CNAV's reduced almanac is a new type of almanac, and the reduced almanacs of four satellites can be broadcasted in one message of type 31. Each set of reduced almanac consists of 31 bits, including 6 bits for satellite PRN number, 8 bits for the reference correction time of the semimajor axis $\delta_A$, 7 bits for reference ascending node longitude $\Omega_0$, and 7 bits for reference time latitude and longitude $\Phi_0$. In addition to the above parameters, the reduced almanac contains the following default reference values: $A_{ref} = 26{,}559{,}710$ m (reference value of $\delta_A$); $e = 0$; $\delta_i = 0.0056\pi$ ($i = 55°$); and $\dot{\Omega}_{ref} = -2.6 \times 10^{-9}\pi/s$.

Compared with an ordinary almanac, the reduced almanac has fewer parameters and less precision, but a smaller number of bits. The reduced almanacs of multiple satellites can be broadcasted simultaneously in one data block. The reduced almanacs of the entire constellation can be broadcasted in a short period of time.

(3) Message type 32: Earth Orientation Parameters (EOPs). The EOP parameters are a new type of parameters that can only be found in the CNAV message, rather than in the NAV message. For the satellite position calculated by the satellite's ephemeris and almanac and given in the Earth's solid coordinate system, the user needs to convert it to the inertial coordinate system if required, by using the EOP parameters.
(4) Message type 33: UTC parameters. These parameters provide the conversion relationship between GPST and UTC, and predict the occurrence time of leap seconds and the number of corrections in leap seconds. The definition of CNAV UTC is similar to that of NAV UTC, except for the added clock drift of GPST relative to UTC.
(5) Message type 34: Differential correction parameters, including Clock Differential Correction (CDC) and Ephemeris Differential Correction (EDC). The CDC has 34 bits in total, while the EDC has 92 bits in total.
(6) Message type 35: Time differential parameters (GGTO) of GPS and other GNSS systems. The GGTO parameters are a new type of parameters that can only be found in the CNAV message, rather than in the NAV message. This shows that the GPS CNAV message has considered interoperability with other GNSS systems.

3-bit system identification IDs are used to identify different GNSS systems, including 001 for the Galileo system and 010 for the GLONASS system. 000 means no data is available, and 011-111 are reserved for other systems.

(7) Message type 36: Text message. The text message is a new type of parameters that can only be found in the CNAV message, rather than in the NAV message. Each message of type 36 can accommodate 18 8-bit ASCII characters (144 bits).
(8) Message type 37: Middle almanac. Each message of type 37 can broadcast the middle almanac of a GPS satellite. The middle almanac in the CNAV message and the almanac in the NAV message have the same set of parameters and the

same user algorithm. Their difference is that the bits occupied by CNAV almanac parameters are fewer.

(4) Message type 12

The message type 12 contains the parameters of reduced almanacs. It differs from the message type 31 in that the message type 31 contains both the clock correction parameters and the reduced almanacs of four satellites, while the message type 12 contains only the reduced almanacs of 7 satellites. The parameters and format of reduced almanacs in this message type are consistent with those in the message type 31. The reduced almanac of each satellite is composed of 31 bits.

(5) Message type 13

The message type 13 contains the Clock Differential Correction (CDC) parameters. It differs from the message type 34 in that the message type 34 contains both a set of CDC parameters and a set of Ephemeris Differential Correction (EDC) parameters, while the message type 13 contains only the CDC parameters for six satellites. With 34 bits for each group, the CDC format of this message type is consistent with that of the message type 34.

(6) Message type 14

The message type 14 contains the Ephemeris Differential Correction (EDC) parameters. It differs from the message type 34 in that the message type 34 contains both a set of CDC parameters and a set of EDC parameters, while the message type 14 contains only the EDC parameters for two satellites. With 92 bits for each group, the EDC format of this message type is consistent with that of the message type 34.

(7) Message type 15

The message type 15 contains the text message. It differs from the message type 36 in that the message type 36 contains both clock correction parameters and text message represented by 18 8-bit ASCII characters (144 bits), while the message type 15 contains only the text message represented by 29 8-bit ASCII characters (232 bits).

(8) Message broadcasting strategy

The data block is broadcasted more flexibly than fixed frame. In principle, the broadcasting of CNAV messages is discretionary without strict sequential limitation. After receiving the message, the user judges the corresponding message type and data content by message type ID. However, in the actual broadcasting scenario, the order and frequency of each message type need to meet the requirement of maximizing the broadcasting interval in order to obtain the best user performance. The length of the broadcasting interval depends on the message importance and real-time requirement. For example, the ephemeris parameters in message type 10 and the clock parameters in message type 11 are the most basic navigation service messages with high real-time requirement, so their longest broadcasting interval in the L2C signal is 48 s, that is, they are broadcasted at least once every 48 s. In contrast, the secondary messages such as the Earth orientation parameters in message type 32 have low real-time requirements, so their longest broadcasting interval in the L2C signal is 30 min, that is, they are broadcasted once every 30 min.

### 3. BDS B-CNAV2 message

The BDS B-CNAV2 message broadcasted by the B2a signal also adopts the data block structure. The basic frame structure of the B2a message is shown in Fig. 4.22. The length of each frame is 600 symbols, the data rate is 100 bps, the symbol rate is 200 sps, and the broadcasting period of one frame is 3 s. The preamble has 24 symbols with the value of 0xE24DE8 in hexadecimal.

The length of B-CNAV2 message frame is 288 bits before encoding, and 576 symbols after LDPC encoding. Its data includes 6-bit PRN, 6-bit Mestype, 18-bit Seconds of Week (SOW), 234-bit message data, and 24-bit CRC. Similar to the GPS CNAV using the "message type ID", the BDS B-CNAV2 also defines a variety of navigation messages by "MesType". So far, 8 types of messages have been defined by B-CNAV2, as shown in Table 4.18.

Similar to the CNAV message, the broadcasting sequence of each type of B-CNAV2 message can also be flexibly configured. After receiving the message, the user judges the specific message type and data content by message type ID. However, in the actual broadcasting scenario, the order and frequency of each message type needs to meet the requirement of maximizing the broadcasting interval in order to obtain the best user performance. Normally, the B2a message types are cyclically broadcasted in the order of "message type 10 - message type 11 - message type 3X - message type 40" to ensure that the interval of broadcasting the ephemeris and satellite clock parameters in the B2a message does not exceed 12 s (3 s $\times$ 4 = 12 s),

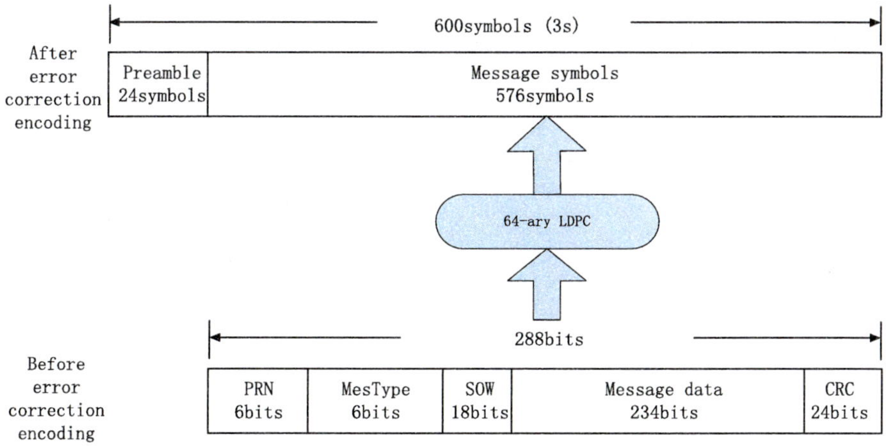


**Fig. 4.22** Structure of B-CNAV2 message frame

**Table 4.18** Definitions of BDS B2a message types

| Message type ID | Message type | Content of broadcasted navigation message |
|---|---|---|
| 10 | Message type 10 | Ephemeris I |
| 11 | Message type 11 | Ephemeris II |
| 30 | Message type 30 | Clock correction parameters, ionospheric delay correction model parameters |
| 31 | Message type 31 | Clock correction parameters, reduced almanacs (for 3 satellites) |
| 32 | Message type 32 | Clock correction parameters, Earth orientation parameters (EOPs) |
| 33 | Message type 33 | Clock correction parameters, BDT-GNSS time offset parameters (BGTO), reduced almanac (1 satellite) |
| 34 | Message type 34 | $SISAI_{oc}$, clock correction parameters, BDT-UTC offset parameters |
| 40 | Message type 40 | $SISAI_{oc}$, Midi almanac |
| | Others | Reserved |

and that the interval of broadcasting the group delay, ionosphere, EOP, BGTO, and other parameters does not exceed 60 s (3 s × 4 × 5 = 60 s), and that the maximum period of broadcasting the middle almanac of the entire constellation does not exceed 756 s (the maximum number of satellites is 63). Actually, the reduced almanac and middle almanac broadcasted by B2a can be configured according to the actual number of on-orbit satellites. When some satellites are not on orbit, their almanacs may not be broadcasted. As a result, the broadcasting frequency of the almanacs of actual on-orbit satellites can be improved. For example, when the number of on-orbit satellites is 30, the actual broadcasting period of middle almanacs will be 360 s.

### *4.3.4 Other Improved Navigation Message Designs*

With the development of satellite navigation systems, the design and format of GNSS navigation messages have been constantly evolving and improving. In addition to fixed frame structure and data block message structure, other improved navigation message structures can also be seen in modern navigation. The main purpose of navigation message structure development is to improve the message efficiency, flexibility, and expandability. This section will briefly introduce several other types of navigation message structures.

1. **GPS CNAV-2 message**

The CNAV-2 navigation message broadcasted by GPS L1C signal adopts an optimized hybrid frame structure based on the fixed frame and data block. It consists of three subframes of different lengths. The subframe 1 broadcasts the time message

(TOI). The subframe 2 broadcasts the ephemeris and satellite clock parameters. The subframe 3 comprises several pages of different types of broadcasting other navigation assistance messages (Fig. 4.23).

The CNAV-2 has the same frame/subframe structure as NAV. Their comparisons are as follows. First, the NAV message consists of 5 subframes of the same length, while the CNAV-2 message consists of 3 subframes of different lengths. Second, either of the NAV subframes 4 and 5 consists of 25 pages, each having the fixed content. The CNAV-2 subframe 3 consists of several expandable pages, each of which has the unfixed content and can be identified by a 6-bit page number.

The data length of CNAV-2 message before encoding is 883 bits, including 9 bits for the subframe 1, 600 bits for the subframe 2 (576 bits for message and 24 bits for CRC), and 274 bits for the subframe 3 (250 bits for message and 24 bits for CRC). The Time Of Interval (TOI) represents the GPS time corresponding to the time epoch of the next frame. The subframe 2 broadcasts the satellite's clock parameters, ephemeris parameters, group delay correction parameters, integrity flags, and Interval Time Of Week (ITOW). The subframe 3 identifies the message type by page number. Currently, six page types have been clearly defined. The CNAV-2 message may be extended by adding new page numbers.

(1) Page 1: GPS-UTC conversion parameters (USNO) and ionospheric parameters (IONO), corresponding to the parameters in CNAV types 33 and type 30.
(2) Page 2: Time Offset (GGTO) parameters of GPS and other GNSS system as well as Earth Orientation Parameters (EOPs), corresponding to the parameters in CNAV type 35 and type 32.

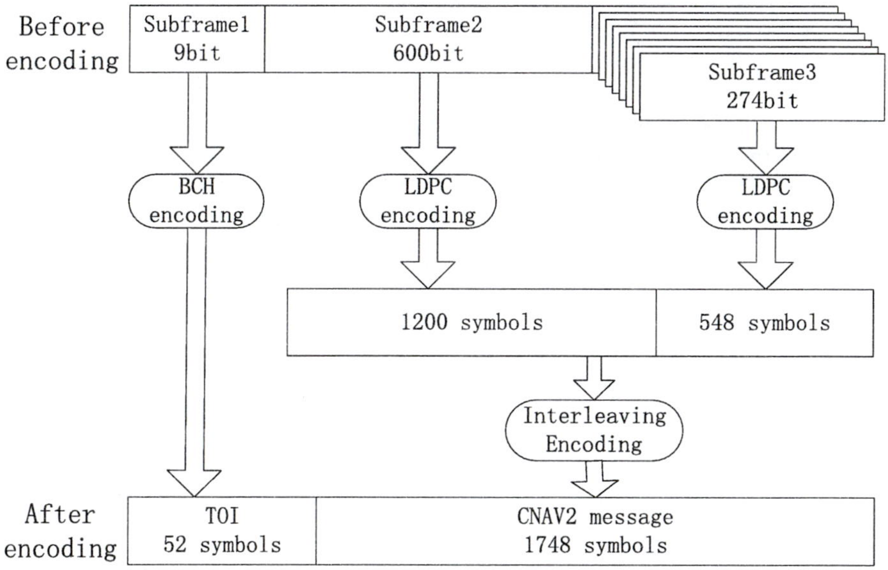


**Fig. 4.23** Structure of GPS CNAV-2 navigation message

(3) Page 3, Page 4: Satellite almanac parameters. Page 3 can provide the reduced almanacs of 6 satellites, and page 4 can provide the Midi almanac of 1 satellite. The definitions of reduced/Midi almanac parameters in the two pages are exactly the same as those in the CNAV message types 31 and 37, respectively.
(4) Page 5: Differential correction parameters. This page can provide a Clock Difference Correction (CDC) parameter and an Ephemeris Differential Correction (EDC) parameter. The 34-bit CDC data block and 92-bit EDC data block in this page have the exactly same contents as those in CNAV type 13 and type 14.
(5) Page 6: Text message. This page can provide 29 ASCII characters, which are exactly the same as those in the CNAV message type 15.

2. **Galileo F/NAV message**

The Galileo E5A-I signal broadcasts the F/NAV message with page flow structure. The F/NAV message structure is shown in Fig. 4.24. In this message, one frame consists of 12 subframes, and one subframe consists of 5 pages. Therefore, page is the basic unit of F/NAV message. One frame lasts for 600 s, one subframe lasts for 50 s, and one page lasts for 10 s. Each page consists of 500 coded symbols, including a 12-symbol sync code used for synchronization pattern, followed by 488 symbols.

Page is the basic unit of F/NAV navigation message and each page corresponds to a certain type. The receiver judges the data content and format by page type, as different types of pages have different navigation message contents and formats. The value of page type is not directly related to the message order, subframe order, and frame order. Therefore, the navigation message can be regarded as a page-by-page or "page flow" structure. The advantage of this arrangement is that it can provide message timely to the user, while the message can be expanded by defining a new page type. So far, 6 page types have been defined in the Galileo F/NAV message, and their corresponding navigation message contents are shown in Table 4.19.

The Galileo F/NAV message is broadcasted in one frame, which is divided into 12 sequentially transmitted subframes. Among the subframes, the page types corresponding to the odd subframes (i.e., subframe 1, subframe 3, subframe 5, subframe 7, subframe 9, subframe 11) are 1, 2, 3, 4, and 5, while the page types corresponding to the even subframes (i.e., subframe 2, subframe 4, subframe 6, subframe 8, subframe 10, subframe 12) are 1, 2, 3, 4, and 6.

The page type 1 mainly provides the clock parameters of the transmitting satellite, and the page types 2–4 jointly provide the ephemeris of the transmitting satellite. The message contents of the above four page types are repeated in each subframe. The parameters provided by the four page types are necessary for positioning. The pages 5 and 6 provide the almanacs of satellites in the constellation, but with different content arrangements. 3 satellite almanacs can be provided in two consecutive subframes, 18 satellite almanacs in one frame, and 36 satellite almanacs in two consecutive frames (i.e., once every 20 min). This design is enough to cover the almanacs of all the 30 satellites in the entire Galileo constellation.

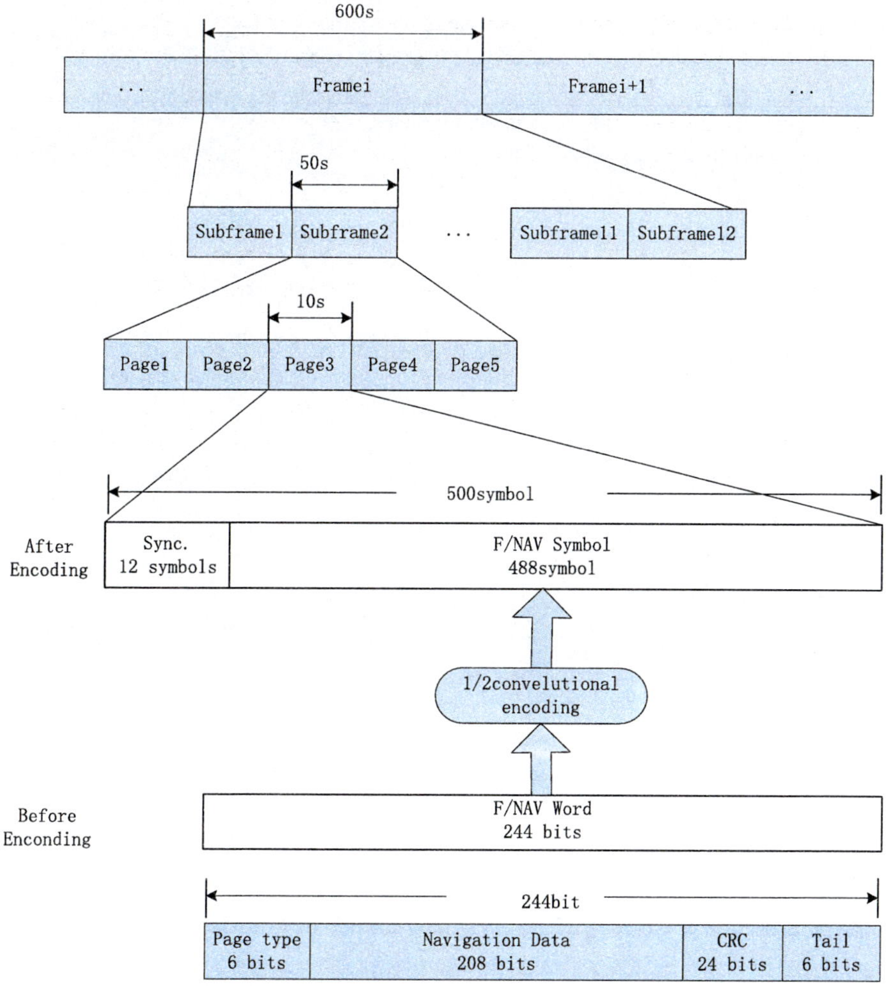


**Fig. 4.24** Structure of Galileo F/NAV navigation message

**Table 4.19** Definitions of Galileo F/NAV message page types

| Page type | Source code | Broadcasted navigation message content |
|---|---|---|
| 1 | 000001 | SVID, clock correction, SISA, ionospheric correction, BGD, signal health and data validity status, GST |
| 2 | 000010 | Ephemeris (1/3), GST |
| 3 | 000011 | Ephemeris (2/3), GST |
| 4 | 000100 | Ephemeris (3/3), GST-UTC conversion, GST-GPST conversion, TOW |
| 5 | 000101 | Almanac ($SVID_k$ and $SVID_{k+1}$(1/2)), Week Type and almanac reference time |
| 6 | 000110 | Almanac ($SVID_{k+1}$(2/2) and $SVID_{k+2}$) |
| Others | Others | Reserved |

# Chapter 5
# Satellite Navigation Inter-satellite Link Technology

Inter-satellite link (or Crosslink) refers to the link between satellites, or the link between spacecraft in a more generalized sense. Its main functions include inter-satellite communication, data transmission, inter-satellite ranging, and inter-satellite TT&C.

The role of inter-satellite link varies with the space system. In the communication satellite constellation, the inter-satellite link can solve nagging problems such as the satellite-ground hops and communication delay. In the reconnaissance formation system, it can increase the aperture of a virtual camera to improve the resolution.

In the navigation satellite constellation, the inter-satellite link can support autonomous operation so as to improve the positioning accuracy. In the relay satellite system, it can increase the TT&C arcs of user satellite. Inter-satellite links are also used in scientific research, for example, in the gravity detection satellite system. They enable multiple satellites to form an organic whole, namely, constellation system, thus improving the operation capability of a single satellite.

This chapter will review the history of inter-satellite link, and address the inter-satellite link frequency selection and signal system, Inter-Satellite Link (ISL) technique, inter-satellite ranging technique, inter-satellite network protocol, and inter-satellite link management technique.

## 5.1 Overview of Inter-satellite Link Technology

### 5.1.1 History of Inter-satellite Link

Inter-satellite links have been used in communication constellations, relay satellites, and satellite formation flights before they are applied in navigation constellation

 
J. Xie et al., *Satellite Navigation Systems and Technologies*, Space Science and Technologies, https://doi.org/10.1007/978-981-15-4863-5_5

systems. In 1983–1995, the United States launched the first generation of relay satellite system, which consisted of seven Tracking and Data Relay Satellites (TDRS) A-G. Each of those satellites was equipped with two sets of 4.9 m single-access parabolic antennas that were shared by S and Ku frequency bands and used for high-speed data exchanging between TDRS and user satellites at the maximum reverse transmission rate of 300 Mb/S. In 1997–2002, the US launched the second generation of relay satellites, namely, TDRS-*H*, *I*, and *J*, whose inter-satellite link was improved by enhancing the *S* band multiple access capacity and increasing the transmission rate to 800 Mb/S or higher. Since 2007, the US has been developing the third generation of TDRS satellites, which implement TT&C communication with the user satellites in the *S* and Ku/Ka frequency bands.

The American iridium satellite system, composed of 66 satellites, is the only large constellation system in the world that establishes the inter-satellite links in the Ka band. This system has 6 polar orbits, each containing 11 satellites and with an altitude of around 780 km. The inter-satellite communication is made through Ka band phased array antennas. Four inter-satellite links are co-established for each iridium satellite, two in the forward and backward directions of the same orbit, and other two on the left and right orbital planes. The maximum inter-satellite data transmission rate is 25 Mb/s, and the working frequency band is 22.55–23.55 GHz. Furthermore, a half-duplex inter-satellite transmission mode is adopted. Iridium satellite system is a polar orbit constellation, where the relative distance between satellites in adjacent orbits and their relative velocity, azimuth, and pitch are changing by a small quantity and rate. Moreover, the inter-satellite topology is a simple structure in fixed connection.

In April 2008, China successfully launched a relay satellite, named "Tianlian 1", which had inter-satellite links with low-orbit spacecraft. "Tianlian 1" has been successfully linked with the spacecrafts in different orbits, such as "Shenzhou 7" and "Yaogan 4", which were spaceship and remote sensing satellite, respectively. In this relay satellite system, the inter-satellite distance is up to 45,000 km, the inter-satellite link in Ka band is a narrow beam antenna with strong directivity, and the dish antenna with a capturing and tracking system is adopted. The inter-satellite links in this system are end-to-end links without any inter-satellite data processing unit. Compared with a navigation constellation, this system has fewer inter-satellite links, no network concept, simpler inter-satellite transparent forwarding, and a smaller antenna rotation angle.

Different from communication satellites, the measurement data of each satellite is required to be collected and updated to support precise orbit determination. In the daily operation of a GNSS, the ground station collects the satellite-ground observation data from domestic satellites for orbit calculation and determination, and updates the ephemeris. For overseas satellites, their orbits cannot be determined because of the failure in directly linking them with ground stations. The inter-satellite link serves as a bridge connecting an overseas satellite with a domestic satellite, making it possible for the overseas satellite to send its observations and data back. Moreover, the radio channel built by an inter-satellite link also facilitates the transmission of satellite measurement/operation control data so that the operation that originally had to wait for the satellite entry becomes a remote operation feasible at any time. The next step is to configure the simplified ground-based orbit determination algorithm on the satellite computers to process the measurements of inter-satellite links, so that the satellites can autonomously generate the navigation message independent of the ground station. Through this development, the autonomous operation capacity of the constellation can be improved, and the ground operation management simplified. Therefore, the establishment of inter-satellite links can greatly improve the control and autonomous operation capacities of navigation constellation. This is of great significance and has been widely studied and adopted by main GNSSs.

GPS is the first system with inter-satellite links in the navigation constellation. Starting from the Block IIR satellite, the GPS satellites have been equipped with the inter-satellite link transceivers with the function of autonomous navigation, thus realizing inter-satellite communication and ranging and autonomous GNSS constellation navigation. The GPS Block IIR and Block IIF satellites are provided with the inter-satellite link transceivers in the UHF band (250–290 MHz). Without the support of the ground control system, the two-way ranging and data exchange between Block IIR satellites is achieved through the Time-Division Multiple Access (TDMA). By processing inter-satellite measurement data through the spaceborne processor unit, the satellite ephemeris and clock correction parameters are autonomously generated, and the basic constellation configuration is autonomously maintained. Within 180 days, the User Range Error (URE) is always less than 6 m, and the navigation and positioning accuracy does not decline sharply. Moreover, the integrity of navigation signals can be monitored to enhance their availability, continuity, and reliability.

The GPS III is a brand-new design mainly featuring high-speed transmission, directional inter-satellite link, and global spot-beam enhancement. To develop the inter-satellite links of GPS III, the areas such as frequency band selection, network topology, routing control, and flow control have been studied. Especially, due to the severe interference and poor measurement accuracy of previous UHF band, the inter-satellite link in the *V* band was finally reported to the ITU after the full demonstration of Ka and *V* bands before 2010. The GPS IIIA will continue to use the inter-satellite link in the UHF band, whereas the future GPS III B will introduce the *V* band link additionally. As a result, the inter-satellite data transmission rate, inter-satellite anti-interference ability, and inter-satellite ranging accuracy will be significantly improved, and the autonomous operation capacity will be further enhanced.

The Galileo system currently has no inter-satellite links, relying mainly on ground monitoring stations distributed around the world for satellite orbit determination. But the next generation of the Galileo system is also expected to use inter-satellite links for mutual measurement and information transmission. Although inter-satellite measurement can achieve the autonomous navigation, an excessively complex dynamic model cannot be adopted due to the limitation of Galileo's spaceborne computing capability. Therefore, the Galileo system comprehensively processes the inter-satellite measurements and satellite-ground measurements to correct the ordinary ephemeris, so as to improve the accuracy of satellite-forecasted ephemeris and spaceborne clock. In addition, Europe will perform the relay communication task on the next generation of Galileo system, that is, to equip each navigation satellite with laser inter-satellite links, to establish the data links between navigation satellites and LEO remote sensing satellites, and to establish the Ka band links between navigation satellites and ground station for high-speed data transmission.

Starting from GLONASS-K, Russia's GLONASS satellite navigation system began to implement the *S* band inter-satellite link in the time-sharing system in order to improve the orbit determination accuracy and autonomous navigation ability of satellites. In addition, the later GLONASS-K satellites also carry two payloads for laser inter-satellite link, which are mainly used for time synchronization and the transmission of a small amount of data.

In China, inter-satellite links are not adopted in the BDS-1 and BDS-2 satellites, but in BDS-3. The BDS-3 inter-satellite links constitute a complex inter-satellite network with high requirements for antenna pointing control, inter-satellite network topology, inter-satellite communication protocol, and inter-satellite ranging accuracy. The overall inter-satellite link technology involves multiple disciplines such as general spacecraft design, structure/mechanism, orbit/orbit determination, antenna/microwave, communication/electronics, computer/network, mathematics, and system simulation. The inter-satellite link scheme based on Ka band phased array antennas has been successfully applied to the launched BDS-3 satellite.

### *5.1.2 Application Requirements for Inter-satellite Links in Navigation Satellite*

Navigation satellite inter-satellite links constitute a dynamic wireless network established among the satellites in the global navigation constellation system and with the functions of precise measurement and data transmission. They can point to the ground targets or spatial expansion user targets as needed at the right time to form a space-ground integrated system. The comprehensive analysis shows that the application requirements of GNSS inter-satellite links are mainly reflected in the following aspects: precise orbit determination and time synchronization, data communication, autonomous operation, and extended services.

1. **Precise orbit determination and time synchronization**

In general, the precise orbit determination and time synchronization of navigation satellites in the GNSS are mainly performed by master ground control station, upload stations, and monitoring stations. The main task of monitoring stations is to continuously observe the navigation satellites and collect data through the receiver, and then preprocess all the measurement data before transmitting them to the master control station. By receiving and processing the data from all the monitoring stations, the master control station can calculate the satellite messages, and update the ephemeris regularly through the upload stations.

Take BDS as an example. The observable orbit arcs of MEO satellites are fewer than half of the total period when using monitoring stations in China and upload stations to track it. When the satellite is in an unobservable arc, the satellite clock error and orbit change cannot be timely monitored and forecasted, leading to a rapid decline in the satellite URE performance and then a sharp decline in the global positioning accuracy.

Therefore, for the satellites in invisible arcs outside China, this problem must be solved by building global stations or inter-satellite links. The inter-satellite links can distribute the satellite ephemerides in the whole network to the constellation, and transmit the observation data and key service data of the whole constellation back to the master control station, so as to increase the observable orbit arcs and supplement the satellite-ground observation data. Besides, using the measurement data continuously provided by inter-satellite links can achieve precise inter-satellite and satellite-ground joint orbit determination and time synchronization, high-precision orbit and clock difference calculation, normal operation of global navigation services, and satisfactory completion of global service performance indicators.

2. **Data communication**

The data communication of GNSS mainly involves global operation control data and global TT&C data. Through inter-satellite links, the TT&C capacity, and data communication coverage can be improved to ensure the continuous monitoring and management of global constellation. In the case that global TT&C stations haven't been established, the MEO satellites in the navigation constellation are not observable in some orbit arcs. The inter-satellite link forwarding can achieve the global coverage of navigation constellation, improve the TT&C command upload, operation control data upload, integrity data transmission, and telemetry download, and expand the communication coverage of dispatching management data.

Furthermore, the telemetry and telecontrol capability of satellites can be improved in emergency through the inter-satellite links. In the absence of inter-satellite links, the satellites in the invisible arcs will be unable to transmit the telemetry data back to the ground in real time, so their emergency failure outside the observed area can't be identified by the ground station. Even if the emergency failure of a satellite is detected in time, it is very likely that effective measures cannot be taken on the satellite because it is in an invisible arc or because it stays too short in the visible arc. Therefore, the construction of inter-satellite links will largely meet the demand for satellite telemetry and telecontrol in emergency.

3. **Autonomous operation**

Autonomous constellation navigation is an implementation process where, without ground system support for a long time, the satellites in the constellation continuously correct the long-time forecasted ephemeris and clock parameters previously uploaded by ground stations through the inter-satellite bidirectional measurement and data exchange as well as spaceborne processing, autonomously generate the navigation message and maintain the basic constellation configuration, in order to meet the user's demand for high-precision navigation and positioning applications. Through the inter-satellite link equipment, the GNSS satellite completes the inter-satellite ranging, obtaining the measurement data such as inter-satellite two-way measurements, the transmitting and receiving time of measurement signal, antenna correction, and other ranging correction information. By means of inter-satellite link equipment, the satellite completes the transmission of autonomous navigation information, such as upload navigation ephemeris, satellite attitude, routing, inter-satellite signal distance and Doppler estimate, satellite health and integrity, as well as satellite operation mode.

Through the autonomous operation, the navigation constellation has stronger autonomous capability and weaker dependence on the ground system, and can meet the requirements of both GNSS intelligentization and navigation warfare in the future. In this sense, the autonomous operation has become the main development direction of GNSS.

4. **Extended services**

Inter-satellite link is essentially a channel for both measurement and data transmission. In some typical user scenarios, the unique capabilities of inter-satellite links can be used to provide the positioning and communication services for some specific extended user targets.

When a near-space vehicle flies at high speed, the user cannot receive traditional *L* band signals for navigation purposes due to the blackout effect of atmosphere. When offensive and defensive weapons and MEO/HEO vehicles carry out the long-distance flight missions, a series of problems will occur, such as insufficient arcs for ground tracking and measurement, difficult long-distance tracking, and difficult acquisition of TT&C data. As a result, the space-based data link service is urgently needed.

In the inter-satellite links, the radio-frequency signals higher than *L* band and capable of penetrating the blackout area are usually adopted. In addition, the inter-satellite link payload is characterized by the global coverage of navigation signals and the compatibility with measurement and data transmission capabilities. The surplus inter-satellite link resources of navigation constellations can provide extended navigation services for MEO/HEO spacecraft, deep-space spacecraft, and high-speed near-space vehicles.

### *5.1.3 Functional Composition of Inter-satellite Links*

The functions of inter-satellite links generally include: antenna pointing and signal acquisition/calculation, antenna signal transmission, inter-satellite signal and information transceiving, and inter-satellite network data routing. The composition of an inter-satellite link is shown in Fig. 5.1.

Antenna pointing control and signal capture & tracking are the basis for the inter-satellite link. The satellites working on orbit have not only the motion relative to other satellites, but also their own platform attitude changes. Therefore, the antenna pointing parameter shall be calculated by antenna pointing algorithm in real time according to the attitude direction of satellite platform and the position of the link targets, and the inter-satellite antenna shall be controlled to achieve mutual alignment. In addition, the high-speed relative motion between satellites will cause the Doppler deviation of signals, the large-range variation of inter-satellite ranges, and

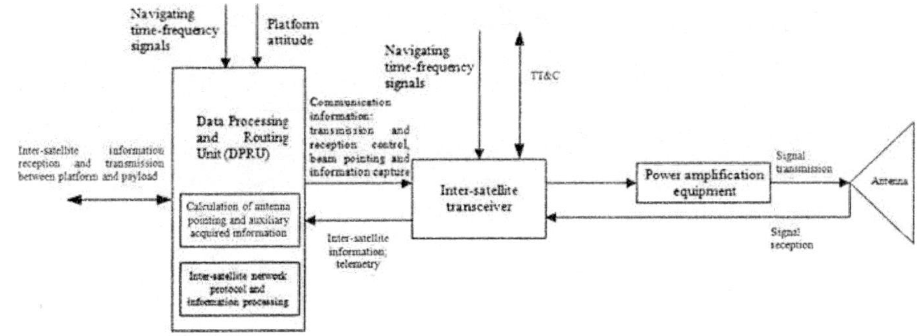


**Fig. 5.1** Composition of the inter-satellite link

the constant changes in signal TOA. Moreover, the fact that independent signal frequencies are adopted at different satellite nodes will lead to the difference in time references. In order to achieve fast signal acquisition, real-time calculation of parameters such as the relative distance, velocity, and clock difference of the two-way link-building objects is required. The antenna pointing and auxiliary information acquisition and calculation software is responsible for calculating the inter-satellite antenna pointing angle and the information parameters acquired by transceivers, and then transmitting the data to relevant equipment to complete the pointing, capturing and tracking of the inter-satellite link. The antenna signal transmission function realizes the long-distance signal exchange through the antenna module and power amplifier.

The inter-satellite signal and information transceiving function is mainly to complete the inter-satellite signal acquisition, demodulation, information coding/decoding, data processing, and ranging through the inter-satellite transceiver. This function involves the selection of inter-satellite link frequency, the design of signal, and many other contents. The key of this function is high-precision measurement. Since the accuracy of inter-satellite measurement is required to be nanosecond level, atomic clock signal should be introduced to the transceiver as time–frequency source, and the pseudocodes or even carrier measurement shall be taken.

The inter-satellite network data routing function is responsible for the management of inter-satellite route planning, and realizes the exchange of inter-satellite ranging values, the operational ephemeris parameters, the satellite commands and the monitoring telemetry data, and the inter-satellite data. The function is based on the inter-satellite data transmission technology and the ground system operation management and collaboration and realized by the inter-satellite network protocol software, implementing the inter-satellite data routing control, and intra-satellite data transmission control.

## 5.2 Frequency Selection and Signal Design of Inter-satellite Link

### 5.2.1 Frequency Selection

1. **Frequency Allocation**

According to the ITU's frequency allocation, the frequency band for relevant Inter-Satellite Service (ISS) is shown in Table 5.1. In this table, the shared service refers to the service where the corresponding frequency band is also allocated to other applications besides the inter-satellite link, and the application region refers to the region to which the shared service is applicable.

According to Table 5.1, the current frequencies allocated by ITU to the inter-satellite links are mainly around 25 GHz (near the K/Ka band), 60 GHz (near the *V* band), and 100 GHZ, and above.

(1) UHF band: 400.15–401 MHz, 410–420 MHz;
(2) *S* band: 2.025–2.11 GHz, 2.20–2.29 GHz;
(3) *C* band: 5.01–5.03 GHz;
(4) Ka band (including the *K* bands in the Table): 22.55–23.55 GHz, 24.45–24.75 GHz, 25.25–27.5 GHz, 32–33 GHz;
(5) *V* band: 54.25–58.2 GHz and 59–71 GHz are allocated for inter-satellite use;
(6) The millimeter wave frequency band: 116–123 GHz, 130–134 GHz, 167–182 GHz, 185–190 GHz, and 191.8–200 GHz are all allocated to inter-satellite use.

In addition, the spectra of laser bands with the wavelengths of 0.8–0.9 4 GHz, 167–182 GHz, 185–190 GHz, and 191.8–200 GHz are all allocated to inter-satellite use or other applications besides inter-sit is difficult to achieve dual-frequency ranging by simply using the *S* and *C* bands. The UHF band used by GPS is legally allocated with a narrow bandwidth. Now the GPS III has gradually replaced it with *V* band. Currently, inter-satellite link frequencies are mostly the bands lower than Ka.

1. **Inter-satellite Frequency Selection**

From the perspective of spaceborne product technology maturity, the antennas with lower frequency in the UHF band are the most mature due to earlier development and have been applied in many satellites. The antennas currently installed on the GPS Block IIR and Block IIRM satellites on orbit are just in the UHF band. The Ka band antennas have been used in many satellites. The moveable spot-beam antenna scheme has been introduced to China's communications satellites, while the *V* band inter-satellite link antenna has been used in the US Iridium satellite.

As far as the design life of spaceborne antennas is concerned, the antenna in the UHF band has a wide beam that frees the antenna from rotation so that the antenna can achieve a long life easily. The spaceborne Ka band antenna has been used all

**Table 5.1** ITU's frequency allocation for inter-satellite links

| Frequency band (GHz) | Shared services | Applicable region |
|---|---|---|
| 22.55–23.55 | Fixed, mobile | ALL |
| 24.45–24.65 | Fixed | 1 |
| | Radio navigation | 2 |
| | Fixed, mobile, radio navigation | 3 |
| 24.65–24.75 | Fixed | 1 |
| | Satellite radio navigation (ground-to-space) | 2 |
| | Fixed, mobile | 3 |
| 25.25–25.5 | Fixed, mobile<br>Satellite standard frequency and time signal (ground-to-space) | ALL |
| 25.5–27 | Earth exploration by satellite (space-to-ground), fixed, mobile, space research (space-to-ground)<br>Satellite standard frequency and time signal (ground-to-space) | ALL |
| 27–27.5 | Fixed, mobile | 1 |
| | Fixed, satellite-fixed (ground-to-space), mobile | 2, 3 |
| 54.25–55.78 | Earth exploration by satellite (passive), space research (passive) | ALL<br>ALL |
| 32.3–33 | Fixed, radio navigation | |
| 55.78–56.9 | Earth exploration by satellite (passive), fixed, mobile, space research (passive) | |
| 56.9–57 | Earth exploration by satellite (passive), fixed, mobile, space research (passive) | |
| 57–58.2 | Earth exploration by satellite (passive), fixed, mobile, space research (passive) | |
| 59–59.3 | Earth exploration by satellite (passive), fixed, mobile, radio positioning, space research (passive) | |
| 59.3–64 | Fixed, mobile, radio positioning | |
| 64–65 | Fixed, mobile (except aviation mobile) | |
| 65–66 | Earth exploration by satellite, fixed, mobile (except aviation mobile), space research | |
| 66–71 | Mobile, satellite mobile, radio navigation, satellite radio navigation | |
| 116–119.98 | Earth exploration by satellite(passive), space research (passive) | |
| 119.98–122.25 | Earth exploration by satellite(passive), space research (passive) | |
| 122.25–123 | Fixed, mobile, service | |
| 130–134 | Earth exploration by satellite(active),fixed, mobile, radio astronomy | |

(continued)

**Table 5.1** (continued)

| Frequency band (GHz) | Shared services | Applicable region |
|---|---|---|
| 167–174.5 | Fixed, satellite fixed (space-to-ground), mobile | |
| 174.5–174.8 | Fixed, mobile | |
| 174.8–182 | Earth exploration by satellite(passive), space research (passive) | |
| 185–190 | Earth exploration by satellite(passive), space research (passive) | |
| 191.8–200 | Fixed, mobile, mobile satellite, radio navigation, satellite radio navigation | |

over the world, and can meet the demand for a long life. In terms of the *V* band, life problem should be studied.

From the perspective of anti-interference capability, the UHF signal beam is wide and vulnerable to the interference signals transmitted by other aircraft in space and by the ground. The *V* band has the narrowest signal beam and the strongest anti-interference capability. From the perspective of the atmospheric attenuation of microwave signal transmission, the Ka and *V* bands are seriously affected by atmospheric attenuation. Especially, 25 and 60 GHz are the frequencies with peak attenuation, so these frequencies can effectively avoid the interference by radio signals from the Earth. However, the UHF band has no such advantage.

From the perspective of control measures, the UHF band antenna is the simplest, without the need for any control measures. The Ka band antenna is more complex, so some control measures should be taken for antenna pointing. The control of *V* band signal beam is the most complex process, and requires accurate antenna pointing and applies to two relatively stationary satellites.

In terms of the available bandwidth, the *V* band has a bandwidth that is up to 15 GHz, the Ka band allocates about 5 GHz for the inter-satellite link, and the UHF band has little available frequency for the inter-satellite link due to its narrow bandwidth. The interference to other satellites and the ground as well as interference from the ground and other satellites must be taken into consideration if the UHF band is selected.

In summary, the UHF, Ka, and *V* bands can be used as the alternate frequency bands for the inter-satellite link. Based on the current technology, the Ka band has comprehensive advantages in terms of maturity, service life, control complexity, and capability resources (see Table 5.2).

**Table 5.2** Comparison of alternative frequencies bands for inter-satellite links

| Frequency band | Technological maturity | Anti-interference ability | Control measures | Available bandwidth |
|---|---|---|---|---|
| UHF band | Mature | Weak | Simple | Few |
| Ka band | Mature | Strong | Complex | 5 GHz |
| V band | To be mature | Strong | Most complex | 15 GHz |

### *5.2.2 Inter-satellite Link Signal Design*

The signal design generally includes multiple-access, duplex, modulation, code spreading, and channel coding. This section will mainly address these aspects related to the current inter-satellite link signal design.

1. **Multiple access**

The multiple access of inter-satellite links may adopt various methods such as SDMA, CDMA, TDMA, and FDMA. The SDMA can adapt to different link-building satellites in different expected directions. But if their directions coincide, different satellites will share a spreading code and thus seriously interfere with each other. The CDMA assigns different code groups to different satellites to ensure that the inter-satellite links are not restricted by airspace. In terms of inter-satellite configuration, the FDMA mode is mainly too complicated to be used in engineering.

There are two TDMA mechanisms: synchronous transmission and Round-Robin transmission. The so-called "synchronous transmission" refers to the mechanism that all the inter-satellite links simultaneously transmit the RF signals in the designated transmission timeslot, and close the transmission channels at the beginning of the designated reception timeslot to begin receiving and processing the RF signals. The "synchronous transmission" system is only suitable for occasions where the inter-satellite clock synchronization is reliable. If the inter-satellite clock difference is large when the link is initially established (the maximum clock difference is 30 ms), it is obviously impossible to adopt the "synchronous transmission" system. The so-called "Round-Robin" refers to the mechanism that each satellite is allocated with a transmission timeslot and then sequentially transmits the RF signals in chronological order. In this mechanism, the launch timeslot of each satellite is not limited by the shortest range of inter-satellite topology, and the requirement for inter-satellite clock synchronization accuracy is not high. This makes up for the defect of "synchronous transmission" mechanism.

2. **Duplex**

The main duplex modes are: Frequency-Division Duplex (FDD) and Time-Division Duplex (TDD). In the FDD mode, signals are transmitted and received at different frequencies, and the reception and transmission channels are separated by guard bands. In the TDD mode, signals are transmitted and received in the same frequency channels but in different time periods, and the reception and transmission channels are separated by guard intervals.

In the case of FDD mode, considering that a loop with odd-numbered side length cannot appear in the inter-satellite network topology, a fixed topology and a simple static routing algorithm are generally adopted. In the TDD mode, the satellite pairing in each timeslot shall be considered. If there is an even number of satellites, each satellite can be linked to a target object in each time slot, otherwise, at least one satellite can't be linked in any time slot.

3. **Modulation and code spreading**

The inter-satellite channel is a power-limited and bandwidth-limited channel, and is also a non-linear channel with certain interference. Therefore, the modulation and demodulation mode selected by the inter-satellite communication system shall work in a concentrated frequency band and have a spectrum featuring fast high roll-off, that is, although the main lobe is big, the side lobes are all small. Therefore, the spectrum spread of the modulation signal becomes as small as possible after the bandwidth limitation and nonlinear processing. In addition, the modulation and demodulation technique for satellite communication shall have a large power gain to compensate for the power loss caused by high-frequency spectrum, and shall have excellent synchronization performance to achieve the synchronization of the inter-satellite communication network.

The modulation and demodulation techniques used in satellite communication are mainly divided into constant envelope modulation technique and nonconstant envelope modulation technique. In general, the nonconstant envelope modulation is not applicable to satellite communication. The constant envelope modulation techniques mainly include BPSK, QPSK, OQPSK, and 8PSK. Currently, QPSK and BPSK are used most frequently in satellite communication. The main digital modulation techniques include: Phase Shift Keying (PSK), Frequency Shift Keying (FSK), and Amplitude Shift Keying (ASK). According to the inter-satellite link budget and spectrum usage requirements, the modulation mode needs to be selected by simulation. Considering the spectrum efficiency and the characteristics of dynamic constellation change, the BPSK or QPSK modulation is generally selected.

Generally, the spread spectra include DSSS, frequency hopping, and time hopping. Considering the requirement of inter-satellite ranging, the DSSS is usually chosen. If using frequency hopping or time hopping, the ranging difficulty will increase. So, the mixture of DSSS and frequency hopping is often adopted.

Combined with the analysis of code-division multiple access, the balanced Gold sequence design can be selected as the spreading sequence. The following is the analysis on the autocorrelation and cross-correlation between the Gold balance code and *m*-sequence. Since the cross-correlation between *m*-sequences is poor and the available sequence of the same length is limited, the performance of the code-division multiple access system is affected to some extent. Gold sequence is a sequence code based on m-sequence, with better autocorrelation and cross-correlation performance than m-sequence to some extent. Since the balance of spread spectrum code has also an impact on the performance of the spread spectrum system, the balanced Gold code shall be selected in the design of the Gold sequence.

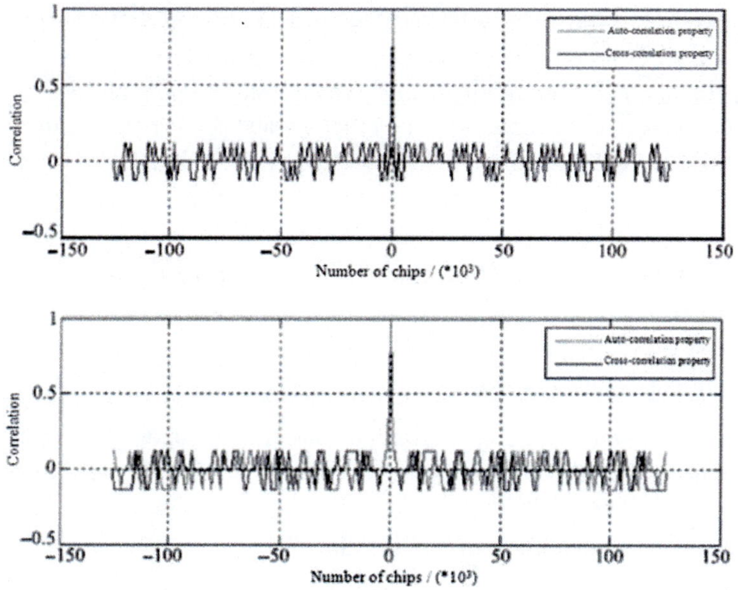


**Fig. 5.2** Normalized autocorrelation and cross-correlation functions of seven-bit m-sequence and balanced gold sequence **a** Normalized autocorrelation and cross-correlation functions for seven-bit m-sequence **b** Normalized autocorrelation and cross-correlation functions for seven-bit balanced gold sequence

As can be seen from Fig. 5.2, the cross-correlation performance of the Gold sequence is slightly better than the m-sequence.

4. **Channel coding**

The inter-satellite link generally requires strong anti-interference ability, so the channel coding technology is needed. Channel coding can increase the operating distance effectively under a certain bit error rate and thus make a contribution to the design of the whole system.

The channel code can be chosen from Turbo code and LDPC code whose decoding performance is close to the Shannon limit. The LDPC code rate can be constructed arbitrarily with greater flexibility, while the Turbo code can only achieve a high code rate through puncturing, in which the puncturing matrix should be considered very carefully, otherwise it will result in a large loss in performance. In addition, the LDPC code has a lower error floor and no basin effect (i.e., the performance of its bit error rate will be no longer significantly improved with the increase of SNR), while the Turbo code will show the basin effect when its bit error rate is $10^{-6}$ to $10^{-7}$.

## 5.3 Network Topology and Routing of Inter-satellite Link

Inter-satellite topology and routing is the basis of inter-satellite links. This section analyzes the constraints of inter-satellite topology and routing design, the network characteristics, and typical inter-satellite links.

### *5.3.1 Constrains of Inter-satellite Topology and Routing Design*

1. **Principle of Topology Design**

In general, the design principles of satellite constellation topology are determined according to the measurement requirements, data transmission requirements, and inter-satellite network topology rules.

(1) Measurement requirement of link-building

Considering the observation geometry requirements in constellation-wide joint orbit determination and autonomous navigation, low-low link-building is required for MEO-MEO satellites. Considering the demand for linking high-orbit GEO/IGSO satellites in autonomous navigation, high-low link-building is required for those satellites. Considering the satellite-to-ground calibration and the anchoring work mode of inter-satellite links, satellite-ground link-building is required. Considering the need for time synchronization, no more than one node shall be linked between overseas and domestic satellites to complete the time transfer of all the nodes in the constellation.

(2) Data transmission requirement of link-building

The inter-satellite link data transmission mainly carries out inter-satellite information distribution and back-transmission such as operation control, remote control upload, and telemetry downlink transmission, and focuses on the information transmission between the satellite nodes on the ground and the invisible overseas satellite nodes. Considering the high real-time requirements of remote control information, the hops in the linking between domestic visible and invisible satellite nodes shall be as few as possible, down to only one hop for transmission.

(3) Inter-satellite network topology rules

The topological relationship of network access for each node is allocated in advance. It determines the linking state in the whole constellation at any time, and directly influences the application service and network transmission performance. The linking relationship between satellites is calculated by the inter-satellite link management department on the ground and is injected into the satellites in the form of topology table. The satellites at different nodes are automatically linked according to the received topology table.

2. **Routing Design Principle**

Route refers to the data transmission path from the source node to destination node. Based on the planning of linking timeslot, a better route can be obtained in accordance with the routing principle. Considering that the inter-satellite link routing algorithm is an off-line algorithm, the shortest routes from one node to all the other nodes should be found to achieve fast data transmission in the multi-node network. The routing principles are as follows:

(1) Matching of routing and link planning

The inter-satellite visibility relationship and connection relationship change dynamically. In order to reduce the difficulty of inter-satellite route design, static routes are generally adopted. The route does not change during the effective period of link planning table, that is, a link-building plan corresponds to a routing table.

(2) Minimum transmission delay

The fastest routes between domestic and overseas nodes shall be chosen to ensure the minimum transmission delay. The uplink instructions and data shall be transmitted from domestic satellites to overseas satellites as soon as possible. The overseas telemetry data shall be transmitted from overseas satellites to domestic satellites as soon as possible. For autonomous navigation data, inter-satellite transmission delay shall ensure that the data can be transmitted between any two satellites within 1 min.

(3) Minimum number of hops

The number of hops shall be optimized under the premise of minimum delay. For the downlink telemetry, uplink instruction, and data information, one-hop transmission between domestic and overseas nodes shall be realized, and the hops of autonomous navigation information transmission between satellites shall be minimized in case of multiple hops.

### *5.3.2 Analysis of Geometric Characteristics in Inter-satellite Measurements*

Suppose that the MEO constellation in the GNSS is a Walker 24/3/1 configuration, with the orbital height of 21,528 km, the orbital inclination of 55°, and the eccentricity of zero. As shown in Fig. 5.3, the three orbital planes are denoted as *A*, *B*, and,*C* respectively, and the 8 satellites in each orbital plane are numbered as 1–8 in sequence, such as A1–A8. At a certain epoch moment, the right ascensions at the ascending points of orbital planes A/B/C are 0°, 120°, and 240°, respectively. The mean anomaly of the satellite $A1$ is 0°, the mean anomaly of the satellite $A2$ is 45°, the mean anomaly of the satellite $B1$ is 15°, and the mean anomaly of the satellite $C1$ is 30°, and so on.

Considering the symmetry of the constellation configuration, this section takes one satellite (named as the node satellite) as the object to analyze, designated as

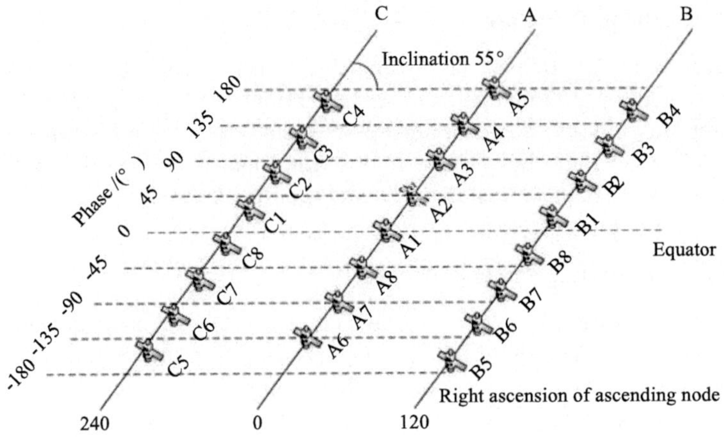


**Fig. 5.3** Walker 24/3/1 navigation constellation planar phases

satellite A, and the satellites (named as the linked satellites) linked with satellite A are denoted as $B_i (i = 1, 2, 3, \ldots)$. In the geocentric inertial coordinate system, $A(x_0, y_0, z_0)$ and $(x_i, y_i, z_i), i = 1, 2, 3, \ldots$ are used to represent the coordinates of these satellites. The distance between satellite A and the link satellite $B_i$ is recorded as $\rho_i$, then:

$$\rho_i = \sqrt{(x_i - x_0)^2 + (y_i - y_0)^2 + (z_i - z_0)^2}, \quad i = 1, 2, 3, \ldots \tag{5.1}$$

The direction cosine of the node satellite $A$ and the linked satellite $B_i$ can be represented as

$$\hat{\rho}_i = \left( \frac{x_i - x_0}{\rho_i} \; \frac{y_i - y_0}{\rho_i} \; \frac{z_i - z_0}{\rho_i} \right) \tag{5.2}$$

Denote the matrix as

$$C = \begin{bmatrix} \frac{x_1 - x_0}{\rho_1} & \frac{y_1 - y_0}{\rho_1} & \frac{z_1 - z_0}{\rho_1} \\ \frac{x_2 - x_0}{\rho_2} & \frac{y_2 - y_0}{\rho_2} & \frac{z_2 - z_0}{\rho_2} \\ \frac{x_3 - x_0}{\rho_3} & \frac{y_3 - y_0}{\rho_3} & \frac{z_3 - z_0}{\rho_3} \\ \frac{x_4 - x_0}{\rho_4} & \frac{y_4 - y_0}{\rho_4} & \frac{z_4 - z_0}{\rho_4} \end{bmatrix} \tag{5.3}$$

$$P = [C^T \cdot C]^{-1} \tag{5.4}$$

Define the inter-satellite link geometric accuracy factor as Crosslink Position Dilution of Presision (CPDOP), to characterize the spatial performance of the inter-satellite link measurement:

$$CPDOP = \sqrt{P_{1,1} + P_{2,2} + P_{3,3}} \tag{5.5}$$

Considering the reciprocity principle of inter-satellite linking, the CPDOP performance of 24 satellites in the constellation is consistent and periodic, in the same as the satellite orbit period. Take any satellite as an analysis object. Denote its orbital parameters as $(a, e, i, \Omega, \omega, M)$. Then its on-orbit position vector can be represented as

$$\begin{aligned} \boldsymbol{r} = a(\cos E - e)\begin{bmatrix} \cos\Omega\cos\omega - \sin\Omega\sin\omega\cos i \\ \sin\Omega\cos\omega + \cos\Omega\sin\omega\cos i \\ \sin\omega\sin i \end{bmatrix} \\ + a\sqrt{1-e^2}\sin E\begin{bmatrix} -\cos 2\sin\omega - \sin\Omega\cos\omega\cos i \\ -\sin\Omega\sin\omega + \cos\Omega\cos\omega\cos i \\ \cos\omega\sin i \end{bmatrix} \end{aligned} \tag{5.6}$$

For all satellites in the constellation, let $e = 0$, $\omega = 0$, $E = M$, and $M = M_0 + n(t - t_0)$, then the above formula can be simplified as

$$\boldsymbol{r} = a\cos M\begin{bmatrix} \cos\Omega \\ \sin\Omega \\ 0 \end{bmatrix} + a\sin M\begin{bmatrix} -\sin\Omega\cos i \\ \cos\Omega\cos i \\ \sin i \end{bmatrix} \tag{5.7}$$

If $a$=1, the mean angular velocity $n = a^{-3/2} = 1$, and the orbit period $T = 2\pi$, then the above formula can be simplified as

$$\boldsymbol{r} = \cos(M_0 + t)\begin{bmatrix} \cos\Omega \\ \sin\Omega \\ 0 \end{bmatrix} + \sin(M_0 + t)\begin{bmatrix} -\sin\Omega\cos i \\ \cos\Omega\cos i \\ \sin i \end{bmatrix}, \tag{5.8}$$

where $M_0$ is the initial mean anomaly of the satellite, $t \in [0, 2\pi]$.

Substituting (5.8) into (5.1)–(5.5) gives the change of CPDOP index in one orbital period.

### 5.3.3 Analysis of Data Communication Linking Characteristics

The common routing algorithms include diffusion routing and inter-satellite minimum-forwarding routing.

(1) Diffusion routing

The diffusion routing is also named as flooding. In the diffusion routing algorithm, each satellite sends a data packet to all visible satellites. To avoid large number of repeated packets, each satellite sets a serial number on the data packet sent out, which increases progressively with each new packet. Each satellite will restore a serial number for each source router. When a new packet comes, if the serial number is less than or equal to the maximum recorded serial number, indicating that the packet is repeated which shall be discarded. If the serial number is greater than the maximum serial number currently recorded, indicating that the packet has not been read, which shall be forwarded to all visible satellites. In order to reduce the number of repeated packets, the number of hops experienced by the packet can also be limited. Packets with a number of hops larger than the limit shall be discarded. However, the diameter of the network shall be known in advance to set such a limit. If the diameter of the network is not known, the limit can be set as a larger value, for example, the number of network nodes. Then the network diameter can be obtained through simulation or actual measurement.

(2) Inter-satellite minimum-forwarding routing

To reduce the repeating packets and further improve the information efficiency, some judgments can be added to the packet processing process. For example, the packet domain can indicate which satellite is the forwarding satellite, so other satellites shall directly discard the packet after receiving it. After the source satellite sends out a packet, only one of the directly visible $N$ satellites receiving the packet will undertake the forwarding task, so that the total number of packets caused by such processing will be reduced from the original $N^2$ to 2N. Since only one satellite undertakes the forwarding task, this method is named herein as the minimum-forwarding routing in this book. However, with only one satellite to undertake the task of inter-satellite information forwarding, this method is less reliable than diffusion routing. This method is applicable to the telemetry data with a large data size and low-reliability requirement. For the telecontrol data and upload data with a small data size and high-reliability requirement, the diffusion routing with limited hops can be pursued.

### *5.3.4 Typical Network Topology Routing Design*

1. **Wide beam system topology and routing**

The inter-satellite network topology and routing of the wide beam system is relatively simple. GPS satellites have an inter-satellite link in the UHF band and TDMA mode. Each satellite is allocated a time slot of 1.5 s. Each of the 24 navigation satellites on orbit is allocated with a different time slot, and all satellites in the constellation are polled within 36 s, which is defined as a frame. There are 25 frames in inter-satellite link transmission, including 1 ranging frame and 24 data frames. Each satellite can broadcast all the data received last time and set the maximum number of rebroadcasts. The counter will be reduced once whenever forwarding, down to 0, where the rebroadcasting will not be continued. Generally, the setting of 5 times can ensure that the information is broadcasted throughout the network. The coverage of a wide beam antenna is as shown in Fig. 5.4.

The UHF band dual-frequency one-way ranging method is used in the GPS Block IIR satellites. During the ranging frame, the ranging mode is used, any of the Block IIR satellite sends ranging signals in the assigned time slot, and all the other visible navigation satellites receive and process the ranging signals. The pseudorange measurement is accomplished by eliminating the effects of delay caused by the ionospheric delay using two different frequencies. During the data frame, the data transmission mode is adopted, and each satellite transmits its own data parameters in its allocated time slot, mainly including the calculation result of the pseudorange measurement, the estimated position of the satellite, the clock parameter, and the corresponding estimated variance.

2. **Narrow beam system topology and routing**

Due to the connectivity and symmetry, the network topology and routing of fixed connection and continuous link has two modes: ringlike co-orbital link and starlike co-orbital link.

(1) Ringlike co-orbital link

In the ringlike co-orbit linking system, each satellite is linked with the satellites in front of and behind it on the same orbital plane, respectively, via two sets of antennas, into a connected ring topology, as shown in Fig. 5.5.

(2) Starlike co-orbital link

In the starlike co-orbit linking system, each satellite is linked with the third satellites in front of and behind it on the same orbital plane, respectively, via two sets of antennas, into a connected starlike topology, as shown in Fig. 5.6.

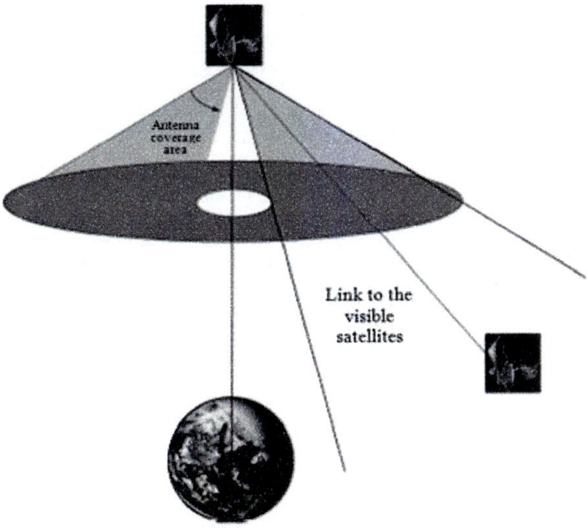


**Fig. 5.4** Wide-beam antenna coverage

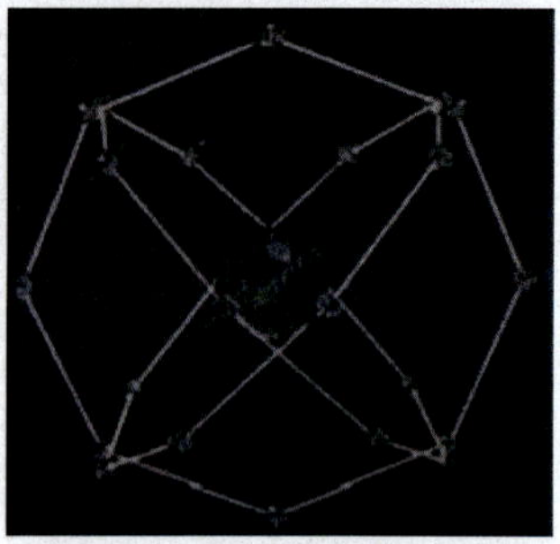

Fig. 5.5 MEO satellite ringlike co-orbital link

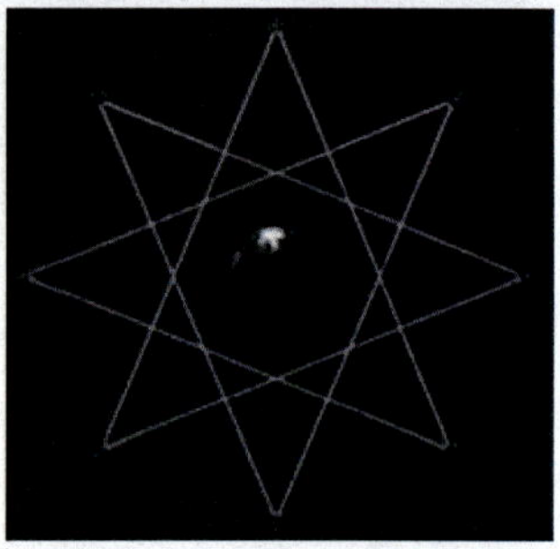

Fig. 5.6 MEO satellite starlike co-orbital link

In order to meet the needs of time synchronization measurement of inter-satellite links at ordinary times, overseas satellites are required to reach the domestic ground station through no more than one hop, that is, through two links "overseas satellites—domestic satellites—domestic ground station". To ensure that the inter-satellite link supports the rapid transmission of the global integrity information and overseas TT&C information, it is required that the time delay of inter-satellite data transmission shall be as small as possible (within 3 s). Therefore, starlike co-orbit topology is usually selected as one of the slow-varying beam topologies used for linking the co-orbit satellites in the navigation constellation.

When linking the satellites on different orbital planes, the deployment of the third slow-varying beam antenna can be considered to establish the inter-satellite links crossing different orbits. A typical topological configuration of cross-orbit inter-satellite links is shown as Fig. 5.7.

In this case, 3 hops (no more than 5 hops) are needed to transmit the inter-satellite link information from the source satellite to any other satellite in a different orbit. This is closely related to constellation configuration and topology selection.

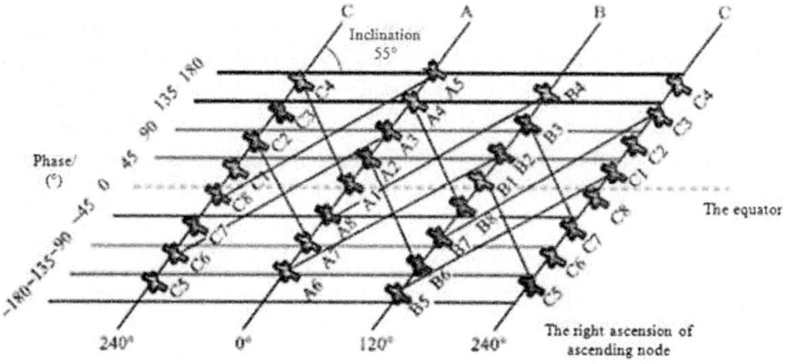


**Fig. 5.7** MEO satellite link topology between different orbits

## 5.4 Implementation of Inter-satellite Link

The implementation of initer-satellite links needs to deal with a series of engineering problems. Generally, satellites need to build an inter-satellite link through antenna beam pointing and other technologies. And the inter-satellite link ranging and information transmission also need to complete the special requirements of the inter-satellite link through protocol design and product design. In addition, the operation and management of the whole inter-satellite link also involve complex processes. This section introduces the implementation technology of an inter-satellite link based on BDS inter-satellite link design.

### *5.4.1 Construction of Inter-satellite Link*

#### 1. **Link-building Methods**

There are many inter-satellite linking methods to establish a network in the navigation constellation, and typical ones are the time-division linking method and the frequency-division continuous linking method. Frequency-division continuous linking is an inter-satellite linking mode featuring the slow change of wireless beam pointing and continuous linking. Each inter-satellite link is established by two fixed satellites and maintained for a long time (e.g., more than 1 h). The resulting inter-satellite network topology for the whole constellation is static, called "static inter-satellite network". The time-division linking method is a mode with fast-changing antenna beam pointing and time-division linking. Each inter-satellite link only exists for a short time (e.g., 3 s). The resulting inter-satellite network topology for the whole constellation is dynamic, called "dynamic inter-satellite network". The two linking modes will be presented, respectively, below.

To achieve the linking with time-division it is required to first specify a work period for the constellation, in which an evenly distributed work timeslot is allocated to each satellite node, and a corresponding linking satellite node is allocated to form a linking satellite pair. In a particular work timeslot, the satellite node in the constellation is only linked with the allocated satellite node. If different satellite nodes are allocated to different work timeslots, reasonable planning shall be made so that in the whole period, each satellite node can be linked with other satellite nodes in the constellation one by one without conflict. The advantage of time-division linking is that the satellite can be linked with as many satellite nodes as possible via a set of antennas. As long as the time of reception and transmission is distinguished in the linking process, a single frequency point can meet the requirements of two-way communication and measurement. The disadvantage is that the continuous measurement communication cannot be achieved, and the linking complexity will increase with the increasing of constellation satellite nodes. The constellation nodes need to maintain high-precision time synchronization to ensure that each node works complying with the time schedule.

The frequency-division continuous linking method adopts different signal frequencies for the signal transmitting and receiving, and keeps frequency isolation to realize simultaneous work of transmitting and receiving. To maintain continuous linking with different satellite nodes, if it is distinguished by adding different signal frequencies for different satellite nodes, the signal will increase with the increase of the number of satellites, resulting that the acquisition of frequency resources will be more difficult, the equipment shall be more complex and the consistency cannot be realized. Since the navigation satellites are widely distributed in space, multiple links that work independently can be connected to the satellite nodes in different space domains, to realize continuous linking. In the working cycle of each constellation, the time-division linking method will build a link with specified satellite nodes. The greatest advantage of the frequency-division continuous linking method is that the link works continuously, without waiting for the work cycle. This is crucial to real-time measurement and data transmission. Its disadvantage is that constrained by the development cost and satellite platform capability, the number of links cannot be increased indefinitely, and needs to be reasonably selected by considering both the linking demand and the satellite resources.

2. **Antenna beam pointing**

For the inter-satellite links based on point beam antenna, the basic linking precondition is that the link antennas beams of two satellites to be linked point to each other. The satellite software calculates the input data sources of antenna pointing, including the orbital position information given by the two satellites in the form of ephemeris data, the satellite attitude data collected by the attitude sensor of the satellite control platform, and the yaw angle of maneuvering target given by the spaceborne computer.

In general, for the relative position relation of the two satellites at a certain moment, the ephemeris data reflects the position vector of one satellite in the Earth-fixed coordinate system. The difference between the two position vectors is just the relative position vector of the two satellites. By shifting this relative position vector to the coordinate system of one satellite through the coordinate system transformation, the antenna pointing angle of a specific antenna can be calculated. In the process of coordinate system transformation, the influence of yaw attitude maneuver control and three-axis attitude shall be considered. In addition, if a satellite-ground link is established, the antenna beam pointing shall be calculated by taking the position coordinates of the ground station in the Earth-fixed coordinate system as the target.

If a link is established between the satellites (as the reference) and target point *P*, the centerline of antenna beam points to the target point. This target point can be a satellite to be linked with other satellites or the ground station to be linked with satellites, as shown in Fig. 5.8.

Central angle θ: the angle between the centerline of the antenna beam and the + *z* axis of satellite;

Azimuth angle Φ: the angle between the projection of the centerline of antenna beam in the *xOy* plane and the +*x* axis of satellite

Denote the unit direction vector of the relative position in the satellite coordinate system as $[X, Y, Z]^T$. From the definition of antenna pointing angle, the following relations can be derived:

$$\begin{bmatrix} X \\ Y \\ Z \end{bmatrix} = \begin{bmatrix} \sin\theta\cos\phi \\ \sin\theta\sin\phi \\ \cos\theta \end{bmatrix} \tag{5.9}$$

To determine the antenna pointing angle and azimuth angle, the main work is to calculate the unit direction vector as $[X,Y,Z]^T$. If the ephemeris information of both the transmitting satellite and the receiving satellite is obtained, their position vectors in the Earth-fixed coordinate system can be calculated separately. The vector difference is just the relative position vector $\left[x_2 y_0 z_0\right]$ of the two linked satellites.

Mainly influenced by Earth rotation, the transformation matrix from the Earth-fixed coordinate system to the equatorial inertial system is $R_1$. The transformation matrix of converting the relative position vector from the geocentric equatorial inertial system to a satellite orbit coordinate system is $R_2$. Considering the three-axis attitude information and rotation sequence of the satellite, the transformation matrix of converting the unit direction vector from the satellite orbit coordinate system to the satellite coordinate system is $R_3$. To sum up, the whole transformation process can be described as follows:

**Fig. 5.8** Diagram of antenna pointing angle

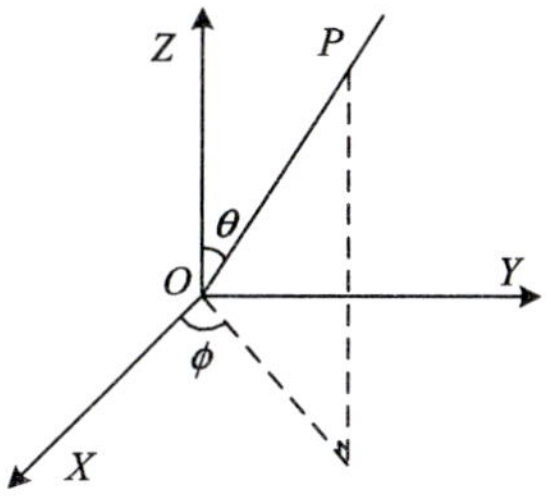

$$\begin{bmatrix} X \\ Y \\ Z \end{bmatrix} = \boldsymbol{R}_1, \boldsymbol{R}_2, \boldsymbol{R}_3, \begin{bmatrix} x_0 \\ y_0 \\ z_0 \end{bmatrix} \tag{5.10}$$

3. **Signal acquisition and tracking**

Signal acquisition is the prerequisite of inter-satellite linking. Signals are usually captured in two ways. One is "blind search" carried out in the whole airspace without priori information, the other is to search the target in a smaller predetermined airspace according to the known approximate position of the target. For the navigation constellations with time-division linking, the objects to be linked shall be frequently switched, the signals from transmitter and receiver shall be captured rapidly, accurately and correctly, and more importantly, high-precision forecast information shall be provided as aid to meet the requirements of the signal search and acquisition and Doppler frequency shift correction in the inter-satellite linking process. As far as the inter-satellite information acquisition is concerned, the spatial position of satellite can be determined through the ephemeris parameters, so the ephemeris can be used to find the way of information acquisition in advance so as to shorten the signal acquisition time of transceiver.

The calculation of acquisition information includes the prediction of inter-satellite TOA, inter-satellite relative velocity, and inter-satellite relative acceleration. Inter-satellite arrival time is defined as the time when the signal arrives at the transceiver of this satellite minus the time when the signal is transmitted by the transceiver of another satellite. Its influencing factors include signal propagation delay, signal reception/transmission delay, relative satellite motion, and relative clock motion. Relative velocity refers to the relative velocity and acceleration required for a transceiver to assist in capturing information, and is defined as the relative velocity and acceleration between transceivers. The relative velocity will cause the TOA signal to deviate from the designed frequency reference and thus invite the Doppler frequency shift.

## 5.4.2 *Inter-satellite Ranging*

1. **Principle of inter-satellite ranging**

In the GNSS, the two-way ranging frame of non-coherent spread spectrum measurement mechanism is generally used in the inter-satellite ranging and the clock difference measurement. The specific principle and sequential relationship are shown in Fig. 5.9.

In Fig. 5.9, satellite $A$ and satellite $B$ transmit ranging signals in their, respectively, assigned time slots based on their own clocks. Due to the time difference between the two satellites, there is also a time difference $\Delta t$ between the ranging frames. In satellite $A$, the transmission delay $\rho_A$ can be obtained by capturing and tracking

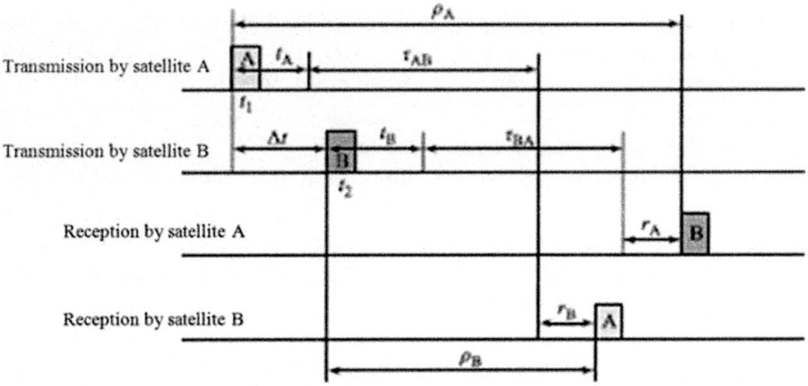


**Fig. 5.9** Principle and sequential relationship of the ranging and clock difference measurement by the two-way ranging frame

the ranging signal transmitted by satellite *B*. The delay include the electromagnetic wave propagation delay $\tau_{BA}$ between the antennas of satellite *A* and satellite *B*, the transmitting equipment delay $t_B$ of satellite *B*, the receiving equipment delay *rA* of satellite *A*, the clock difference $\Delta t$ between the two satellites, $\Delta t = t_2 - t_1$, and the measurement error, whose relationship is

$$\rho_A = \Delta t + t_B + \tau_{BA} + r_A + \varepsilon_A \tag{5.11}$$

Similarly, the time delay of satellite *B*, denoted as $\rho_B$, can be measured in accordance with the following relation:

$$\rho_B = (-\Delta t) + t_A + \tau_{AB} + r_B + \varepsilon_B \tag{5.12}$$

It shall be noted that if the ranging frame sent by satellite *B* is regarded as an electric wave scale between satellite *A* and satellite *B*, the epoch of transmission frame will be the electric ruler scale that can measure the geometrical distance between the two satellites at a certain moment. Different from the measurement of the distance between two objects under normal static conditions, the relative motion always exists between two satellites and between satellite and electric ruler scale. Therefore, an ideal "sampling shutter" is needed to be required to simultaneously extract the scale of this electric ruler in satellites *A* and *B*, respectively, at a moment, so as to calculate the geometrical distance of the two satellites in an inertial reference time system.

2. **Accuracy of inter-satellite ranging**

The space ranging performance of the inter-satellite links depends on the specific type of ranging terminal. For example, the phase center and channel delay of a highly directive dish antenna suitable for static inter-satellite networking are related to the antenna structure, and the position of the phase center in the satellite coordinate system varies with the antenna rotation. The phased array antenna suitable for dynamic inter-satellite networking has no moving parts, but its phase center is not fixed under the control of beam controller. Besides, its channel delay will change with the beam pointing direction. In addition, the time delay characteristics of other devices in the inter-satellite link channel, together with different temperature conditions, will also affect the ranging accuracy.

The object measured by the inter-satellite link is the true distance between the satellite centroid (i.e., the true distance). The direct measurement value is the pseudorange calculated by the inter-satellite transceiver, which shall be deducted by the channel delay zero value, corrective phase center variation, and corrective clock difference. The relation is as follows:

$$\tau_{ij} = \tau_i^{\text{clk}} + \tau_i^0 \underbrace{-\tau_i^{\text{pcv}} + \tau_{ij}^{\text{true}} - \tau_j^{\text{pev}}}_{\tau_{ij}^{pc}} + \tau_j^0 - \tau_j^{\text{clk}} \tag{5.13}$$

where the subscripts *i* and *j* represent satellite numbers, the superscript *true* indicates the true distance between the two satellites, the superscript *pc* indicates the phase center distance between the two satellites, the superscript 0 indicates the channel delay zero value, the superscript clk indicates the satellite clock difference, and the superscript pck indicates the phase center variation.

As seen from the above equation, inter-satellite ranging accuracy is just the accuracy of the pseudorange $\tau_{ij}$. The factors that affect the ranging accuracy are mainly the delay zero accuracy representing the time delay characteristic of ranging channel and the phase center variation accuracy representing the phase center characteristic of the antenna. Ranging channels are the ranging signal transmission channels including transceiver and antenna. On one end of the link, the time delay zero of ranging channel starts from the baseband processing of ranging signal and ends at the antenna phase center. Based on the structural correlation between antenna phase center and satellite, the phase center variation (pcv) refers to the projection length of the line connecting phase center and satellite centroid in the ranging direction. In practical application, the time delay zero is assumed to be a constant and the phase center change to be a variable. Then they can be predicted by using the satellite software.

### 5.4.3 *Inter-satellite Network Protocol*

1. **Protocol of Consultative Committee for Space Data Systems (CCSDS)**

The inter-satellite link between navigation satellites establishes a data transmission and measurement relationship among multiple linked points. It has a distinct network characteristic. Therefore, in addition to a physical layer and a data link layer, the inter-satellite network protocol also contains a network layer, a transport layer, and an application layer in order to meet the needs for inter-satellite data transmission, error control, and different services as well as applications.

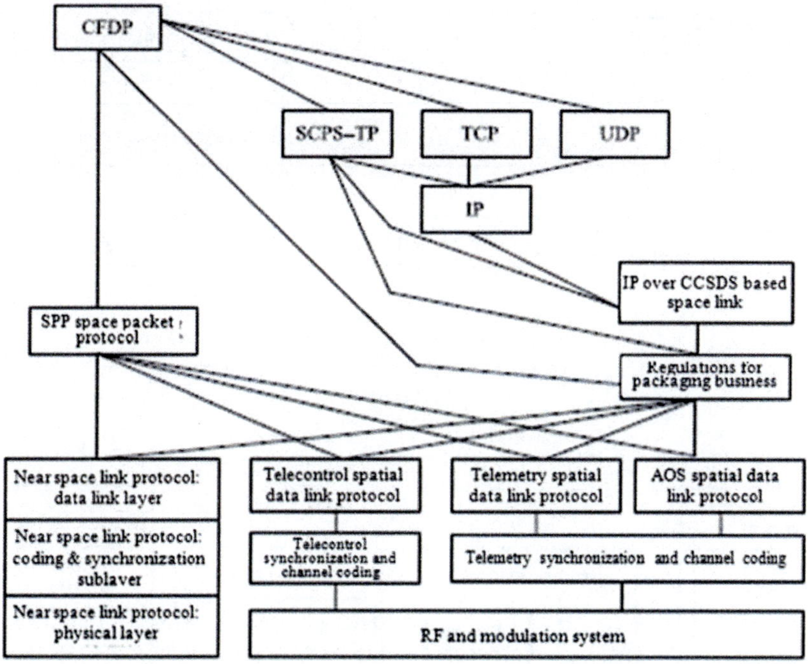


**Fig. 5.10** CCSDS hierarchical protocol model

The CCSDS protocol is a layered reference model established by CCSDS for space data system, including a physical layer, a data link layer, a network layer, a transport layer, and an application layer. It has been widely applied and promoted in spacecraft developed by NASA and ESA. This model is shown in Fig. 5.10.

2. **Inter-satellite protocol design**

Since the data transmission of inter-satellite links is characterized by relatively simple data services and no multimedia services, simple protocols can be introduced to reduce the implementation complexity. The protocols in each layer are as follows.

a. The AOS protocol or the Proximity-1 protocol is used in the data link layer;
b. The space packet protocol or the reduced SCPS-TP (Spatial Communication Protocol Specification-Transportation Protocol) protocol is used in the network layer to solve the routing problems;
c. No transport layer is adopted.
d. The specified application protocol is used in the application layer to solve the problems such as time synchronization, track extrapolation, ephemeris autonomous update, and navigation message autonomous generation.

The resulting layered protocol configuration is shown in Fig. 5.11.

The communication protocol configuration between any two satellites is shown in Fig. 5.12.

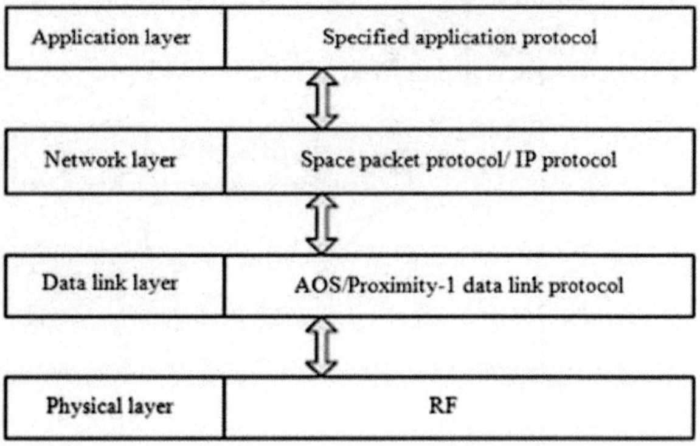


**Fig. 5.11** Layered protocol configuration of navigation inter-satellite protocol

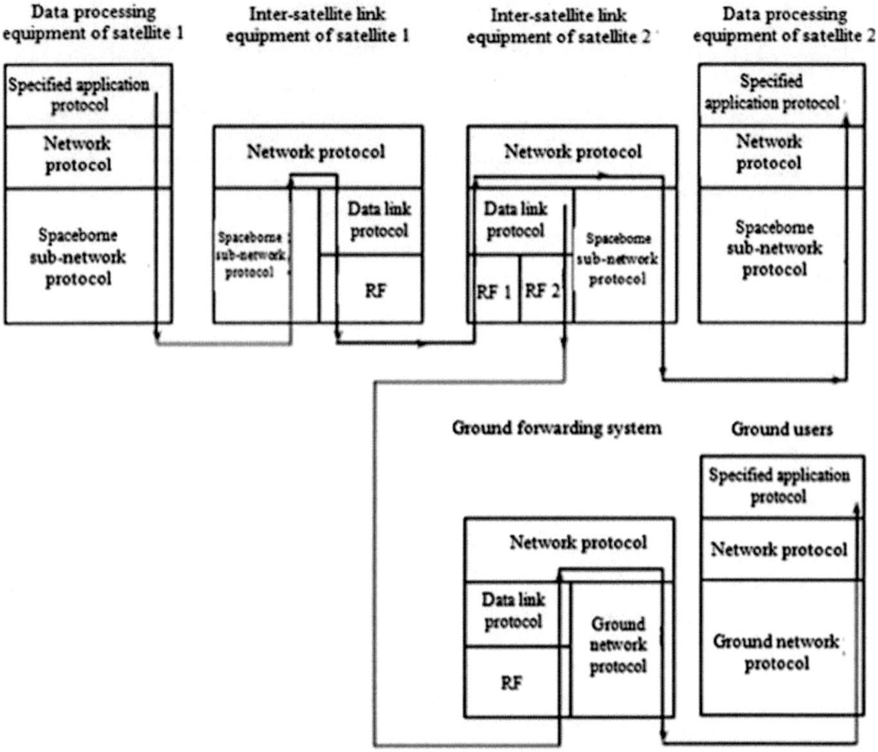


**Fig. 5.12** Communication protocol configurations between satellites

## 3. Inter-satellite transmission strategy

The services provided by the inter-satellite network layer are routing services, including the end-to-end transmission of data from one point to another, and the broadcast transmission of data from one point to many points.

(1) End-to-end transmission

End-to-end transmission, also called end-to-end communication, refers to the communication from one node to any other node in the network, possibly through multiple links and nodes.

Take the satellite $A1$ as an example. The satellites linked to $A1$ include $A2$, $A8$, and $C4$. The data sent by $A1$ can reach $A3$ and $B6$, both linked to $A2$, through $A2$, or reach $A7$ and $B4$, both linked to $A8$, through $A8$, or reach $C3$ and $C5$, both linked to $C4$, through $C4$... and so on. Thus, it can be concluded that the data sent by $A1$ can reach all the linked satellites in the same or different orbit (s). The data generated by the source node finally reach the target node through relay nodes and by route forwarding. The end-to-end routing is as shown in Fig. 5.13.

(2) Broadcast transmission

Broadcast transmission, also called broadcast routing, refers to the transmission of data from one node to all nodes in the network. The link channel manages the inter-satellite data uploaded by the ground system with the number of hops. When the access satellite broadcasts the data to other satellites, SCID is defined as the broadcast address in the inter-satellite data frame. The hop count is the life cycle of broadcast frame, and can be set by ground command, as shown in Fig. 5.14.

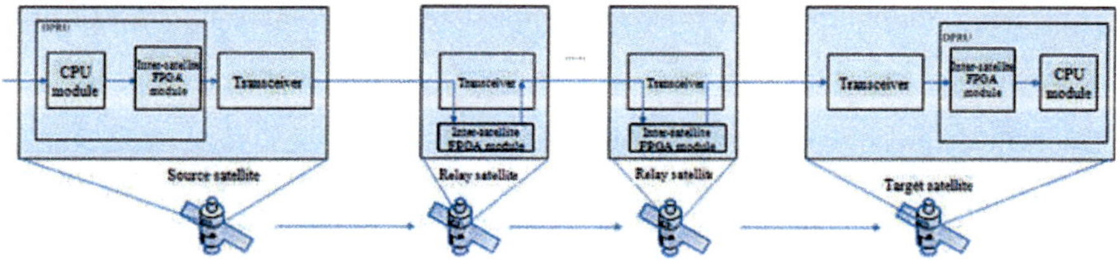


**Fig. 5.13** End-to-end data transmission path

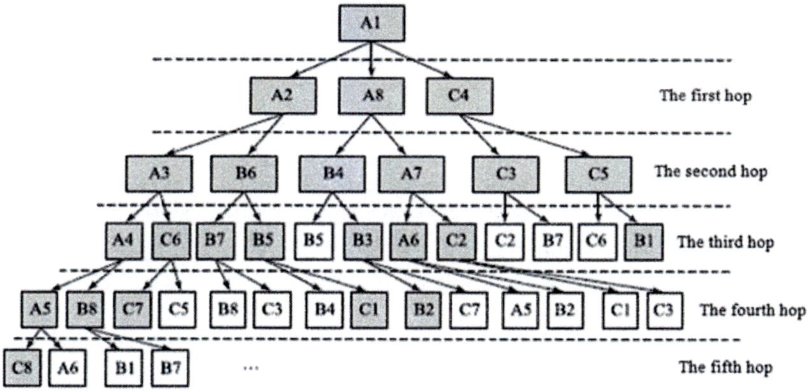


**Fig. 5.14** Transmission paths of broadcast data

Whenever a network node is passed through, the hop count will be subtracted by 1. The data is continuously broadcasted to each node in the network until the hop count becomes 0.

Take the broadcast data sent by the ground through $A1$ as an example to describe the data broadcasting route, as shown in Fig. 5.14. The gray box means that a packet arrives at this satellite for the first time, and then is forwarded to the other two satellites linked with this satellite. The white box means that a repeated packet is received by the satellite. Starting from $A1$ for broadcasting, it takes up to 5 hops for all the satellites to receive data provided that all the 24 MEO links are functioning normally. To control the flooding process, the life cycles of data shall be specified. Whenever the data is forwarded, its life cycles will be subtracted by 1. If the number of life cycles becomes 0, the data won't be forwarded.

### 5.4.4 Inter-satellite Operation Management

Inter-satellite link operation management is the process that through the calculation, generation and transmission of link control instructions (such as the topology table and routing table for each node) and based on various user requirements, carries out the antenna pointing control and information transmission routing control, the management and allocation of inter-satellite link resources, the status monitoring and operation maintenance of inter-satellite links, and handles the anomalies, so that the inter-satellite links can complete various tasks like precision measurement and data transmission and extended tasks efficiently and stably.

The use of inter-satellite links, including all kinds of satellites and all the ground nodes, by all users on the ground and in space, shall be subject to inter-satellite link control and management, whose framework is shown in Fig. 5.15.

According to the application requirements and modes of the inter-satellite links, navigation inter-satellite links are designed with five operation modes: system debugging, ground support, semiautonomous operation, fully autonomous operation, and extended application, as shown in Fig. 5.16.

In the five operation modes of the constellation network system, the inter-satellite links have different tasks, roles, and implementations. They are described as follows:

(1) system debugging mode

In the initial stage of satellite navigation system construction, the system is in the process of testing and debugging, and performance is still not up to the designed requirements, due to the insufficient number of satellites on orbit and inter-satellite links, or due to a certain number of non-operational links.

In the system debugging mode, the inter-satellite link assumes no normal network services, and is under the to-ground test status or the pointing calibration process.

(2) ground support mode

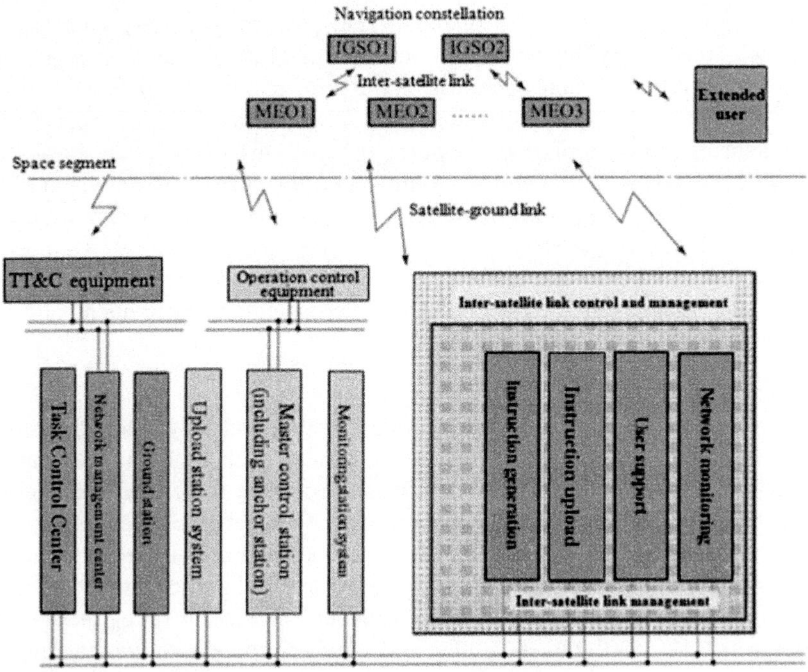


Fig. 5.15 Control and management framework of inter-satellite link

Fig. 5.16 Operation modes of constellation network system

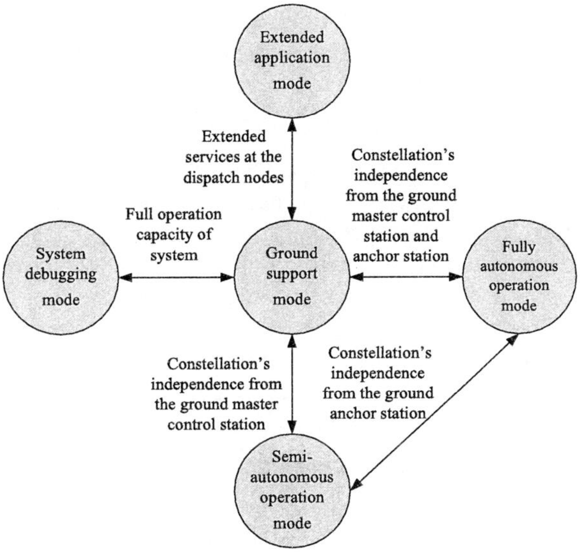

Ground support mode is the normal operating state of the system. The ground segment is responsible for the maintenance and management of all the satellites and ground stations. Through the ground planning and control of inter-satellite link network topology and pointing and routing strategies, the system can fully achieve the required performance of precise orbit determination, time synchronization and autonomous navigation, meet the requirements of information transmission capacity and timeliness in the TT&C and operation control services, and provide the signals and services satisfying the navigation users.

In the ground support mode, the inter-satellite links are oriented to the above services and assume all the functions of inter-satellite measurement and data transmission.

(3) semiautonomous operation mode

Semiautonomous operation is between ground support and fully autonomous operation. The satellite system is only linked with ground anchor station, and maintain the inter-satellite and satellite-ground networks using the forecasted ephemerides, the constellation topology and the parameters of spaceborne inter-satellite link equipment, all uploaded by the management system. Furthermore, the satellite autonomously generates high-precision ephemerides, realizes long-time ephemeris maintenance, and provides users with the navigation services meeting the accuracy requirement, based on the inter-satellite and satellite-ground two-way ranging and range and parameter exchange.

Since the ground anchorage station uses the inter-satellite links pointing to ground as its satellite-ground channels, the inter-satellite link network in the semiautonomous operation mode shall not only assume all the functions of ground support mode and fully autonomous operation mode, but also allocate some links pointing to ground to serve as satellite-ground channels.

(4) fully autonomous operation mode

In the autonomous operation, the satellite system runs independently from the ground stations. The inter-satellite network is maintained by the forecasted ephemerides uploaded by the management system, the constellation topology and the parameters of spaceborne inter-satellite link equipment. Moreover, the satellite autonomously generates high-precision ephemerides, realizes long-time ephemeris maintenance, and provides users with the navigation services meeting the accuracy requirement, based only on the inter-satellite two-way ranging and range and parameter exchange.

In this mode, the inter-satellite links are free from ground services and only assume the functions of autonomous navigation measurement and data transmission.

(5) extended application mode

The extended application of inter-satellite links in the satellite navigation system means that through planning and allocation, the system allocates some nodes and link resources to support deep-space navigation, specific spacecraft communication and measurement and other services. In this mode, the ground system will assign some inter-satellite link to provide inter-satellite measurement and communication services to the extended users.

## 5.5 Inter-satellite Link of BDS-3

As one of the four global satellite navigation systems, the BDS-3 system has established a unique inter-satellite link system in order to meet the requirements of global measurement and control, satellite orbit determination, autonomous navigation, and extended application,

### 5.5.1 Constellation and Satellite Configuration

The BDS-3 constellation has three types of satellite nodes: GEO, IGSO, and MEO. All the satellites are equipped with inter-satellite devices. 24 MEO satellites formed a Walker 24/3/1 MEO satellite constellation. The six high-altitude satellites include three GEO satellites and three inclined IGSO satellites. The inter-satellite link Ka antenna has a beam with a half transmitting angle of 60° with the satellite $+Z$ axis.

The test satellite of BDS-3 is equipped with the inter-satellite link payload devices. The ranging and data transmission experiments have been demonstrated for inter-satellite links. In the experiments, a large amount of on-orbit application data has been accumulated, laying a technical foundation for the application of inter-satellite links.

### 5.5.2 Inter-satellite Link Performance

The inter-satellite link of the BeiDou satellite navigation system has played a significant role in improving the orbit determination accuracy of overseas satellites, implementing data transmission with overseas satellites, and the constellation autonomous navigation.

In the test satellite phase of BDS-3, the ranging performance of terminal type $A$ and terminal type B was tested on orbit. Without true inter-satellite range as a reference, it is difficult to measure the ranging performance of a link separately. For the link network with two experiment satellites, because the inter-satellite ranging values of terminals $A$ and $B$ are available at the same time, the relative ranging changes of the two links can be compared, and the residual error can be analyzed by using the telemetry data down-linked by the experiment satellites.

The downlink period of telemetry parameters was selected as 0:00 on December 19, 2016 to 0:00 on December 21, 2016 (Beijing time). The links between two experiment satellites (satellite 1 and satellite 2) and with terminals $B$ and $A$ were taken as analysis objects.

By inquiring the ground test records according to the working condition of on-board equipments, the time delay zero of each channel was obtained. The pseudorange and phase center variation were directly obtained from the downlink telemetry

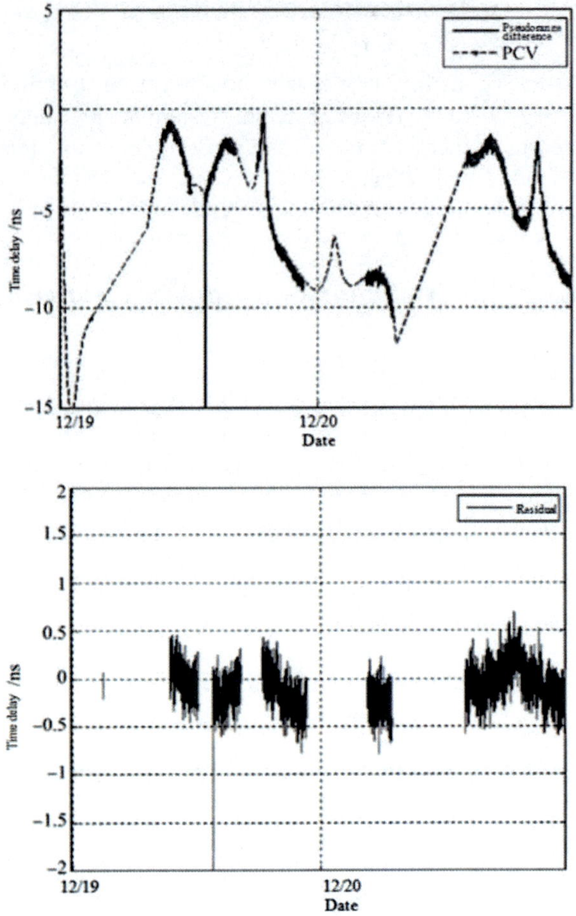


**Fig. 5.17** Inter-satellite pseudo range difference and residual

data by the satellites. Each pseudorange value generated a corresponding BDS time reference. Calculate the pseudorange and phase center change in the same time reference to obtain the residual curve, as shown in Fig. 5.17. The fluctuation range of pseudorange error was ±5 ns. After being corrected by phase center, the random error (RMS) was 0.21 ns.

### *5.5.3 Prospect for Application*

As far as the inter-satellite measurement of navigation constellation is concerned, a dynamic inter-satellite network has better spatial measurement performance and more links than a static one. In terms of data transmission, the highly directive dish antenna or laser link device in the static inter-satellite network has a potential to increase the data transmission rate, and the static network has the advantages of simple routing strategy and large data transmission capacity.

Looking forward to the development of inter-satellite links in the BDS, the hybrid inter-satellite link system is a combination of excellent measurement performance and large data transmission capacity. Its spaceborne equipment is of a moderate scale and has good implementability. A global satellite navigation system composed of a dynamic inter-satellite network made by one set of phased array antenna, and a static inter-satellite network made by two or three sets of highly directive dish antennas or laser link equipment can be build to achieve excellent performance and high-level global operational control management under the condition of regional (rather than global) station distribution.

# Chapter 6
# Time-Frequency System for Satellite Navigation

As the core of a navigation satellite system, the time-frequency system directly determines the accuracy of satellite navigation timing. The work of this system mainly includes the establishment of system time, the generation and maintenance of satellite time, the intra-system time synchronization, the tracing of system time, the monitoring and prediction of GNSS time offset, and other work.

The establishment and maintenance of system time is the core technology of navigation satellite system, while the generation and maintenance of satellite time is the basis for transmitting the system time data to users. Through the intra-system time synchronization, the satellite time is synchronized with the ground system time. The synchronization between navigation system time and the UTC is realized by tracing the system time, while the compatibility and interoperability between system times are achieved by monitoring, predicting, and evaluating the time offset between GNSS systems.

In this chapter, the time-frequency system for satellite navigation is addressed from the perspectives of time-frequency system, spaceborne time-frequency system, and atomic clock technology.

## 6.1 Overview of Time-Frequency System

The navigation performance of a navigation satellite system relies primarily on the accurate time provided by a precise time-frequency reference signal. The time-frequency reference signal with high accuracy is the basis for performing the functions such as navigation positioning, timing, and velocity measurement, so it should be highly stable, accurate, and continuous.

The generalized time-frequency system is a system that provides users with the signals and information of time frequency. It mainly consists of three subsystems:



J. Xie et al., *Satellite Navigation Systems and Technologies*, Space Science and Technologies, https://doi.org/10.1007/978-981-15-4863-5_6

a time-keeping subsystem for generating, establishing, and maintaining a standard time frequency, a time transportation subsystem for broadcasting a standard time frequency, and a time-using subsystem for extracting time-frequency information from a broadcast signal and converting it into a time-frequency signal available for users. At present, the GPS, GLONASS, Galileo, and BDS have established their relatively independent time-frequency systems.

### 6.1.1 Composition and Function of Time-Frequency System

In Fig. 6.1, the navigation time-frequency systems are illustrated by a block diagram, mainly including a satellite time-frequency system, a ground time-frequency system, and a user time-frequency system. The three time-frequency systems provide time-frequency standard signals and synchronization services to the satellite, the ground,

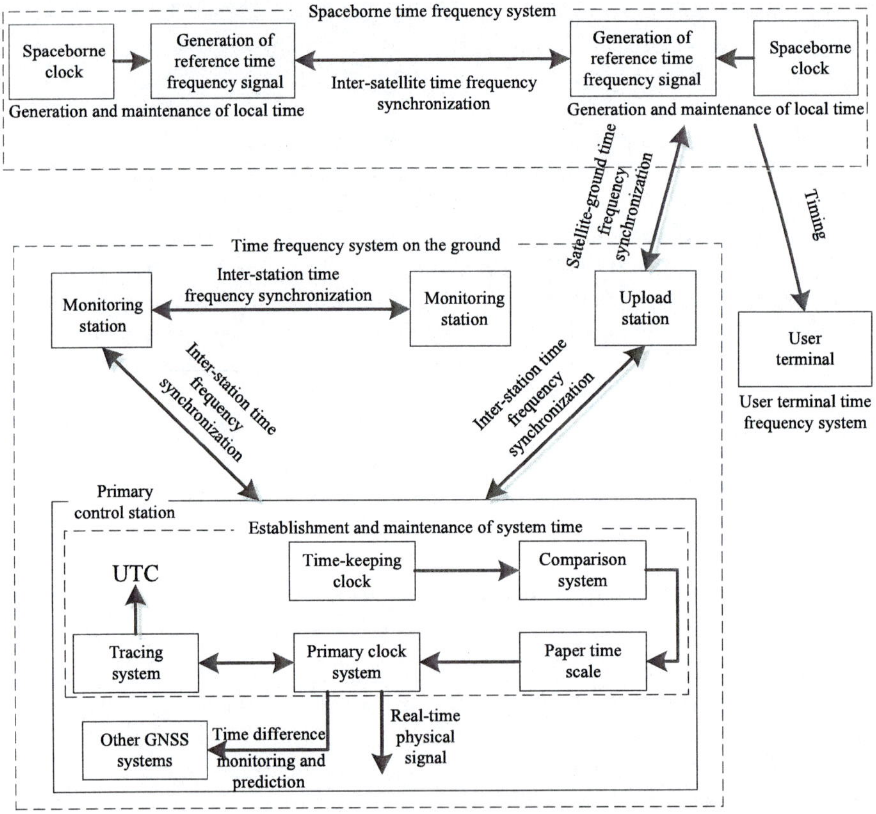


**Fig. 6.1** Block diagram of time-frequency system

and the user in the navigation satellite system, respectively, thus enabling them to work together.

The satellite time-frequency system is mainly to provide reference time-frequency signals for a satellite. Usually, a satellite is equipped with multiple high-precision atomic clocks, which cooperate with the generation unit of reference time-frequency signal to generate the onboard local time. In the BDS, the difference between satellite time and system time is measured by a satellite-ground comparison link, and the onboard local time is controlled and broadcasted by labeling frequency and phase adjustment information under the command of the ground control system.

The ground time-frequency system provides reference time and reference frequency signals for all the equipment in the ground system. The generation process of system time in the navigation satellite system is as follows. A time-keeping clock group composed of multiple high-precision atomic clocks delivers a paper time scale via a time scale algorithm according to the comparison results of ground-based atomic clocks. In the system operation process, the working clock substantializes the system time and generates the system-level reference time frequency, which serves as the local time of master control station in the ground system. The local time of other monitoring/control stations in the ground system is synchronized with the time of master control station through technical means such as common-view/bidirectional time comparison.

The user time-frequency system provides the user with the required time frequency, which is generated, distributed, and synchronized by the user receiver.

The time-frequency system in a navigation satellite mainly performs the following functions:

(1) Time generation

The time generation function includes the following aspects: the establishment and maintenance of system time; the establishment and maintenance of autonomous satellite time system; and the measurement, control, and forecasting of time difference through spaceborne clocks, master control stations, monitoring stations, upload stations, wide-area difference stations, and user terminal clocks.

(2) Time synchronization

The time synchronization function is mainly to use inter-station, satellite-ground, and inter-satellite time-frequency synchronization links to realize intra-system time synchronization.

(3) Time tracing

The time-tracing function is mainly to realize the synchronization between navigation system time and UTC time so as to ensure the time compatibility between GNSS systems.

(4) Timing

The timing function is mainly to provide the user with time services such as one-way timing, two-way timing, and long-distance time transfer (such as common-view and all-in-view).

(5) Time difference monitoring and prediction

The time difference monitoring and prediction is mainly to measure the time difference between GNSS systems through the GNSS time difference monitoring system and to release the forecast information on the changing trend.

As shown in Fig. 6.1, the time synchronization links in the navigation satellite system are mainly inter-station, satellite-ground, and inter-satellite time-frequency synchronization links that synchronize the equipment time with signal time in the whole system. The time difference monitoring and prediction is mainly to measure the time difference between GNSS systems through the GNSS time difference monitoring system. The system time tracing is to keep the system time synchronized with UTC through the time-tracing link. The system time-frequency transmission is mainly to transmit the system time to the user via a satellite.

### *6.1.2 System Time Generation and Tracing*

GNSS system time is a mathematical and physical basis for describing the satellite motion, processing the observed data, and expressing the observations. It serves as a reference basis for satellite ephemeris calculation, satellite clock difference prediction and timing, and also as the reference time for the coordination between space segment and ground segment.

The ground system time is indirectly linked with UTC through a system time-tracing/comparison link to synchronize the system time with UTC. The clock difference between local satellite time and system time can be measured by spaceborne atomic clock. The local time signal is generated by the reference time-frequency signal generation unit via a satellite-ground time comparison link. The clock difference parameters are input into the satellite via the uploaded messages to synchronize the local satellite time with system time.

1. **Generation of BDS Time (BDT)**

BDT is implemented by "paper time", and generated and maintained by the working clock group in the master control station of the ground system. The atomic clocks used for time services are mainly hydrogen atomic clocks, along with some cesium atomic clocks. A clock group includes more than 10 atomic clocks. The system time algorithm is conducted through the optimized paper time with excellent performance. In the algorithm, the frequency shift, frequency drift, and frequency stability of atomic clocks are fully considered. The atomic time is mainly weighted by Allan variance minus clock slope and the maximum weight limit. To keep the BDT consistent with

UTC as much as possible, the shift is adjusted within the limit of $5 \times 10^{-15}$ at the appropriate time using a frequency control method.

2. **Generation of GPS Time**

The GPS time (GPST) is given by the combination of atomic clocks in the master control station, in the monitoring stations, and in the satellite. GPST is based on the UTC (USNO) of the United States Naval Observatory. The time shift between GPST and UTC (USNO) is basically less than 50 Ns (module 1 s). At present, the USNO's time-keeping system is a clock group that consists of 69 cesium clocks, 24 hydrogen clocks, and 2 cesium fountain clocks, as well as huge comparison equipment and a data processing system. The ground master control station averages the weight of onboard clock group unequally with reference to the short-term stability of various atomic clocks by using a Kalman filtering algorithm. The basic principle to select clocks is an excellent running state without periodic phase changes. Therefore, the resulting clock combination has high stability and good geometric distribution, without being affected by a single point of failure. The weight of the cesium atomic clock of the ground master station is the maximum, followed by the weight of the atomic clock of the ground monitoring station and the weight of spaceborne atomic clock in turn. Then the GPST can be derived from the clock weights.

3. **Generation of Galileo Time**

The Galileo time system generates the Galileo System Time (GST) by using synthetic atomic time, while generating real-time reference signals under master clock control. The master control station is equipped with two hydrogen clocks and four cesium atomic clocks with high performance, which constitute a precise timing system. Such a small system mainly takes advantage of the short-term stability of hydrogen clocks and the long-term stability of cesium clocks. The two hydrogen atomic clocks include a master clock and a redundant clock. Their real-time measurements are compared to realize seamless switching. The real-time measurements of the four cesium atomic clocks, which constitute the GST free time, are also compared periodically. After being measured and compared with the UTC(k) time provided by major European laboratories, the time of the hydrogen atomic clocks is corrected to obtain true GST.

4. **Generation of GLONASS Time**

The Time and Frequency Service Center of VNIIFTRI (All-Russian Scientific Research Institute for Physical-Engineering and Radiotechnical Metrology) produces and maintains Russian independent local atomic time TA(SU) and Universal Time Coordinated UTC(SU). As the national standard time of Russia, UTC(SU) is a reference source of such time service systems as GLONASS, ground-based radio timing system, TV, and network. At present, the UTC(SU) system is composed of nine Russian-made hydrogen atomic clocks. The calculation of the TA(SU) is based on all the hydrogen atomic clocks participating in time services, under the cooperation with international atomic time (TAI). Then the UTC(SU) time is transferred to GLONASS ground master control station, the reference source of GLONASS.

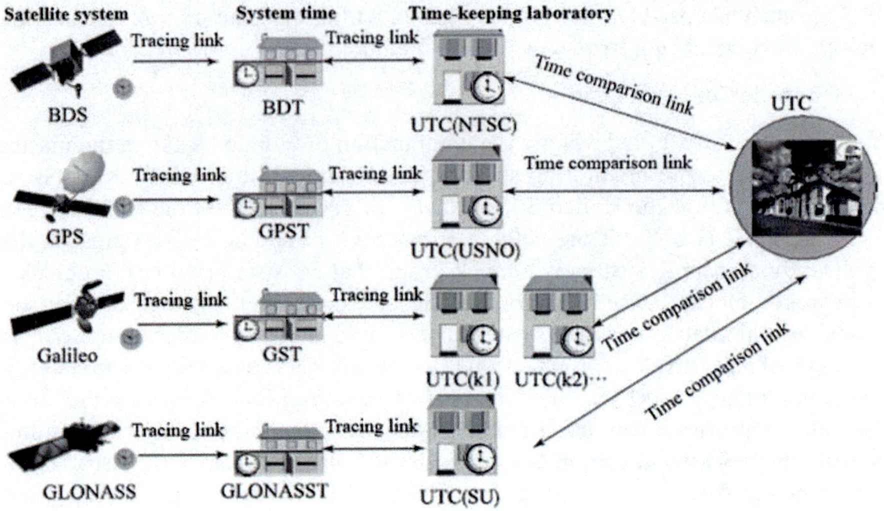


**Fig. 6.2** Tracing link of GNSS navigation system

5. **Tracing of GNSS Time**

In general, the GNSS system time is traced back to UTC(K) and then to UCT by establishing a tracing relationship with the national standard time UTC(k). The tracing of the system time is expressed by a tracing model in navigation messages. Users can know the time difference between the local time and the national standard time UTC(k) by receiving the tracing data from this system and correcting GNSST into UTC(k). To improve the time precision, UTC(k) can be corrected into UTC by using the data published in the T Bullet of Bureau International des Poids et Mesures (BIPM), thus achieving complete traceability.

As shown in Fig. 6.2, different navigation systems have different system time and tracing links. GPST is traced back to the UTC(USNO) maintained by the US Naval Observatory, GLONASS back to the UTC(SU) maintained by VNIIFTRI, and BDT back to the UTC(NTSC) maintained by the National Time Service Center of China. The time reference of GST is calculated by using the UTC time provided by several European time laboratories.

## *6.1.3 Time Transfer*

In a navigation satellite system, independent and reliable system time that continuously operates shall be established. The difference between the time of each atomic clock and the system time shall be accurately known through the synchronization of atomic clocks between ground stations and between ground station and satellite.

In order to realize the synchronization, the satellite-ground, inter-station, and inter-satellite time transfer links are established so that the satellite time and ground time can be synchronized with the system time.

1. **Satellite-Ground Time Transfer**

The navigation satellites are all equipped with high-precision atomic clocks that are synchronized with ground-based working clocks. In the satellite-ground time transfer, the satellite-generated time (1 pps) is synchronized with the time of ground master control station mainly through satellite-ground two-way radio ranging, laser pseudoranging, and reverse positioning

In terms of satellite-ground two-way radio ranging, the ground station receives the navigation signal from a satellite in the visible arc and measures the pseudorange. In the meantime, the satellite receives an uploaded signal and also measures the pseudorange. The satellite sends the measured pseudorange value back to the ground station through a downlink signal to calculate the time difference between the satellite clock and ground clock, thus realizing the satellite-ground time transfer. This method is a conventional satellite-ground comparison method with higher accuracy which can eliminate the path delays such as ionosphere delay and troposphere delay.

In terms of laser pseudoranging, the time difference between satellite clock and ground clock is calculated based on the propagation delay of satellite transmission signal measured by the ground station and the bidirectional propagation delay of the laser traveling from the synchronization station to the satellite. The laser time synchronization has extremely high accuracy, but is greatly affected by the weather. It is generally used as a calibration method for radio devices.

In terms of reverse positioning, four ground stations are used as a basis to calculate the position and clock difference of the satellite by referring to the receiver positioning, thus achieving the satellite-ground synchronization.

2. **Inter-station Time Transfer**

In general, the ground stations are distributed separately. The inter-station time transfer mainly includes satellite two-way time transfer, GNSS time transfer, and optical fiber time transfer.

Satellite two-way time transfer refers to the time transfer method that two ground stations send their own time information to each other through the same satellite (usually a GEO satellite) and receive time information from each other at the same time. Then the clock difference between the two stations can be calculated with reference to the correction of the Sagnac effect caused by Earth rotation. Because the signal transmission paths of the two stations are approximately symmetric, the effect of path delay can be ignored. Therefore, the time transmission accuracy is high, up to 1 ns. The satellite two-way time transfer is also an important means for the BIPM to calculate the TAI and the UTC.

GNSS time transfer is usually a common-view method. In this method, two ground stations observe the same GNSS satellite at the same time. The time difference between the ground receiver clock A and the spaceborne clock, and the time difference between the receiver clock B and the spaceborne clock can be determined according to the GNSS system time contained in the navigation message. The time difference between the clock A and the clock B can be determined by subtracting the time of the two clocks. The common-view time transfer can eliminate not only the effect of the spaceborne clock, but also part of the effects of satellite position, ionosphere, and troposphere. This time transfer method has become an important means to calculate the TAI and the UTC. The accuracy of GNSS time transfer can be further improved by the so-called "all-in-view" method (i.e., observing multiple satellites) and PPP method. However, those methods need accurate orbital data and clock data from an IGS so that a delay will be produced.

High-precision optical fiber time transfer is achieved mainly in two ways: loop-back and two-way comparison. The loop-back method is to measure the time delay of the whole transmission loop of a time signal from the central station A to the terminal station B and then back to the station A along the original route, so as to calculate the one-way time delay between the station A and the station B, compensate for the time signal received by the station B with the calculated time delay, and achieve the time signal synchronization between the station B and the station A. The two-way comparison method is similar to the satellite two-way time transfer method. The two stations participating in the comparison transmit time signals to each other and receive time signals from each other. Either of the two stations measures the time difference between its own signal and the other station's signal, and then calculates the atomic clock difference between them. Optical fiber has the advantages of large communication capacity, small temperature coefficient, long relay distance, low loss, and strong anti-interference ability.

3. **Inter-satellite Time-Frequency Transfer**

The inter-satellite time-frequency transfer techniques mainly include radio two-way time-frequency transfer and laser two-way time-frequency transfer. At present, the radio technique is more popular. The spaceborne processors continuously update the satellite parameters through inter-satellite two-way pseudocode ranging and data communication, thus achieving the time-frequency synchronization among satellites. The inter-satellite laser time-frequency transfer is mainly to synchronize different satellite time frequencies through laser pulse ranging or optical carrier radiofrequency. Its principle is similar to that of radio two-way ranging.

### *6.1.4 Time-Frequency System Monitoring and Forecasting*

After the establishment of a satellite time-frequency standard, a clock difference will be generated between satellite time and system time due to the aging shift of atomic frequency standard. If accumulated to a certain level, the clock difference will affect the navigation positioning and timing accuracy. Therefore, it is necessary to forecast and compensate for the satellite clock difference and eliminate the influence of clock difference on the system.

In China's BDS system, satellite clock difference is just the difference between spaceborne clock time and BDT. The ground stations of BDS are located within the territory of China. Therefore, when a satellite is flying on an orbital plane, the ground monitoring station can only monitor part of its orbit. When the satellite is in the orbit arc that can be monitored, the ground station can calculate the satellite clock difference by a certain algorithm. However, when the satellite is in the unobservable orbit arc, the ground station cannot obtain the satellite clock difference directly, but through an inter-satellite link. The clock correction value broadcast by the satellite to the user is the forecasted satellite clock difference. In order to avoid the degradation of accuracy due to the long forecasting time, it is necessary to upload the latest forecast value of clock difference at intervals. To acquire more accurate satellite clock difference, a more accurate clock difference prediction model shall be established.

The prediction accuracy of the satellite clock difference mainly depends on the frequency stability of onboard clock, the prediction time duration, the prediction model, and the quality of observations used for prediction. The current studies on clock difference prediction usually rely on the following three models: linear model, quadratic polynomial model, and grey model.

1. **Quadratic polynomial model**

In the quadratic polynomial model, a power function is used as the fitting function to fit the satellite clock difference data with equal time intervals and then is extrapolated to forecast the clock difference. The relationship between satellite clock time and system time is expressed as

$$\Delta t = a_0 + a_1(t - t_0) + a_1(t - t_0)^2 + v \tag{6.1}$$

where $\Delta t$ is the satellite clock difference, $t$ is the system time, $a_0$, $a_1$ and $a_2$ are the clock difference, clock velocity, and half acceleration of the atomic clock, respectively, at the time $t_0$ relative to the system time $t$. $v$ is the error of observation. If the duration of clock difference is uniform, then $a_2$ is 0.

If $\hat{a}_0$, $\hat{a}_1$ and $\hat{a}_2$ are the estimated values of the satellite clock parameters $a_0$, $a_1$, and $a_2$, then

$$\Delta\hat{t} = \hat{a}_0 + \hat{a}_1(t - t_0) + \hat{a}_2(t - t_0)^2 \tag{6.2}$$

It is known that $x_1, x_2, \ldots, x_n$ are the clock difference data relative to the equal-interval time $t_1, t_2, \ldots, t_n$. Then the optimal parameters $\hat{a}_0$, $\hat{a}_1$ and $\hat{a}_2$ can be fit through least-square estimation, i.e.:

$$\hat{a} = \begin{bmatrix} \hat{a}_0 \\ \hat{a}_1 \\ \hat{a}_2 \end{bmatrix}, \quad A = \begin{bmatrix} 1 & t_1 - t_0 & (t_1 - t_0)^2 \\ 1 & t_2 - t_0 & (t_2 - t_0)^2 \\ \cdots & \cdots & \cdots \\ 1 & t_n - t_0 & (t_n - t_0)^2 \end{bmatrix}, \quad B = \begin{bmatrix} \Delta\hat{t}_1 \\ \Delta\hat{t}_2 \\ \cdots \\ \Delta\hat{t}_n \end{bmatrix} \tag{6.3}$$

So the estimated value is

$$\hat{a} = \left(A^T A\right)^{-1} A^T B \tag{6.4}$$

When the parameter $\hat{a}$ is obtained, this quadratic polynomial prediction model can be fit, and the satellite clock difference in the required time can be forecasted.

2. **Grey model**

The Grey System Theory, which was first proposed and founded by professor Deng Julong in China, is a system engineering methodology based on mathematical theory. This theory is specialized in analyzing data with little information, few samples, and a small amount of calculation. Generally, it directly extracts information from the confronted data rather than requiring the data to obey a certain distribution law.

GM (1,1) (1 order and 1 variable gray model) is the most commonly used gray system model. The original data sequence of satellite clock difference at equal time intervals is defined as $X^{(0)} = \left\{X^{(0)}(1), X^{(0)}(2), \ldots, X^{(0)}(n)\right\}$, where $n$ is the number of the originally observed data selected to establish the model.

Firstly, the sequence $X^{(1)} = \left\{X^{(1)}(1), X^{(1)}(2), \ldots, X^{(1)}(n)\right\}$ can be obtained through one cumulation of original clock difference sequence, where $\mathrm{X}^{(1)}$ satisfies $X^1(k) = \sum_{i=1}^{k} X^0(i), \quad k = 1, 2, \ldots, n$. Then, a differential equation of the new sequence $X^{(1)}$can be derived, as shown in Eq. (6.5):

$$\frac{\mathrm{d}X^{(1)}}{\mathrm{d}t} + \mathrm{a}X^{(1)} = b \tag{6.5}$$

where both $a$ and $b$ are model parameters that can be estimated and calculated through the least-square calculation of cumulated clock difference observations.

Given that $\mathrm{Z}^{(1)}(k+1) = \left[X^{(1)}(k+1) + X^{(1)}(k)\right]/2, k = 1, 2, \ldots, n-1$, then the differential equation can be written as

$$\begin{cases} X^{(0)}(2) + aZ^{(1)}(2) = b \\ X^{(0)}(3) + aZ^{(1)}(3) = b \\ \quad \cdots \\ X^{(0)}(n) + aZ^{(1)}(n) = b \end{cases} \tag{6.6}$$

The estimated values of the parameter $\hat{a}$ and the two matrixes are, respectively, given by

$$\hat{a} = \begin{bmatrix} a \\ b \end{bmatrix}, \quad Y = \begin{bmatrix} X^{(0)}(2) \\ X^{(0)}(3) \\ \cdots \\ X^{(0)}(n) \end{bmatrix}, \quad B = \begin{bmatrix} -\mathrm{Z}^{(1)}(2) & 1 \\ \mathrm{Z}^{(1)}(3) & 1 \\ \cdots & \cdots \\ -\mathrm{Z}^{(1)}(n) & 1 \end{bmatrix} \tag{6.7}$$

According to the least-square estimation, the estimated value of the parameter $\hat{a}$ can be yielded:

$$\hat{a} = \left(B^T B\right)^{-1} B^T Y \tag{6.8}$$

By substituting the estimated value of $\hat{a}$ into the differential equation, we can obtain

$$X^{(1)}(k+1) = \left(X^0(1) - \frac{b}{a}\right)\mathrm{e}^{-\mathrm{ak}} + \frac{b}{a}, \quad k = 0, 1, \ldots, n-1, \ldots \tag{6.9}$$

where $k$ represents the number of the original data used for prediction.

The data used to establish the prediction model should be the data obtained after one cumulation rather than the original clock difference observations. Therefore, it is necessary to regain the original clock difference observations through subtraction.

The prediction model after the subtraction is as follows:

$$\hat{X}^{(0)}(k+1) = \hat{X}^{(1)}(k+1) - \hat{X}^{(1)}(k) = \left(1 - \mathrm{e}^{a}\right)\left[X^{(0)}(1) - \frac{b}{a}\right]\mathrm{e}^{-\mathrm{ak}},$$
$$k = 1, \ldots, n-1, \ldots \tag{6.10}$$

From the above equation, the general form of GM (1,1) prediction model after cumulation can be derived:

$$\hat{X}^{(0)}(k+p) = \left(1 - \mathrm{e}^{a}\right)\left[X^{(0)}(1) - \frac{b}{a}\right]\mathrm{e}^{-a(k+p-1)} \tag{6.11}$$

where $p$ is the number of clock difference prediction points, $p \geq 1$. When subtraction is used for data processing, its forecasting model can be derived in a similar way.

The satellite clock difference prediction based on the GM (1,1) model has made great achievements in the real navigation system. In the long-term prediction of satellite clock difference, microsecond precision can be achieved. However, in the short-term prediction, the index coefficient in the Eq. (6.11) is only related to the number of epochs (the time corresponding to the obtained data), so the prediction precision is relatively poor.

3. **Kalman model**

Kalman Filtering is a computer-based real-time recursive algorithm proposed by Kalman (R. E. Kalman) in 1960. With the aim of processing random data and signals, this algorithm utilizes the statistical characteristics of system noise and observation noise, takes the observed quantity of the system as input and the value to be estimated as output, and correlates the input with the output by the algorithms of time updating and observation updating.

Owing to its optimality, the Kalman prediction algorithm has been widely used in the prediction of atomic clock time difference and GNSS system clock difference.

Let the start time be 0. The satellite clock difference at the time $t$ can be expressed by

$$\Delta t = a_0 + a_1 t + a_2 t^2 + w_1 + w_2 \tag{6.12}$$

where $a_0$, $a_1$ and $a_2$ are the parameters of the quadratic polynomial model, $w_1$ is the phase noise, and $w_2$ is the measurement noise. Its Kalman prediction models are as follows:

$$X(t) = X_1(t) + \varepsilon_4(t) + \varepsilon_1(t) \tag{6.13}$$

$$X_1(t) = X_2(t) + \varepsilon_2(t) \tag{6.14}$$

$$X_2(t) = X_3(t) + \varepsilon_3(t) \tag{6.15}$$

$$X_3(t) = 0 \tag{6.16}$$

where $X(t)$ is the satellite clock difference, i.e., the $\Delta t$ in Eq. (6.12); $\varepsilon_1(t)$ represents the phase white noise in $w_1$; $\varepsilon_2(t)$ represents the frequency white noise in $w_1$; and $\varepsilon_3(t)$ represents the frequency random walk noise in $w_2$. The variances of the noises $\varepsilon_1(t)$, $\varepsilon_2(t)$ and $\varepsilon_3(t)$ are expressed as $\sigma_1^2$, $\sigma_2^2$ and $\sigma_3^2$, respectively, all of which can be obtained by specific calculation formulas.

The state equation and observation equation of satellite clock in Kalman filtering can be expressed as

$$\begin{bmatrix} X_1(k+1) \\ X_2(k+1) \\ X_3(k+1) \end{bmatrix} = \begin{bmatrix} 1 & \tau & \frac{1}{2}\tau^2 \\ 0 & 1 & \tau \\ 0 & 0 & 1 \end{bmatrix} \begin{bmatrix} X_1(k) \\ X_2(k) \\ X_3(k) \end{bmatrix} + \begin{bmatrix} u_1(k) \\ u_2(k) \\ 0 \end{bmatrix} \tag{6.17}$$

$$X(t) = X_1(t) + \varepsilon_1(t) + \varepsilon_4(t) \tag{6.18}$$

where $\varepsilon_4(t)$ is the measurement white noise. The noise variance matrix $Q$ is

$$Q = \begin{bmatrix} \tau\sigma_2^2 + \frac{1}{3}\tau^2\sigma_3^2 & \frac{1}{2}\tau^2\sigma_3^2 & 0 \\ \frac{1}{2}\tau^2\sigma_3^2 & \tau\sigma_3^2 & 0 \\ 0 & 0 & 0 \end{bmatrix} \tag{6.19}$$

Suppose that the interval between the current prediction time and the time of the last observation is $n\tau$. Then the predicted value of clock difference is

$$\Delta t((k+n)\times\tau) = \begin{bmatrix} 1 & 0 & 0 \end{bmatrix} \begin{bmatrix} 1 & \tau & \frac{1}{2}\tau^2 \\ 0 & 1 & \tau \\ 0 & 0 & 1 \end{bmatrix}^n \begin{bmatrix} X_1(k) \\ X_2(k) \\ X_3(k) \end{bmatrix} \tag{6.20}$$

It can be seen from Eqs. (6.19) and (6.20) that the noise equation matrix in this algorithm is determined by $\sigma_1^2$, $\sigma_2^2$, $\sigma_3^2$ and $\sigma_4^2$, where $\sigma_4^2$ can be derived from the data of the observation station, while $\sigma_1^2$, $\sigma_2^2$ and $\sigma_3^2$ can be derived from the number of the observation stations through Allan variance inversion.

The advantage of Kalman filtering is that with the input of a new measured value, the predicted value in Kalman prediction will change accordingly. When the data volume is large enough, the overall data trend can be analyzed. The state equation and observation equation can be established against the influences of different noises, thus predicting the short-term clock difference.

## 6.2 Spaceborne Time-Frequency System

As an important part of a navigation satellite system, the spaceborne time-frequency system provides a high-precision satellite time-frequency standard that is the basis for the whole navigation satellite system to realize navigation, positioning, timing, and velocity measurement. Therefore, it should have good stability, accuracy, and reliability. The time-frequency standard is derived from the high-precision atomic frequency standard on the satellite. The frequency conversion, high-precision measurement, clock difference adjustment, and other precise control functions are implemented by a time-frequency generation and maintenance system, thus providing a stable and reliable time-frequency reference and contributing to the excellent performance of satellite navigation payloads.

### *6.2.1 Composition and Function of Spaceborne Time-Frequency System*

The block diagram of the spaceborne time-frequency system is shown in Fig. 6.3. Unlike the output frequency of a spaceborne atomic clock, the reference frequency of a navigation satellite is usually 10.23 MHz, the conversion result of the output frequency of the atomic clock. The frequency conversion is usually realized by a reference frequency generation and maintenance unit (time-keeping unit for short). The onboard time-frequency system is just composed of an atomic clock and a time-keeping unit to generate and maintain a time-frequency standard signal and obtain the atomic time without introducing any leap seconds. The satellite time-frequency standard can participate in the generation of navigation system time and synchronize the generated navigation system time with the satellite time precisely by timing.

The time-keeping unit locks the output frequency of a crystal oscillator to the atomic frequency standard by comparing the spaceborne atomic clock with a high-stability crystal oscillator to obtain the spaceborne time-frequency standard signal. In order to ensure the system reliability, the spaceborne time-frequency system is

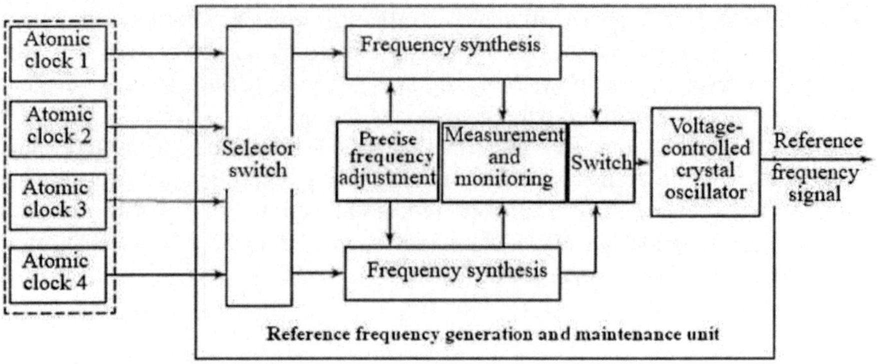


**Fig. 6.3** Block diagram of spaceborne time-frequency system

usually equipped with multiple atomic clocks, including one main atomic clock used for generating the reference frequency and several backup atomic clocks. The time-keeping unit usually measures the phase difference between the working clock and the backup clock, and sends this measurement result to the ground as the basis for evaluating the backup clock. As a result, the health state of onboard clocks can be monitored. In case of the failure of any atomic clock, an alarm will be given or autonomous processing can be implemented. The spaceborne time-frequency system can accurately synchronize the generated satellite time with the system time, monitor and manage the integrity of spaceborne atomic clocks, and provide a stable and reliable time-frequency reference for the generation of navigation signals.

The main functions of spaceborne time-frequency system are as follows:

(1) Generation of spaceborne time-frequency reference

The spaceborne time-frequency reference usually refers to a standard time-frequency signal (10 MHz) generated by spaceborne atomic clocks. It is usually provided by multiple spaceborne atomic clocks, one of which serves as the working clock and the others of which as the backup clocks. The navigation satellite uses the signal of atomic clocks to generate the time-frequency reference signal as needed.

The spaceborne time-frequency reference is generated by a synthesizer. It inherits not only excellent medium-term and long-term stability from atomic clocks, but also high short-term stability from a high-stability crystal oscillator. The spaceborne time reference is defined by a 1 pps signal. With the performance consistent with the atomic clock, this signal is usually obtained by the direct frequency division of a time-frequency reference signal.

(2) Generation of spaceborne reference frequency

The spaceborne reference frequency signal is usually generated by a reference frequency generation and maintenance unit. The spaceborne reference frequency provides a reference frequency for navigation task unit, and supports the code and frame synchronization during the onboard signal reception and transmission. Without a high-precision clock frequency, high-precision navigation, positioning, and timing cannot be implemented. The spaceborne reference frequency is usually designed with a backup to ensure reliability, by outputting one selected signal through a switch.

(3) Fault detection, alarm, and repair

The fault detection, alarm, and repair are mainly implemented by the fault detection module in the reference frequency generation and maintenance unit. The fault detection module can set the fault mode and parameter failure threshold for spaceborne clocks in advance, and detect the satellite clock fault and alarm in time by using a high-precision frequency/phase measurement technique and an appropriate detection algorithm. When a failure in satellite clock is detected, it can be handled autonomously by the satellite, or adjusted and compensated for by the commands sent by the ground.

(4) Time-frequency synchronization among spaceborne clocks

The satellite time-frequency synchronization is mainly implemented by the reference frequency generation and maintenance unit after receiving a clock difference adjustment command from the ground and adjusting the clock difference parameter through a phase regulator. Meanwhile, the working clock can also be synchronized with the backup clock. The precise clock difference between the working clock and the backup clock can be determined through high-precision frequency measurement and phase comparison. When the clock difference reaches the preset threshold, the time generated by the backup atomic clock can be adjusted by the phase regulator to realize the synchronization with the time of the working clock.

(5) Time-frequency signal allocation

The time-frequency signal allocation, including the allocation of a 1 pps spaceborne time-frequency reference signal and the allocation of a reference frequency signal, is mainly implemented by the frequency standard allocation unit. Through the signal allocation, multiple low-noise stable signals are generated for other equipment on the satellite. In some special applications, the time delay of the signals needs to be allocated equally or can be adjusted.

1. **Overview of time-frequency system in GPS satellite**

In 1985, the time and interface expansion unit was introduced to the Block I satellite. This is the earliest GPS time-frequency generation device, which uses spaceborne rubidium atomic clocks and cesium atomic clocks as a reference to provide time-frequency signals for payloads. The time and interface expansion unit has been further developed into a frequency synthesis and allocation unit. In the Block IIR satellite, the unit was equipped with three rubidium atomic clocks, including one working clock and two backup clocks. The reference frequency generation and maintenance unit in this satellite is called a time standard device. By taking rubidium clock as a reference, the output signal frequency is 10.23 MHz, the frequency adjustment step is 1 μHz, the frequency adjustment range is 10 Hz, and the measurement precision of phase difference is 1.67 ns

The functional block diagram of reference frequency generation and maintenance unit in the GPS Block IIR is shown in Fig. 6.4. The output frequency signals of reference frequency source (rubidium clock) and output frequency source (crystal oscillator) are divided by a reference epoch generator and a system epoch generator, respectively, to generate a reference time signal (a period signal of 1.5 s) and a system time signal. The phase difference between reference frequency source and output frequency signal, including the phase difference between the output signal of spaceborne rubidium atomic clock and the signal of the voltage-controlled crystal oscillator, is obtained by a phase comparator. Based on the phase error, the processing unit calculates the control amount of voltage-controlled crystal oscillator according to a certain algorithm and mathematical model, and converts the signal, through a digital-to-analog converter, into the oscillator's adjusted voltage to be acted on the system output frequency, thereby ensuring that the 10.23 MHz output frequency of the system is locked onto the reference frequency source.

The reference frequency generation and maintenance unit of GPS Block IIR satellite can detect and judge whether the output of rubidium atomic frequency standard is normal, whether the output of voltage-controlled crystal oscillator is normal, whether the rubidium atomic frequency standard is locked, and whether the modulation signal

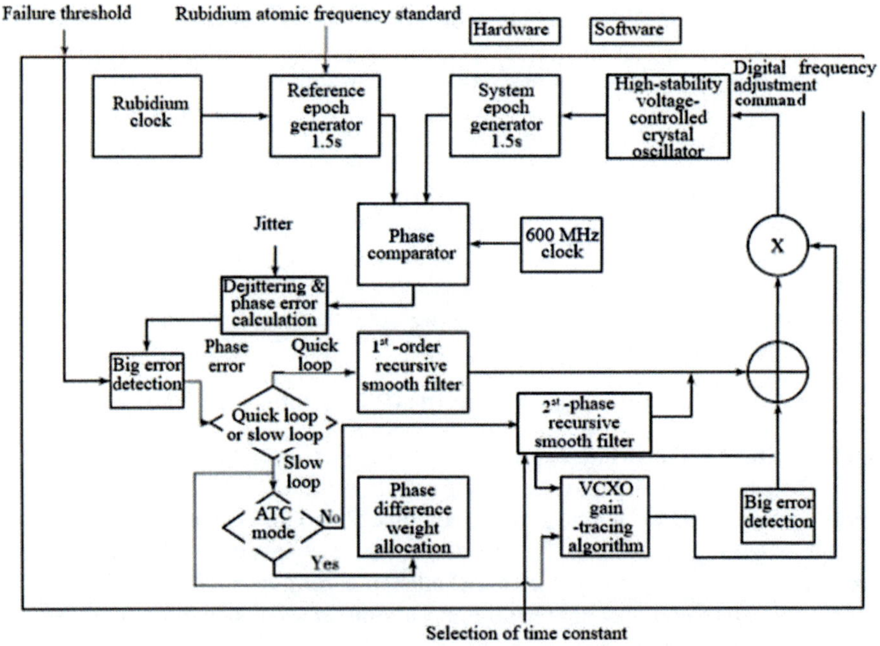


**Fig. 6.4** Functional block diagram of reference frequency generation and maintenance unit of GPS block IIR satellite

level of rubidium atomic frequency standard and the state of other monitors are normal. Therefore, the failure of time-keeping system can be monitored and judged at any time on the basis of the above detection functions. The time-keeping system in the GPS Block IIR satellite can also adjust the phase, frequency, and frequency drift of the 10.23 MHz output signal, so that the frequency accuracy and frequency drift of a spaceborne clock with respect to the working clock are lower than $3 \times 10^{-12}$ and $2 \times 10^{-14}$ per day, respectively.

Each GPS Block IIF satellite is equipped with 2 rubidium clock and 1 cesium clock. On this satellite, the time standard device is upgraded into a satellite time-keeping system (TKS), and the phase difference measurement accuracy of time-frequency standard signal is improved to 0.8 ps.

## 2. Overview of time-frequency system in GLONASS satellite

The time-frequency system in a GLONASS satellite includes three atomic clocks and a Spaceborne Time-Frequency Standard (STFS). Its main functions include generating high-precision frequency and time signals with reference to spaceborne atomic clock, and ensuring the stability and reliability of time-frequency signals. In the GLONASS-M, the STFS provides a 5 MHz time-frequency standard signal for the satellite, and the spaceborne atomic frequency standard is redundantly configured in a one-working-two-standby mode. The basic principle of STFS is to connect the reference frequency source and output frequency source of the system with a phase-locked loop in order to lock the output frequency of the system to the reference frequency source. The working principle of STFS is shown in Fig. 6.5.

The GLONASS system uses a spaceborne cesium atomic clock as the reference frequency source, whose output frequency value is 5 MHz. With the nominal value of 5 MHz, the output frequency of the high-stability voltage-controlled crystal oscillator used in the system can be adjusted within a certain range. The frequency signal

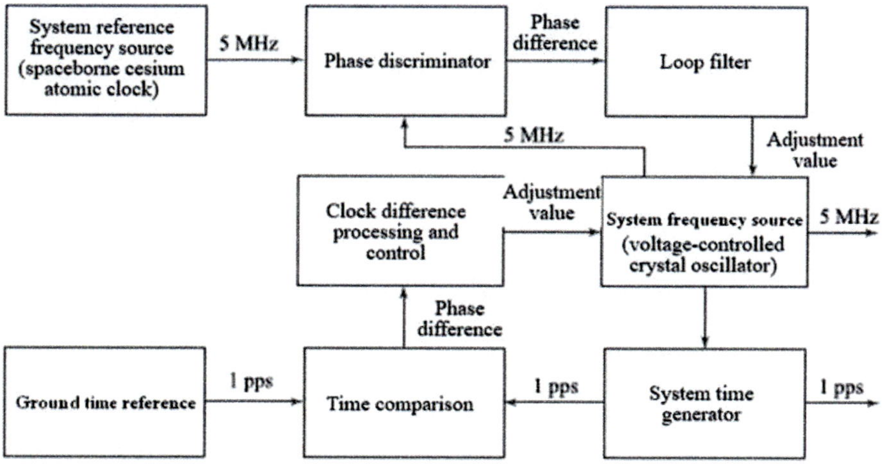


**Fig. 6.5** Functional block diagram of GLONAS STFS

output by spaceborne cesium atomic clock and the signal output by crystal oscillator are subjected to phase discrimination by a phase discriminator to obtain their phase difference, which, in turn, is converted into the control amount of the crystal oscillator by a loop filter circuit. Then the output frequency of the crystal oscillator is adjusted according to the control amount, and the crystal oscillator is locked to the frequency of spaceborne cesium atomic clock.

With the frequency output of the crystal oscillator, the system time generator generates a 1 pps signal, which is compared with the time reference signal of the ground system (1 pps) to obtain the phase difference between the two pulses. Then the phase difference is converted into the adjustment value of the crystal oscillator by a phase difference processing module according to a certain algorithm and mathematical model, in order to adjust the output signal of the crystal oscillator and synchronize the satellite time with the ground system time.

3. **Overview of Time-frequency System in Galileo Satellite**

The time-keeping unit in the time-frequency system of the Galileo navigation satellite is called Clock Monitoring and Control Unit (CMCU). By referring to spaceborne rubidium atomic clocks and hydrogen atomic clocks, the CMCU synthesizes the time reference and frequency reference required by onboard payloads. The structure of the corresponding time-frequency system is shown in Fig. 6.6.

The time-frequency system in a Galileo satellite has an output frequency of 10.23 MHz, a frequency adjustment step of 0.056 μHz, a frequency adjustment range of 10 Hz. The CMCU's frequency reference is based on four 10 MHz

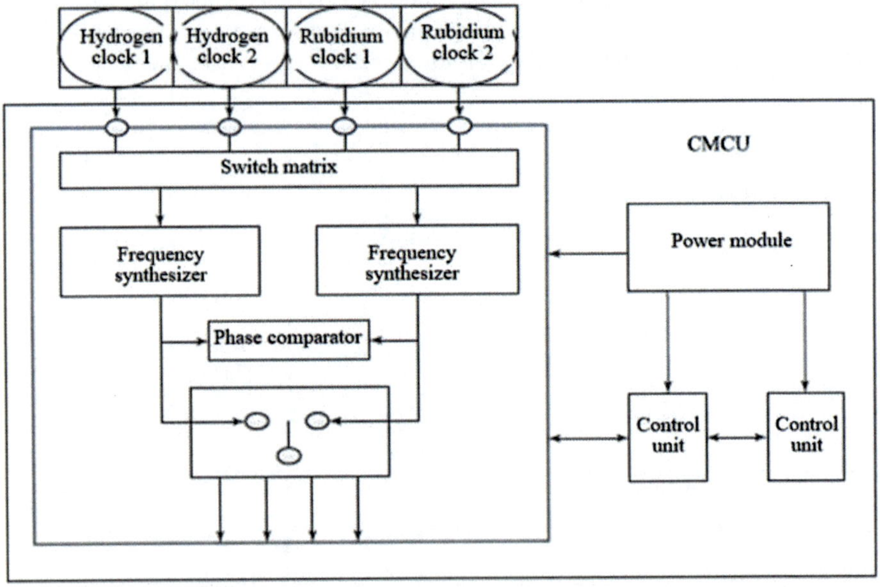


**Fig. 6.6** Block diagram of time-frequency system in Galileo Satellite

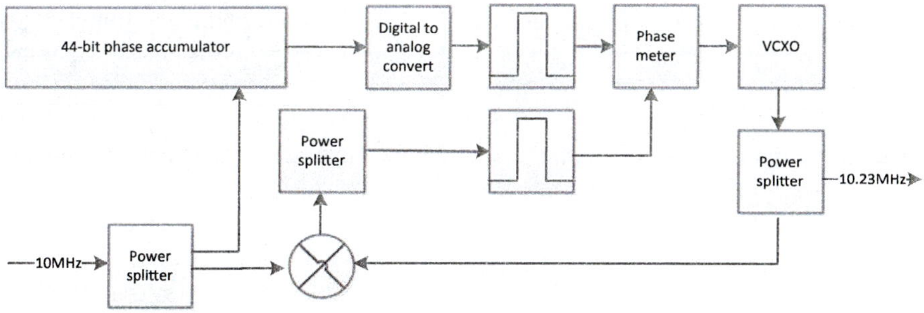


**Fig. 6.7** Functional block diagram of frequency synthesizer in Galileo Satellite

atomic frequency standards. The four frequency standards are selected from a switching matrix, including two frequency standards provided to the CMCU's two active frequency synthesizers (one is working and the other is redundant). The two frequency synthesizers generate the working-standby 10.23 MHz signals. The phase difference between them is obtained through a phase comparator, and then sent to the ground as a basis for monitoring the satellite clocks. One of the 10.23 MHz signals output by the two frequency synthesizers is selected by an electronic switch to lock the crystal oscillator so that the 10.23 MHz signal output by the crystal oscillator has good long-term and short-term stability.

As the core circuit of CMCU, the frequency synthesizer uses the 10 MHz signal of spaceborne atomic clock as a reference source and adopts the Direct Digital Synthesizer (DDS) technique to generate and output a 230 kHz signal with high-frequency resolution to a phase discriminator. Its block diagram is shown in Fig. 6.7. The other input of the phase discriminator is generated by the down-conversion of atomic frequency standard (10 MHz) and crystal oscillator signal (10.23 MHz). After the two 230 kHz signals are processed by a digital algorithm, the phase difference between atomic clock signal and crystal oscillator signal is obtained so that the Voltage-Controlled Crystal Oscillator (VCXO) can be locked and controlled.

### *6.2.2 Time-Frequency System in BDS Satellite*

1. **System Overview**

With the development of BeiDou navigation satellite system, the time-frequency system in BDS satellite has been gradually established step by step according to the "three-step" development strategy.

In the BDS-1 satellite, transparent repeaters are used for positioning and timing by forwarding the signals from ground stations and users. The design of a unified frequency source is adopted for the spaceborne time-frequency signal and implemented through the redundant configuration of the rubidium clock and high-stability crystal oscillator. The time-frequency signal is provided by a frequency-time signal generation and allocation device. The time-frequency system in the BDS-1 satellite is shown in Fig. 6.8.

The time-frequency system in BDS-2 Satellite is China's first generation of spaceborne time-frequency generation and maintenance device. It is responsible for providing stable and continuous reference frequency signals and system time for payloads and other devices in the satellite, and for realizing time synchronization with ground reference clock in coordination with the ground system.

The time-frequency system in BDS-2 satellite is shown in Fig. 6.9. Four Rubidium Atomic Frequency Standards (AFSs) are mounted in the satellite as a frequency source.

The time-frequency system in BDS-2 satellite consists of 4 atomic clocks, a high isolation switch, a reference frequency synthesizer, a command processing unit,

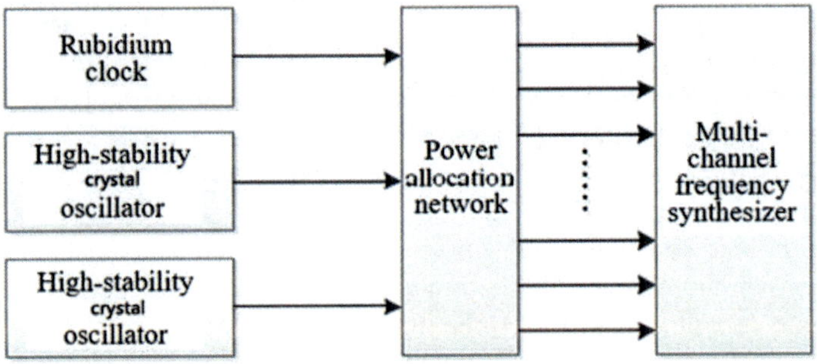


**Fig. 6.8** Block diagram of time-frequency system in BDS-1 Satellite

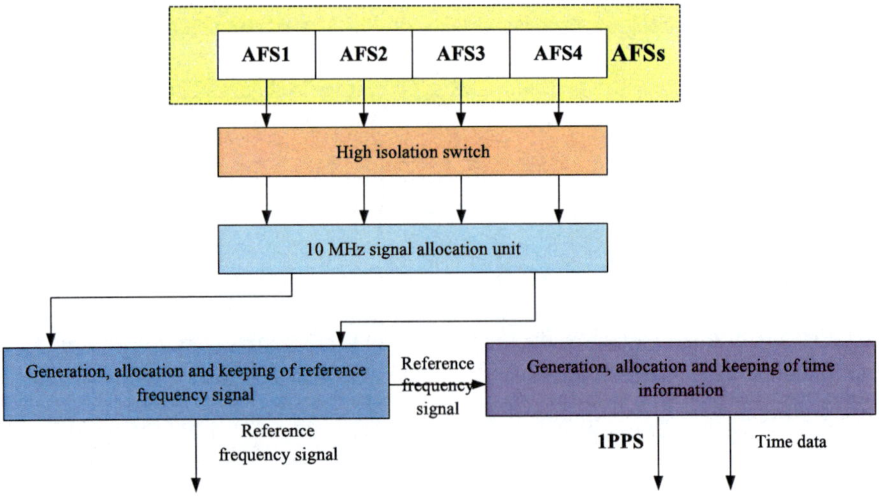


**Fig. 6.9** Block diagram of time-frequency system in BDS-2 Satellite

and other devices. The reference frequency synthesizer receives 10 MHz signals from two rubidium atomic clocks (a working clock and a redundant clock), and generates two 10.23 MHz signals (a working signal and a redundant signal) through two 10.23 MHz frequency synthesization circuits. The phase meter measures the phase difference between the two 10.23 MHz signals to monitor the health state of spaceborne clocks. The two signals are, respectively, locked to two high-stability voltage-controlled crystal oscillators with a nominal value of 10.23 MHz through a phase-locked loop. One of the output signals of the two crystal oscillators is selected by the switch as the frequency standard for system output. The characteristics of phase-locked loops determine the characteristics of the output signals. The phase-locked loops are required to inherit not only excellent long-term stability from atomic clock, but also good short-term stability from a high-stability crystal oscillator. The two phase-locked loops serve as the backup for each other, thus improving the system reliability. The structure of reference frequency generation and maintenance unit in BDS-2 Satellite is shown in Fig. 6.10.

In addition to synthesizing the 10.23 MHz reference frequency, the reference frequency synthesizer of BDS-2 satellite can also adjust and monitor the clock difference parameters of time-frequency standard by fine-tuning the 10.23 MHz

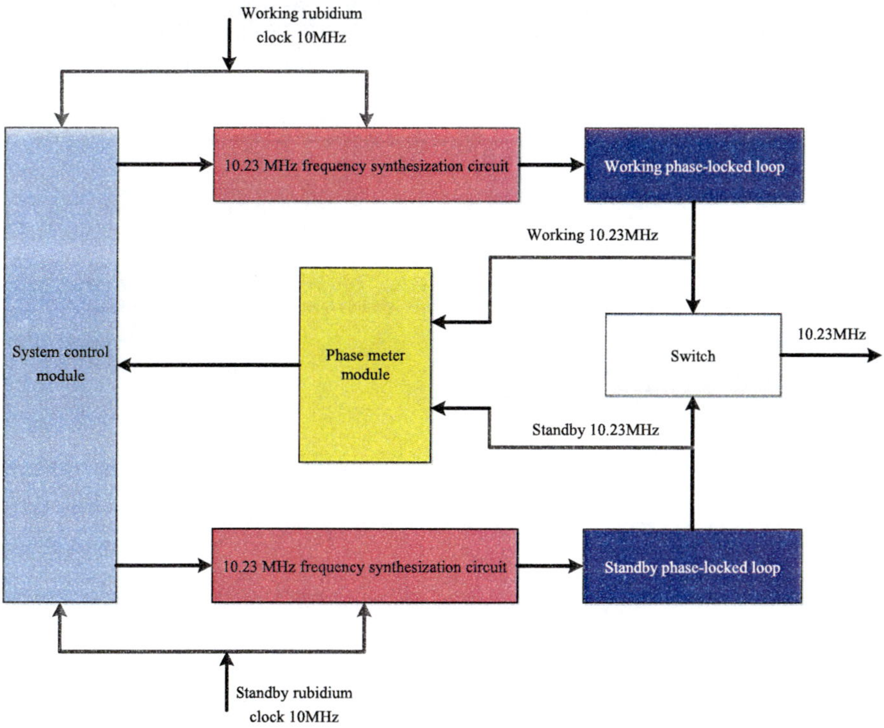


**Fig. 6.10** Block diagram of time-frequency generation and maintenance unit in Beidou-2 Satellite

reference frequency with a frequency control accuracy better than $5 \times 10^{-11}$ under the frequency adjustment command sent by the ground, thus achieving the synchronization between satellite time and system time. The reference frequency synthesizer monitors the phase of the 10.23 MHz signals. If the working reference frequency signal is found faulty, it can be switched to the redundant signal autonomously to ensure that the system has a certain autonomous operation capability.

The reference frequency generation and maintenance unit in the BDS-3 satellite has been improved in the following aspects on the basis of BDS-2. The stability of 10.23 MHz signals is greatly improved. The clock difference parameter can be accurately adjusted. The autonomous operation capability is achieved through a hardware upgrade and finer software algorithm control. The frequency, phase, and shift rate of the reference frequency can be compensated for by ground upload. The phase of the standby 10.23 MHz (signal) can be locked to the working 10.23 MHz (signal) by using a working-backup clock synchronization technique. When the working 10.23 MHz circuit fails, the system can be automatically and smoothly switched to the standby signal, and the frequency and phase jump of the output signals before and after switching can be confined to a certain range.

Figure 6.11 shows the typical stability test results of the 10.23 MHz reference frequency, whose stability in 10,000 s is $3.24 \times 10^{-14}$. In the test, the signal of the rubidium clock is used as the input signal of reference frequency synthesizer, and the hydrogen maser signal with a high-stability index is used as the reference signal of input frequency stability tester. The 10.23 MHz signal output by reference frequency

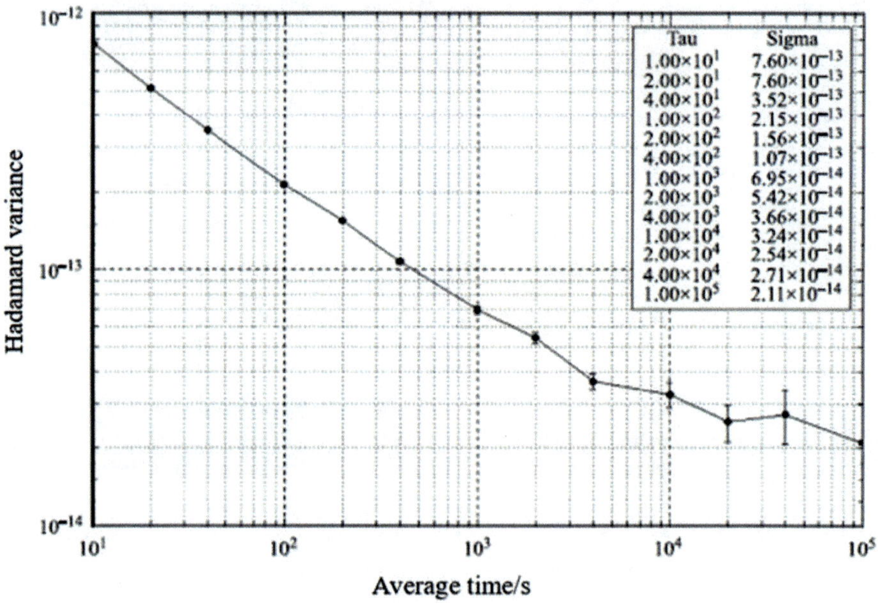


**Fig. 6.11** Typical stability test data of 10.23 MHz reference frequency

synthesizer is tested and input to the frequency stability tester with a sampling period of 10 s, in order to obtain a set of frequency difference data. The Hadamard variance of frequency difference data is obtained by the stability data minus the shift rate.

2. **Precision Adjustment of Spaceborne Time-Frequency System**

The parameter adjustment for 10.23 MHz signals includes the adjustment of frequency, phase, and shift rate. The high-frequency resolution of the DDS circuit is used to adjust the frequency of the 10.23 MHz output signal. Meanwhile, to make full use of the phase noise characteristics of VCXO, a suitable PLL loop bandwidth is selected. The noises in the loop bandwidth mainly include the 230 kHz phase noises generated by 10 MHz atomic clock signal and DDS circuit, and the additional phase noises generated by the PLL loop filter and phase discriminator. The output signal inherits excellent phase noise characteristics from VCXO while maintaining the frequency stability of the atomic clock signal by filtering, through PLL, the spurious noises generated by DDS circuit. The structure of the DDS circuit is shown in Fig. 6.12.

The N-bit phase accumulator, the phase adder as well as the secondary accumulator for adjusting the frequency drift of the DDS have been optimized on the basis of traditional DDS circuit. By superimposing pseudorandom codes on the phase accumulator, the signal power is allocated over a wider frequency band to reduce spurious output signals.

When detecting that the offset parameters of a satellite reference signal (including frequency, phase, frequency shift rate, etc.) are large enough to be adjusted, the ground will send a parameter adjustment command to the DDS circuit for compensation.

After receiving a frequency adjustment command, the frequencies are accumulated in the DDS frequency accumulator 1 in a certain time sequence and then superimposed in the phase accumulator to complete the satellite clock frequency adjustment. After receiving a phase adjustment command, the phase accumulator is controlled in a certain time sequence to complete the satellite clock phase adjustment. And after receiving a frequency drift adjustment command, the frequency drift accumulator is controlled in a certain time sequence, so that the frequency control word is changed at a certain frequency by a minimum step. The frequency adjustment degree depends on the drift adjustment value, and the frequency drift is compensated for indirectly by controlling the frequency adjustment degree.

The frequency drift adjustment function of the time-frequency system in the BeiDou satellite is verified as follows. A high-precision rubidium clock is used as the source of the input signal for testing, whose drift is known to be about $-3.9 \times 10^{-13}$. The frequency of the 10.23 MHz reference signal is tested by a frequency stability tester PicoTime with a sampling interval of 1 s.

The total testing time, when both the rubidium clock and the test instrument work stably, is nearly 7 days, covering 600,000 points. Among them, the measured drift of the top 180,000 points without a shift rate adjustment command is $-3.88 \times 10^{-13}$. After the 180,000 points, a reference drift adjustment command is sent according to the measured drift, and the test is continued to the 600,000th point, i.e.,

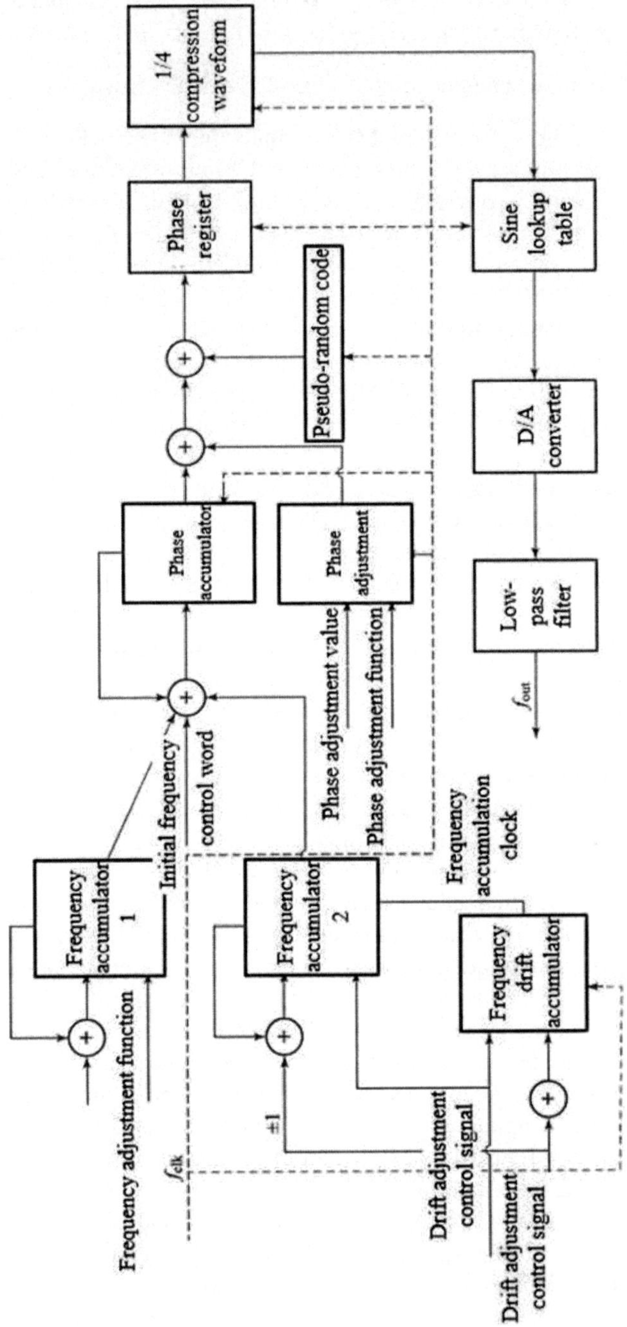


**Fig. 6.12** Structure of adjustment circuit of time-frequency reference signal parameter

from 180,000 points to 600,000 points. The drift of the 10.23 MHz signal under the reference control is measured at this time to be $1.45 \times 10^{-14}$, down by an order of magnitude from that before the adjustment, as shown in Fig. 6.13.

### 3. Smooth Switching of Spaceborne Time-Frequency System

Another important function of precision time-frequency signal management is to manage and switch the working and backup clocks according to the navigation satellite requirements for the continuity and reliability of the 10.23 MHz signal. When the spaceborne atomic clock or frequency synthesizer fails, discontinuous situations will occur to the 10.23 MHz signal, including signal interruption, frequency jump, and phase jump. These abnormal phenomena will interrupt the navigation services or bring down the positioning accuracy.

To solve this problem, the method of redundant backup and switching is employed by the reference frequency synthesizer of the time-frequency generation and maintenance device in the BeiDou satellite. When the working circuit signal is abnormal, the fault detection module in the system will detect the time abnormality and switch to the output frequency of the standby circuit. To ensure a smooth signal transition before and after switching, two problems need to be solved, One problem is that the frequency and phase of the redundant clock must be consistent with the working clock before switching. The other problem is that the switching process must be as fast as possible to reduce the time for getting out of the abnormality.

The design structure of redundant switching is shown in Fig. 6.14. A synchronization control module is the key to realize smooth switching. It synchronizes the

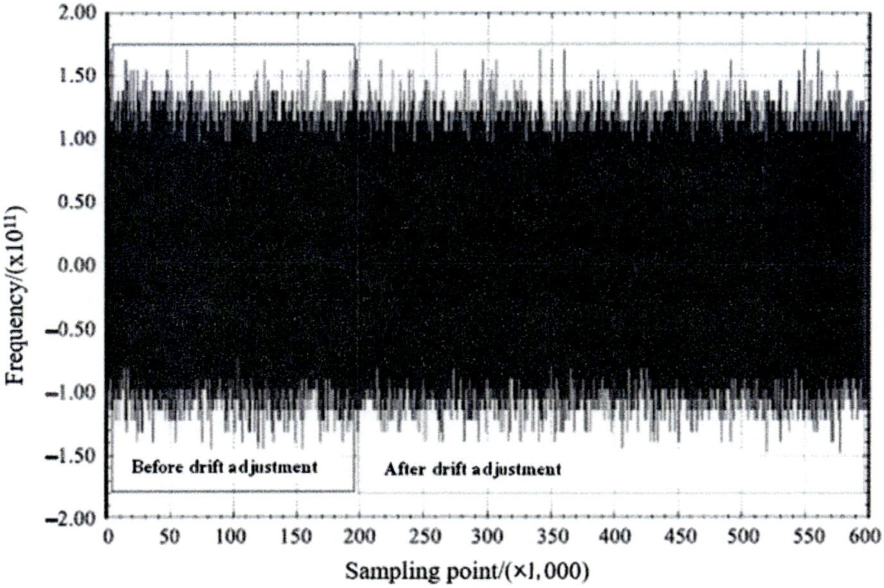


**Fig. 6.13** Test data of drift adjustment

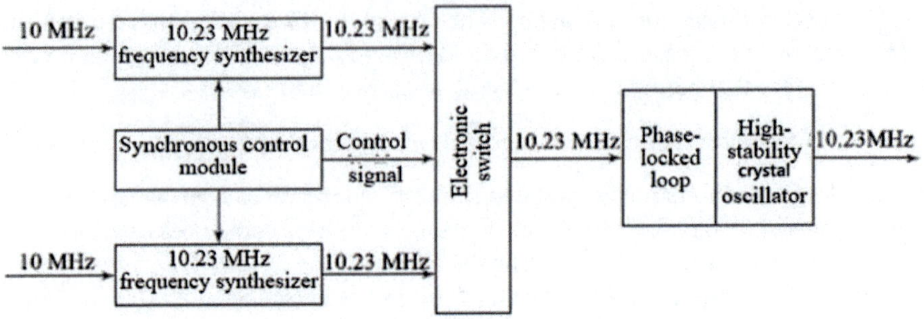


**Fig. 6.14** Switching structure of time-frequency system in BeiDou Satellite

frequency and phase of the 10.23 MHz signal between the standby circuit and the working circuit by a precision tracking algorithm to keep the phase synchronized at time. Supported by rapid detection and switching logic, the 10.23 MHz signal output by the complete system can be quickly switched to the standby circuit after the working clock fails and goes beyond the set threshold. Since the working clock signal has already been tracked by the redundant clock, the frequency and phase of the output signal can remain unchanged after being switched.

In order to realize smooth switching, the BDS satellite samples the phase difference between the working clock and the backup clock at a 61.374 MHz sampling frequency. After receiving a remote switching command from the ground, the single-phase difference value measured in real time can be used as the basis for adjusting the phase of the standby circuit. To achieve autonomous switching of the satellite, the output signal needs to be switched to the standby circuit autonomously when the working circuit signal is abnormal. When the working clock is abnormal, the measurement result of the phase difference will be wrong. To overcome this problem, the real-time phase difference measurements will be buffered within a certain period. The data buffered in a period of time is used as the basis for adjusting the phase of the standby circuit during autonomous switching.

The frequency difference between the working clock and the backup clock is calculated based on the variation of the phase difference between the two clocks in the period of N. The difference between every two consecutive phase difference measurements is calculated and then accumulated for (N-1) measurements in total. The frequency difference between the working signal and the standby signal is calculated by Eq. (6.23), where $\Delta p_i$ is the $i$th phase difference measurement, and $f_0$ is 10.23 MHz.

$$\Delta f = \frac{\sum_{i=1}^{N-1}(\Delta p_{i+1} - \Delta p_i)}{N} \times f_0 \tag{6.21}$$

If there is a frequency difference between the working signal and the standby signal, the phase difference between the working clock and the standby clock will change periodically. As shown in Fig. 6.15, after the 9th measurement reaches the

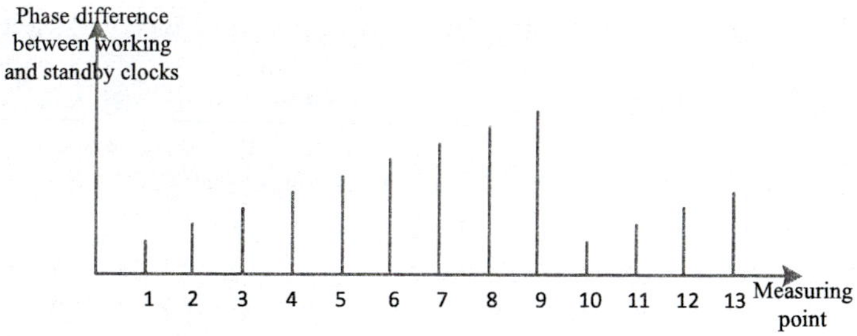


**Fig. 6.15** Measured phase difference between working and standby clocks

peak, the measured value will start to cycle from a smaller value. When the difference between the 9th and 10th measuring points is calculated by the Eq. (6.15), an error will occur.

To avoid such an error, a cross-period processing algorithm for phase difference data has been designed to solve this problem by calculating the relative difference—rather than the absolute difference—of two consecutive phase difference measurements.

The frequency and phase jump values of the 10.23 MHz signal before and after switching are checked using a frequency stability tester and a time interval tester, respectively. Two hydrogen clocks are used as the input signal sources of the reference frequency synthesizer. Upon testing, the minimum frequency jump before and after switching can be better than $1 \times 10^{-14}$, and the phase jump better than 0.1 ns.

### *6.2.3 Management of Spaceborne Clock Group*

With the development of navigation satellites, the spaceborne clock group under management will provide higher reliability and stronger autonomous operation capability. In the future, the clock management should be introduced to spaceborne time-frequency reference signal generation and maintenance system.

Compared with traditional time-keeping system, the output of the spaceborne clock group will no longer depend on the performance of a single atomic clock, but integrate the advantages of multiple satellite carrier frequency standards. For example, although a hydrogen clock is significantly better than a rubidium clock in stability, the latter has obvious advantages in mass, size, cost, and life. By using a combined clock algorithm based on the advantages of hydrogen clock and rubidium clock, a time-frequency reference more stable than the rubidium clock and cheaper than the hydrogen clock can be generated..

Another advantage of clock group management is that when some clock in the atomic clock group fails (including excessive frequency shift, unexpected frequency jump, unexpected phase jump, etc.), the system can detect this abnormality in time and remove the faulty clock from clock group calculation, thus ensuring that the time-frequency reference signal output by the system is not affected. As a result, the continuity and reliability of the system can be improved.

For example, the clock group management technique can be applied to the 10.23 MHz reference signal output by four spaceborne atomic clocks. The time-frequency system based on atomic clock group management can be established, as shown in Fig. 6.16.

The basic principle of time-frequency system is as follows. The signals of four atomic clocks are converted into digital signals by analog-to-digital converters, respectively. The phases of the four 10 MHz signals and the 10.23 MHz signal of crystal oscillator are compared pairwise in a digital domain. The phase comparison results are used as the input for signal processing and clock group algorithm. A time scale algorithm is used to calculate the clock difference data in order to obtain the optimal clock difference, which is just the output signal of the module. Then the controlled voltage signal output by the module is fed back to control the crystal oscillator so as to obtain the spaceborne 10.23 MHz reference signal.

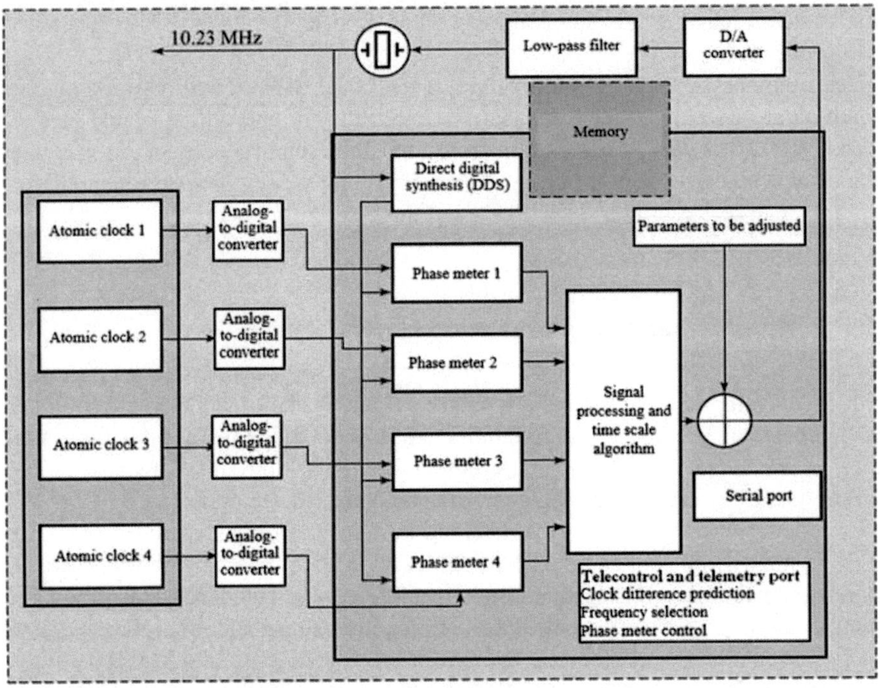


**Fig. 6.16** Block diagram of the clock group management

## 6.3 Atomic Clock Technology

Atomic clock is a highly integrated product of quantum physics and electronics. It is one of the most outstanding achievements in the application field of spectroscopy and atomic-molecular physics. The atomic clock uses the internal stable transition frequency between atoms (molecules, ions) as a reference to achieve accurate and stable frequency signal generation and keeping by locking the frequency of crystal oscillator or the laser frequency.

With the emergence and technological advances of atomic clocks, the accuracy and stability of time-frequency measurements have been greatly improved. The atomic clock plays an important or even irreplaceable role in many important areas of human social activities and daily life, including communications, electricity, network, transportation, and satellite navigation. In the field of scientific research, the accurate measurement of some physical constants and the verification of relativistic theory also requires high-precision time-frequency standards. As one of the seven international basic units, time (frequency) is the physical quantity with the highest measurement accuracy that can be achieved so far.

### *6.3.1 Development of Atomic Clock*

In the early days, human beings took the movement of celestial bodies (such as the sun, moon, and stars) as a tool for timing. With the development of productivity and the progress of science and technology, people began to turn attention to time accuracy, as it was difficult for the original timing method to meet the needs of these applications. With the development of quantum physics and spectroscopy, scientists have recognized that if the highly stable transition level inside an atom or molecule is used as a frequency reference, a time-frequency standard with higher precision can be obtained.

In 1936, I. Rabi put forward the theory of atomic and molecular beam resonance, and carried out the corresponding experiments. He concluded that the transition frequency of an atom depended only on its inherent characteristics, thus revealing the possibility of frequency control by the quantum transition. In 1945, I. Rabi initially announced that the atomic frequency standard could be implemented by a magnetic resonance technique. Subsequently, N. Ramsey proposed the separation field oscillation technique to effectively narrow the spectral line width.

On this basis, the first atomic clock in the world, namely absorption-type ammonia molecular clock, was born in 1948. It was developed by Lyons at the National Institute of Standards and Technology. In 1953, American scientist Townes and Chinese scientist Wang Tianjuan successfully developed an ammonia maser oscillator.

In 1952, Sherwood and Lyon et al. developed the first cesium beam frequency standard. In recent years, with the development of optically pumped cesium beam frequency standard, the frequency stability of the cesium beam frequency standard has been further improved.

In the early 1950s, A. Kastler invented the optical pumping technology. Later, Dicke et al. discovered that the Doppler spectral line broadening could be effectively reduced by using buffer gas to "trap" alkali metal atoms. The combination of the two key technologies contributed to the successful development of a bulb rubidium atom frequency standard. After decades of development, the rubidium atomic frequency standard has been widely used in communications, satellite navigation, and other fields owing to its high reliability, small size, low cost, and excellent frequency stability.

In the 1960s, the storage bulb technology invented by n Ramsey et al. contributed to the successful development of hydrogen atomic frequency standard. At present, hydrogen maser has high-frequency stability, with the stability per second of $1 \times 10^{-13}$ and the stability per day of $10^{-16}$. It is frequently used as a reference source for frequency testing. Because the size and weight of hydrogen maser are too large, some international companies and entities have developed passive hydrogen clocks with a relatively small size and weight. The passive hydrogen clock developed by Spectratime Company in Switzerland has been successfully applied to the European Galileo satellite navigation system.

With the increasing maturity of semiconductor lasers as well as laser cooling and atom trapping technologies, the cold atomic fountain clock has emerged. Compared with the traditional atomic frequency standard, this clock has extremely high frequency accuracy and stability. The accuracy of cesium atomic fountain clock can be as high as $2 \times 10^{-16}$ and its stability per second up to $1.6 \times 10^{-14}$.

Scientists have found that by choosing an optical transition frequency higher than microwave transition frequency by 4–5 orders of magnitude, the performance of the atomic frequency standard is expected to be greatly improved. At present, the best optical clock is the Sr atomic optical clock developed by NIST in the United States, with the frequency uncertainty of $10^{-18}$.

### 6.3.2 *Principle of Atomic Clock*

The working process of the atomic clock includes four steps: atom state preparation, atomic clock frequency excitation, clock transition signal detection, and clock locking.

1. **Atom state preparation**

The process of changing the relative number of atoms at two levels is called "state preparation". The purpose of state preparation is to prepare the atoms into appropriate energy states by some method so as to observe the net effect of transitions.

At present, the state preparation methods used in microwave atomic resonators approximately include magnetic state selection and optical pumping. The magnetic state selection is to classify the molecules in different energy states or the atoms with different magnetic moments (just like the magnetic poles in different directions) in a nonuniform electric or magnetic field into different trajectories, and select the atoms in desired states from space and discard unwanted ones. The optical pumping is to "change" all or most of the atoms into those in desired states through a special optical selection function by taking advantage of the atomic characteristic that the atoms in different energy states have different absorption rates for a special frequency and polarized light. The optical pumping has a higher atomic utilization rate.

2. **Atomic Clock Frequency Stimulation**

In order to generate transitions, the atoms must be provided with a "resonant cavity" where the atoms interact with the radiation field and exchange energy with each other. If the frequency of the microwave radiation field in the resonant cavity is exactly equal to the transition resonance frequency of atoms between the two energy states, an electrical signal proportional to the transition effect can be obtained. This signal can be used as a reference standard to control an oscillator (usually a quartz crystal oscillator), thus building a more advanced timing standard - atomic clock.

3. **Clock Transition Signal Detection**

Atomic signals can be detected basically in three ways: atom detection, optical detection, and microwave detection.

(1) Atom detection

The atoms passing through a state selector (which can separate the atoms of two states in space) will transit under the effect of the excitation signal in the microwave cavity. When passing through another state selector, they will be divided into two beams according to their different energy states in space. The atom detector set in the proper position can only receive one of the two beams. The change in the number of atoms received by the detector before and after the transition excitation shows the level transition probability of atoms. In this way, the signal change in the detector reflects the microwave resonance of atoms (Fig. 6.17).

(2) Optical detection

At present, the "clock" transition signal in the atomic frequency standard is detected by two optical methods, namely, linear absorption-transmission light detection and fluorescence detection.

The linear absorption-transmission light detection method is mainly used in bubble atomic clocks. When this method is used to select the states and the pumping light frequency is appropriate, the light irradiating the atomic bubble may excite the optical frequency resonance of atoms so that the level pumping process can be completed quickly. At this time, a downstream photodetector will receive the maximum optical signal. If a microwave interrogation signal is fed into the resonant cavity at the same time at the same frequency as atomic microwave resonance frequency, the atoms will undergo microwave resonance ("clock" transition). The

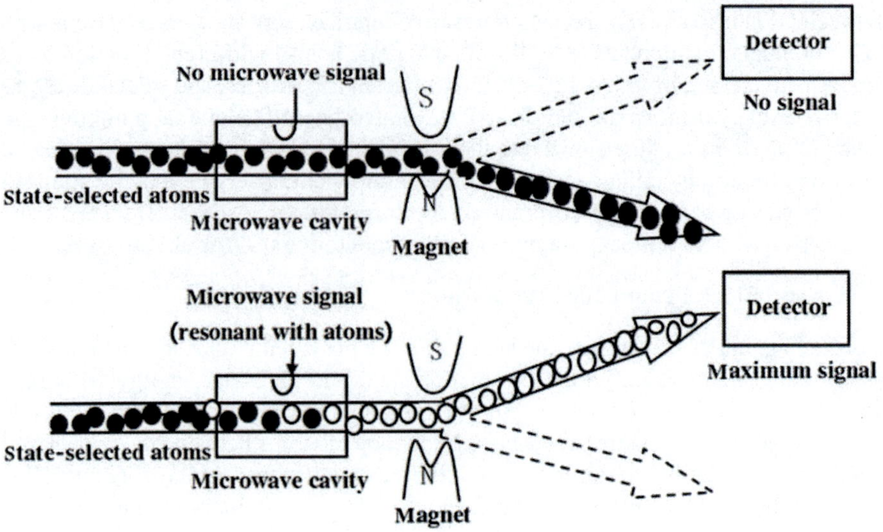


**Fig. 6.17** Block diagram of atom detection

atoms that have been pumped to a high energy state will be transited to low energy state, while the atoms that have been transited to low energy state will absorb the excitation light. As a result, the optical signal received by the downstream photodetector will be minimized. In this way, the signal change in the photodetector reflects exactly whether the microwave resonance of atoms and the frequency of external interrogation signal are correct or not (Fig. 6.18).

Fluorescence detection is a method that uses a photodetector to collect and detect the fluorescence emitted by the atoms returning to ultrafine ground-state levels after the "clock" transition (pumping or cyclical transition that pumps the atoms to the excited state). The magnitude of optical detection signal reflects the probability of microwave transition.

(3) Microwave detection

There is also a microwave detection method used in the microwave atomic frequency standard. It is performed during the interaction of microwave signals and atoms, as shown in Fig. 6.19.

When the frequency of the signal fed into the resonant cavity is equal to the resonant frequency of atoms, the atoms will either increase the signal energy (when the atoms in the cavity are in a high energy state) or decrease the signal energy (when the atoms in the cavity are in a low energy state). Detecting the change of microwave signal level in the microwave cavity is equivalent to detecting the microwave resonance of atoms. When the atoms entering the cavity are in a high energy state and in sufficient quantity, self-sustained oscillation will occur in the cavity so that the energy of the excited radiation is enough to offset the loss of cavity resonance. This

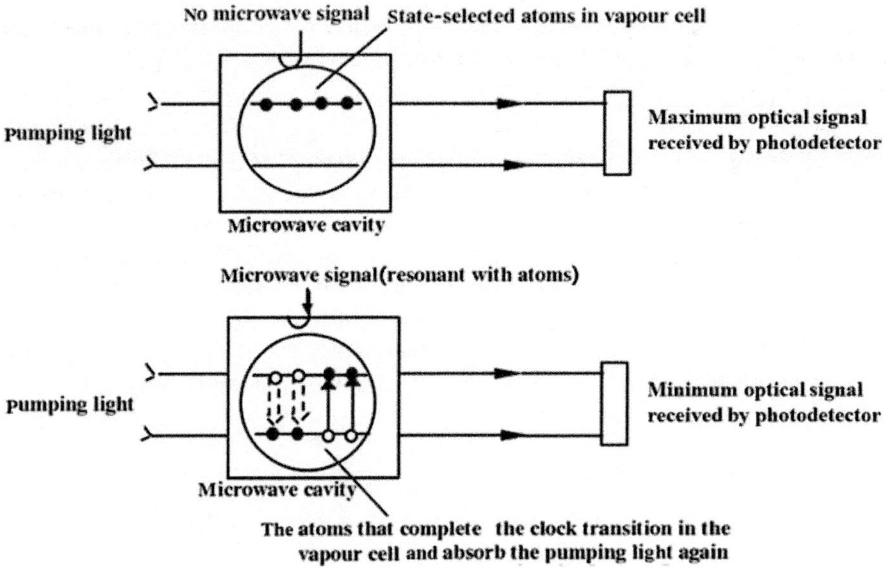


**Fig. 6.18** Block diagram of linear absorption-transmission light detection

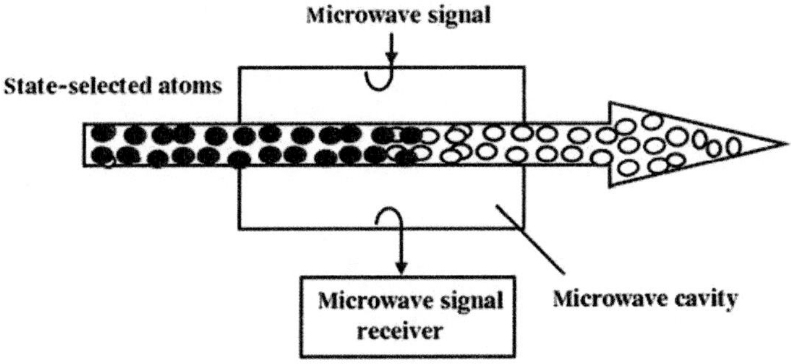


**Fig. 6.19** Block diagram of microwave detection

method is used in the Microwave Amplification by Stimulated Emission of Radiation (MASER).

4. **Clock Locking**

The clock transition detection signal is processed by a phase-locked (or frequency-locked) loop to obtain a correction voltage feedback that can control the oscillator. Then the frequency of the controlled oscillator is locked to the clock transition frequency of the quantum system to obtain a frequency signal output with high accuracy and stability. The clock transition frequency signal can be directly output or used as a frequency discrimination standard. Therefore, the clock locking can be divided into two types: the closed-loop locking of the Phase-Locked Loop (PLL) and the closed-loop locking of Frequency-Locked Loop (FLL).

(1) PLL closed-loop locking

When the energy radiated by high-level atoms in the resonant cavity is strong enough and the electromagnetic field energy generated by atomic induction radiation in the cavity is enough to compensate for the loss of cavity resonance, the quantum system will begin to oscillate. Then an atomic transition frequency signal will be output by the resonant cavity. In general, this signal has very low power and its frequency cannot be used directly. Therefore, the frequency of the controlled oscillator needs to be controlled by a phase-locked loop in order to obtain a high-stability frequency output that can be used directly. In this case, the locking process of an atomic clock is called the closed-loop locking of PLL.

(2) FLL closed-loop locking

When clock transition frequency signal is used as a frequency discrimination standard, the locking process is called the closed-loop locking of FLL. As a frequency discriminator, the quantum system discriminates the microwave excitation signal frequency generated by voltage-controlled oscillator through a frequency multiplication synthesizer. The detection signal output by quantum discriminator is subjected to synchronous phase-sensitive detection through a frequency-locked loop to generate a DC correction voltage feedback that can control the oscillator. Then the frequency of the microwave excitation signal is locked to the center frequency of the clock transition spectrum line, thereby enabling the oscillator to output a frequency signal with high stability.

### 6.3.3 Classification of Atomic Clocks

According to different technical schemes, atomic clocks can be divided into active atomic clocks and passive atomic clocks, whose core modules are atomic oscillator and atomic discriminator, respectively.

In fact, the quantum transition generated by a physical package cannot be directly applied, so can't the output frequency of self-sustained atomic oscillator. This is because it is a fractional frequency with low power, which will bring inconvenience during its direct use. The atomic resonator cannot generate self-sustained oscillation because it has no resonant cavity or the $Q$ value of the resonant cavity is not high. Usually, an atomic oscillator or resonator is used as a reference or frequency discrimination standard to control the voltage-controlled oscillator through a phase-locked (or frequency-locked) loop so as to transfer the high accuracy and stability of the physical package to the voltage-controlled oscillator, which, in turn, provides useful output. In the microwave atomic clock, the oscillator is usually a crystal oscillator. In the optical frequency standard, the oscillator is a frequency-stabilized laser.

1. **Active atomic clock**

Atomic oscillator refers to a resonant cavity that does not need external excitation. The system will start to oscillate when it works under certain positive feedback conditions, the intensity of high-level atomic beam entering the cavity is quite high, and the electromagnetic field energy generated by induction radiation in the cavity is enough to compensate for the loss of cavity resonance. The energy of this self-sustained oscillation comes directly from high-level atoms. This phenomenon is called MASTER when the stimulated radiation frequency is in the microwave band. The principle block diagram of the active atomic clock is shown in Fig. 6.20.

The active atomic clock is essentially a phase-locked loop. A microwave signal with a frequency close to the output frequency of the maser is obtained after being multiplied by the output of the crystal oscillator. Then the signal obtained after the mixed beat of the two is amplified. On the other hand, a signal consistent with the beat frequency is synthesized out of the crystal oscillator, serving as an error signal after the phase detection with the beat signal to control the crystal oscillator.

2. **Passive Atomic Clock**

The passive atomic clock cannot generate self-sustained oscillation because it has no resonant cavity or the $Q$ value of the resonant cavity is not high. Usually, an atomic resonator is used as a frequency discrimination standard to control the oscillator through a frequency-locked loop so as to transfer the high accuracy and stability of atomic transition frequency to the oscillator, which, in turn, provides a useful frequency signal output. In the microwave atomic clock, the oscillator is usually a crystal oscillator.

The atomic discriminator has to obtain the output signal by controlling the XTAL oscillator. A microwave interrogation signal that excites the atomic transition is fed into the resonant cavity. When the microwave frequency in the cavity is adjusted to be exactly equal to the transition frequency of higher and lower levels, the atom

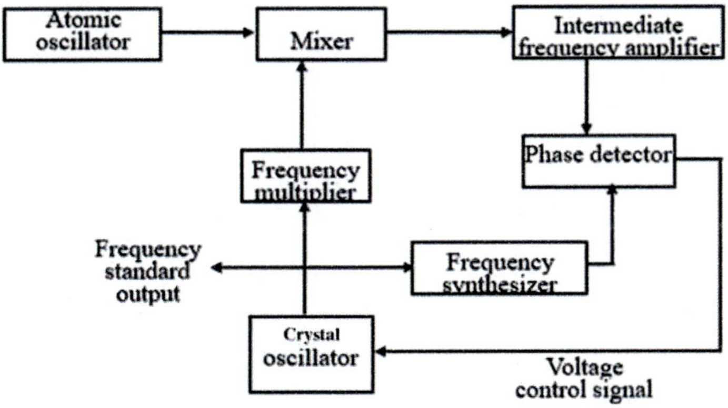


**Fig. 6.20** Principle of active atomic clock

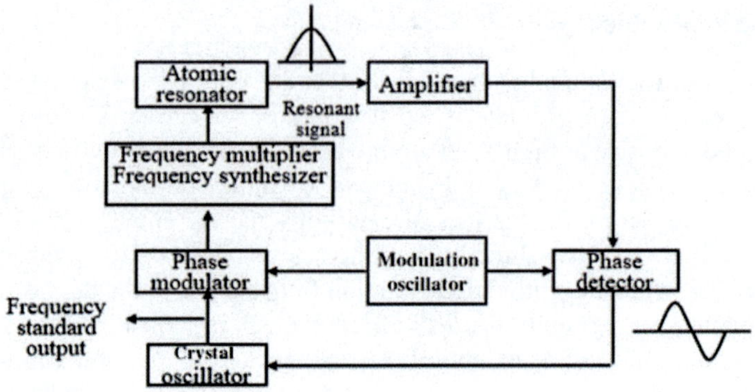


**Fig. 6.21** Principle of passive atomic clock

produces the maximum induced transition. Then this transition signal is detected and fed into the XTAL oscillator to generate an excitation signal to be controlled, so that the frequency of the oscillator can be locked to atomic transition frequency.

The principle block diagram of the passive atomic clock is shown in Fig. 6.21.

The microwave signal that excites the atomic transition in the atomic discriminator is often obtained from the output of the crystal oscillator through a frequency multiplication synthesizer. A microwave interrogation signal close to atomic transition frequency is generated by the output of the crystal oscillator through a frequency synthesizer and a frequency multiplier. This signal is modulated with a low-frequency signal, and then interacts with the atoms to generate a transition signal, which is converted into a DC error signal after the amplification and phase detection to control the crystal oscillator through an appropriate filter.

### *6.3.4 Spaceborne Atomic Clock Technology*

As the key payload of a satellite positioning system, spaceborne atomic clock provides high-precision time-frequency signals for the system as the source of navigation signals and timing signals. Its performance directly determines the accuracy of navigation positioning and timing, and the autonomous operation ability and even service life of navigation satellite. So far, three kinds of clocks have been applied to satellites, including rubidium clock, hydrogen clock, and cesium clock. The rubidium clock is superior in frequency stability, size, weight, power consumption, and reliability, but inferior in accuracy and frequency drift. The hydrogen clock has relatively optimal frequency stability and a small frequency drift, but with a relatively large size, weight, and power consumption. The cesium clock has excellent long-term frequency stability, high accuracy, and extremely low drift, but with relatively poor short-term frequency stability.

So far, the atomic clocks that have been successfully used in orbit include: spectral lamp optically pumped bubble rubidium clock, magnetic state-selection cesium beam clock, and passive hydrogen clock. Motivated by the demand for BDS construction, China has made great progress in the research of spaceborne atomic clock after long-term unremitting efforts. The spaceborne rubidium clocks made in China have reached an internationally advanced level in performance and service life, thus playing an important role in the construction of the BDS system. In 2016, the spaceborne passive hydrogen clock developed by China was successfully applied to BeiDou navigation satellite system (BDS for short) for the first time. Meanwhile, China is also actively developing the spaceborne cesium clock, which is now under key technology development and engineering application.

1. **Characteristics of spaceborne clock**

As the source of spaceborne time-frequency reference signal, the atomic clock is required to meet the performance requirements of high precision, high stability, long-term reliability, continuity, and integrity under the conditions of limited size, weight, and power consumption. Furthermore, it has to adapt to the change of environmental conditions such as space mechanics, heat, atmospheric pressure, electromagnetism, and space radiation. It has to undergo strong vibration and acceleration during the satellite launch. At the same time, it must have the functions of remote control and telemetry.

(1) High Performance

High precision is the basic requirement for spaceborne atomic clocks. Meanwhile, size, weight, and power consumption are also important considerations in the design of spaceborne atomic clocks. It is required that the frequency accuracy, stability, and shift rate of the output signals of atomic clocks meet the design requirements of the satellite system and their size, weight, and power consumption be reduced as much as possible.

(2) Environmental adaptability

The spaceborne atomic clock needs to adapt to the influences and changes of the following environments: the mechanical environment during the ground development and launch of carrier rocket; the thermal environment, space radiation environment, and atmospheric pressure during on-orbit operation; and the electromagnetic environment for ground development and satellite platform.

For the mechanical environment, special and effective technologies should be introduced to the design and fabrication of spaceborne atomic clock to ensure that the physical package of atomic clock, the crystal oscillator, and various circuit boards are stably and reliably supported and connected without displacement and deformation or component damage during various impacts, noises, and vibrations. Before and after the mechanical environment test for the module and equipment, it is necessary to test and check whether the performance of the atomic clock has changed.

For the thermal environment, a special thermal control cabin can be configured for the spaceborne atomic clock after the satellite is put into orbit. The temperature of the cabin baseplate is generally controlled to about 0 °C, changing by less than 1 °C within one orbital period. One of the difficulties in the development of the atomic clock is the temperature difference between ground-based working environment and orbital working environment. To overcome this difficulty, special design measures need to be taken and the clock should be tested under different conditions.

The change of atmospheric pressure in the working environment of the atomic clock will cause the change of the working state of the physical package and the change of heat transfer path inside the clock. When the satellite is working on orbit, there is no way to exchange heat by convection due to the absence of atmosphere so that the thermal environment inside the clock will inevitably change. Therefore, during the design and fabrication of the atomic clock, measures must be taken to ensure that the change of atmospheric pressure will not affect the normal operation of the clock.

The atomic clock should meet a high requirement for electromagnetic interference resistance. As the satellite carries a large number of current cables, high-power microwave amplifiers, digital signal products, and various frequency sources and products with magnetic characteristics, the electromagnetic environment around the clock will change with onboard working condition. In addition, to meet the requirements of high product integration and miniaturization, a comprehensive trade-off must be made among the layout and wiring of various modules when designing an atomic clock, and effective techniques such as electrical isolation and magnetic shielding should be adopted.

After the orbit injection, the spaceborne products will be exposed to the space radiation environment and the changing geomagnetic field environment for a long time, which may cause the components and materials in the products to change and eventually lead to the change of product functions and performance. Therefore, the space radiation environment and its influence on the selected components and materials must be fully considered in the design of the atomic clock. The mass loss coefficient and deflation coefficient of the materials must be controlled to ensure that the derating design and safety margin coefficient meet the requirements.

(3) Strict experiment and assessment test

The atomic clock must satisfy the requirements of ground system joint test and entire satellite test, as well as the requirements for storage and on-orbit service life after the completion of the ground delivery test. The ground test items that need to be conducted in the ground development process of atomic clock include: performance test under normal temperature and pressure, power bias test, mechanical test, magnetic test, high-temperature aging test, high- and low-temperature cyclic test, thermal vacuum test, performance test under thermal vacuum, etc.

(a) Performance test under normal temperature and pressure

The performance test of an atomic clock at room temperature and normal pressure is required to check and verify the basic working state of clock product, the range of telemetry parameters and the implementation of telecommands, and to test and check the performance parameters such as frequency accuracy, phase noise, output power, output signal spectrum, and frequency stability.

(b) High- and low-temperature cyclic test

Under the low-temperature working condition, attention should be paid to the heating and temperature control functions of clock product in order to test and check the temperature control capability of the atomic clock as well as its performance parameters such as frequency accuracy, output signal power, output signal spectrum, and short-term frequency stability.

Under the high-temperature working condition, special attention should be paid to checking the temperature change of key modules inside the atomic clock as well as the parameters and contents mentioned above for low-temperature condition according to the telemetry parameters, so as to ensure that the temperature coefficient of the clock meets the requirement.

(c) Thermal vacuum test

Ground thermal vacuum test is the most important test before the delivery of an atomic clock. Multi-cycle thermal vacuum tests are carried out for the atomic clock according to the test standard of spaceborne products. The data of the preheating process, signal balance process, and aging steady process of spaceborne atomic clock can be fully evaluated by a long-time experiment in the vacuum environment.

For a satellite after entering the orbit, the basic characteristics and predicted values of the atomic clock with high precision can be accurately given by performing a long-term experiment under the thermal vacuum condition. During the multi-cycle thermal vacuum test, the baseplate temperature of the thermal vacuum system needs to be set as a certain temperature value for the atomic clock. In the test process, the frequency accuracy, output power, and frequency stability (10 ms–1000 s) of its output signal are tested and evaluated.

If the spaceborne atomic clock meets the functional and performance requirements after the completion of thermal vacuum test, it is required to continue the test and data acquisition measurement for no less than the time specified for long-term stability (usually two months for rubidium atomic clock). In this process, attention should be paid to analyzing the changing trend of the frequency shift rate (per day) and telemetry parameters of the output signal, so as to ensure that the telemetry parameters of the atomic clock are continuous and stable without a significant change.

2. **Spaceborne Rubidium Clock**

The spaceborne rubidium clock has the advantages of small size, lightweight, low power consumption, long life, high reliability, and excellent frequency stability. It has been applied in the global navigation satellite systems including BDS, GPS, Galileo, and GLONASS, and the regional navigation satellite systems such as NAVIC and QZSS.

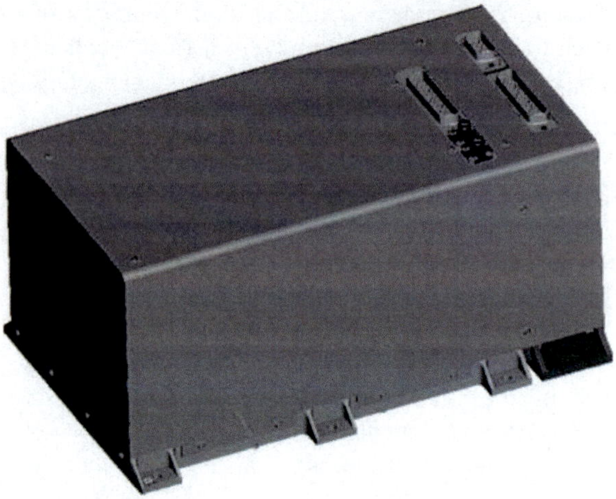

**Fig. 6.22** High-precision rubidium clock product of CAST-Xian

The main research and development entities of rubidium clock include Excelitas in the United States, Spectratime in Switzerland, China Academy of Space Technology in Xian (CAST-Xian), China Aerospace Science and Industry Corporation-203(CASIC-203), and Wuhan Institute of Physics and Mathematics of Chinese Academy of Sciences (CAS-WIPM). The high-precision rubidium clock developed by CAST-Xian is shown in Fig. 6.22.

The existing rubidium clocks used in main GNSS systems are all spectral lamp optically pumped bubble rubidium clocks. The simplified level structure of $^{87}$Rb in the ground state as well as the working principle and physical package structure of the rubidium clock are shown in Fig. 6.23.

The physical package of a rubidium clock is mainly composed of a rubidium spectrum lamp, a filtering cell, an absorption cell, a microwave cavity, a photoelectric detector, and other components. The rubidium bulb, the filtering cell, and the absorption cell are placed in a thermostatic bath to keep their temperature constant. The rubidium spectrum lamp in the left figure mainly includes a rubidium bulb, an exciting circuit, and an oscillating circuit. The rubidium bulb is filled with metal rubidium and exciting gas, and placed in the RF exciting coil of the exciting circuit. Under the combined action of RF exciting and oscillating signals and exciting gas, the $^{87}$Rb atoms in the bulb are excited to emit light in a spectrum that contains ultra-fine spectrum $(a, b)$ light. The unnecessary ray $\alpha$ in the luminescent spectrum of the rubidium bulb is filtered by the filtering cell filled with $^{85}$Rb and buffering substance through isotope filtering. The remaining ray $\beta$ enters the resonance absorption cell in the microwave cavity. The population inversion between the super-fine levels $|F = 1, m_F = 0\rangle$ and $|F = 2, m_F = 0\rangle$ of $^{87}$Rb in the ground state is achieved by the optical pumping of the ray $\beta$. A 6834.6875 MHz microwave signal required by the resonant transition is synthesized by the circuit of the rubidium bulb and then fed into the physical package of the rubidium clock. Under the joint action of pumping light and microwave signal, the $^{87}$Rb atoms in the bulb are absorbed to perform optical

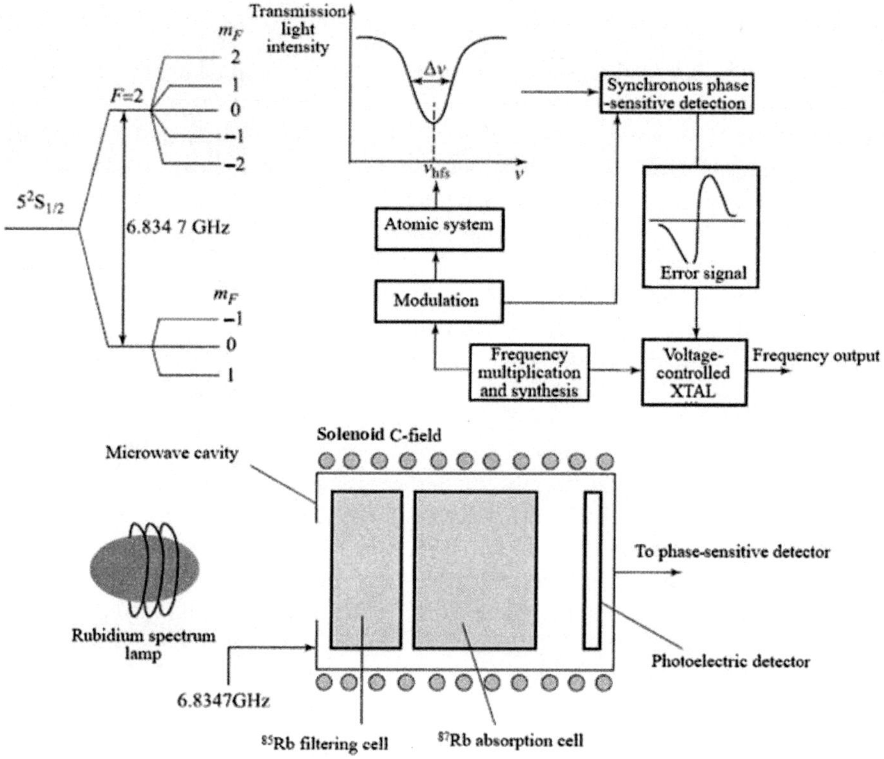


**Fig. 6.23** Simplified level structure of $^{87}$Rb in ground state, working principle, and physical package structure of rubidium clock

microwave double-resonance. The resonant signal is detected by the photodetector at the right end and then an optical detection signal is output. The detection signal is processed by a service circuit to obtain an error signal, which, in turn, is fed back to the voltage control end of the crystal oscillator to output a stable frequency signal.

The rubidium spectrum lamp of the optically pumped rubidium clock has a wide spectral line and low pumping efficiency. In addition, its optical pumping and microwave interrogation will not bring an optical frequency shift effect to time and space domains separately. For spaceborne rubidium clock, these factors have restricted further improvement of its frequency stability.

An important development direction of the spaceborne rubidium clock is to introduce Pulsed Optically Pumped (POP) technique to this clock, i.e., to use a semiconductor laser instead of rubidium spectrum lamp and carry out optical pumping and microwave interrogation in a time-sharing manner through time service control. This technique can greatly improve the pumping efficiency, reduce the background light current noise, enhance the signal-to-noise ratio of frequency discrimination signal, and further improve the frequency stability of the rubidium clock.

3. **Spaceborne Hydrogen Clock**

The existing research institutes and companies that are committed to developing and commercializing the hydrogen atomic clock technology include: Symmetricom of the United States, Vremya-CH and Kvarz in Russia, Spectratime and T4 Science in Switzerland, CASIC-203 in China, and Shanghai Astronomical Observatory of Chinese Academy of Sciences (CAS-SAO) in China. Among them, the Vremya-CH, the Spectratime, the CASIC-203, and the CAS-SAO have developed and produced passive hydrogen clocks.

The passive hydrogen clock product installed on the European Galileo satellite was developed by Spectratime in Switzerland.

Both the CASIC-203 and the CAS-SAO are developing active and passive hydrogen clocks. In the late 1970 s, the CAS-SAO started the development of hydrogen clocks. Its hydrogen clocks have been widely used in major national scientific projects (such as "VLBI network"), ground stations, and the satellite positioning systems for longwave navigation. In 2015, the passive hydrogen clocks developed by CASIC-203 and CAS-SAO were successfully applied to the BDS-3 test satellite on orbit for the first time.

The $\sigma$ transition of hydrogen atom from $F = 1, m_F = 0$ to $F = 0, m_F = 0$ in the ground state is applied in the hydrogen atomic clock with a transition frequency of 1.42 GHz. The structural block diagram of the physical package of passive hydrogen clock is shown in Fig. 6.24. Its physical package is mainly composed of a hydrogen source, a magnetic state selector, a storage cell, a microwave cavity, and a C-field magnetic shield. The hydrogen molecules emitted from a hydrogen storage bottle are firstly purified by a palladium tube, then dissociated into hydrogen atoms in an ionizing cell, and finally screened by a multipole state selector. The hydrogen atoms in the $F = 1, m_F = 0$ state are firstly biased toward the central axis in a strong uneven magnetic field under the action of a magnetic moment. Then they converge

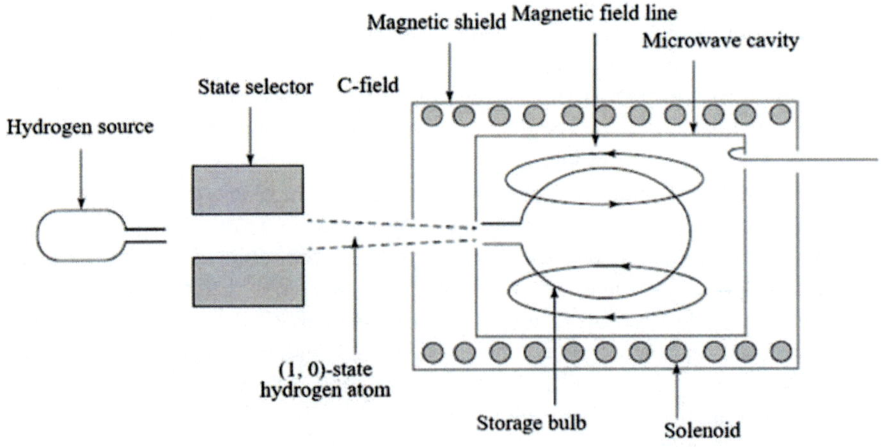


**Fig. 6.24** Working principle of the physical package of passive hydrogen clock

and enter the storage cell, and later interact with the microwave field in the microwave cavity. The storage cell technique greatly prolongs the action time between hydrogen atoms and microwave radiation field, thus reducing the width of the spectral line and increasing its $Q$ value in the clock transition.

The spaceborne passive hydrogen clock is based on a magnetic separation and storage cell structure. Its key technologies include: the purification and dissociation of hydrogen atoms, the maintenance of high vacuum system, the narrow line width under storage cell pressure, and the automatic frequency tuning of microwave cavity with a high $Q$ value. However, the palladium poisoning in the hydrogen purifier and the low ionization efficiency of the ionizing cell will reduce the generated hydrogen atoms, thus weakening the clock transition amplitude and affecting the overall clock performance. Meanwhile, the bulb-wall frequency shift effect, the cavity pulling effect, and the cell aging caused by the collision of hydrogen atoms with the storage cell will restrict the improvement of frequency accuracy and long-term frequency stability of the whole clock.

Passive hydrogen clock is now developing toward small size, low mass, and high performance. The key is the small size design of cavity cell system and the suppression of bulb-wall frequency shift effect and cavity pulling effect.

4. **Spaceborne Cesium Clock**

As the primary frequency standard, the cesium atomic clock has the advantages of high-frequency accuracy, excellent medium and long-term frequency stability, and extremely low-frequency drift. Both the GPS and GLONASS satellites have been successfully equipped with cesium clocks—to be specific, magnetic state-selected cesium beam clocks. The cesium clock has a very small frequency shift and outstanding long-term stability, thus reducing the communication capacity occupied by atomic clock calibration in constellation networking and weakening the dependence of navigation system on ground stations. It is of great significance to guarantee the autonomous operation of navigation satellites.

In 1977, the NTS-2 satellite carrying two cesium beam atomic clocks designed by Symmetricom was launched, marking the first use of cesium clock in space. At present, the cesium clock carried by GPS BLOCK IIF is the FTS4410 digital magnetic state-selected cesium beam clock developed by Symmetricom.

The existing entities developing the cesium clocks in China mainly include CASIC-203, Peking University, Hubei Hanguang Electricity Factory (formerly Hubei 4404 Factory), CAST-510 and Chengdu Spaceon Electronics Co., Ltd. The cesium clocks developed by them include magnetic state-selected cesium clocks and optically pumped cesium clocks. The Peking University and the 4404 Factory are co-developing the engineering prototype of a new cesium clock type based on a unique "magnetic state-selection—storage cell—laser detection" design. This new design can keep the sample state of "beam" atoms, ensure high accuracy, replace electron multiplier with optical detector, and introduce the storage cell technique of hydrogen clock to cesium clock so as to improve the signal-to-noise ratio.

In recent years, the CAST-510 and its partners have jointly developed a ground-based high-performance single-beam magnetic state-selected cesium clock with

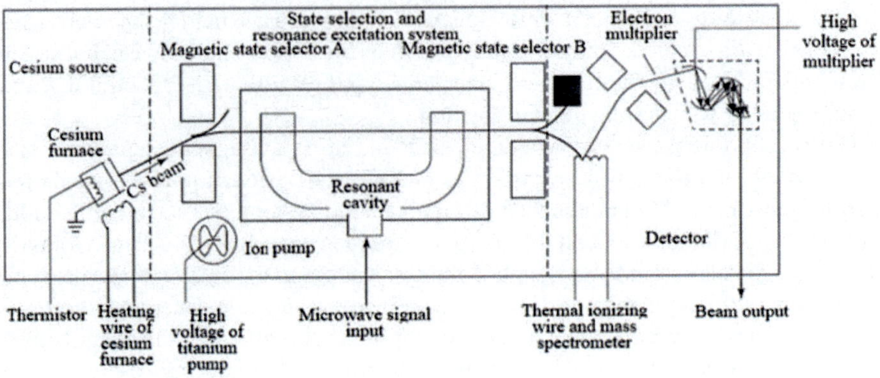


**Fig. 6.25** Basic composition of the magnetic state-selected cesium beam tube

higher adaptability to space environment attributed to the lightweight design and component reinforcement. Their goal is to develop a magnetic state-selected cesium clock that can be applied to BDS satellites.

The $\sigma$ transition of cesium atom from $F = 4, m_F = 0$ to $F = 3, m_F = 0$ in the ground state is applied in the cesium atomic clock with a transition frequency of 9.193 GHz. The structure of the magnetic state-selected cesium atom frequency standard quantum system is shown in Fig. 6.25. The entire system can be roughly divided into three parts: a cesium beam source, a state selection, and resonance excitation system, and a detector.

The cesium furnace is filled with metal cesium. When the cesium furnace is heated to a certain temperature (generally about 100 °C), the cesium steam atoms are sprayed out toward a certain direction through a collimator to form a cesium beam. When the beam passes through a magnetic state selector A, the cesium atoms at the super-fine level $F = 4, m_F = 0$ in the ground state enter the U-shaped Ramsey microwave resonant cavity due to the effect of nonuniform magnetic field. After passing through the two resonance regions in the Ramsey cavity with a resonance frequency of 9.193 3 GHz, the cesium atoms in the $F = 4, m_F = 0$ state will be transited to the $F = 3, m_F = 0$ state, i.e., the clock transition will occur. After passing through another magnetic state selector B, the outgoing atoms will undergo the clock transition. Then they are spatially separated from the cesium atoms that are sprayed on the detector but have not undergone the transition. The detector is made up of hot wires and can ionize the incoming cesium atoms. The mass spectrometer directs the cesium ions to an electron multiplier, where an electron current is generated and then amplified for a servo circuit to process.

The oscillating-field separation technique of Ramsey microwave resonant cavity is the key to greatly improve the performance of cesium atomic frequency standard. Compared with the previous single-cavity technique of Rabi, this technique can effectively increase the interaction time $\tau$ between the atoms and the cavity in a limited space. It can also effectively reduce the width of atomic spectral line $\Delta\nu = \frac{1}{\tau}$.

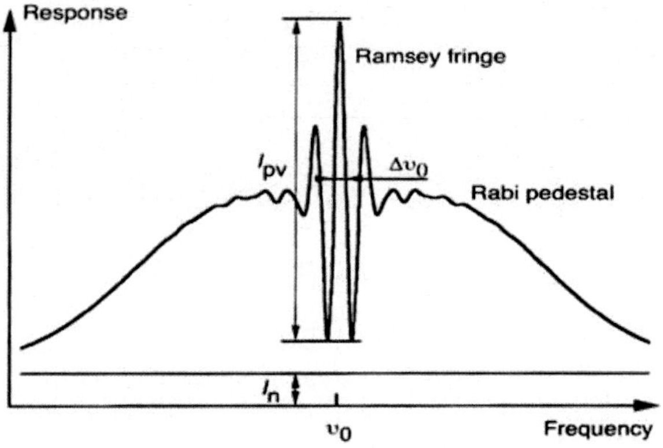


**Fig. 6.26** Ramsey spectral line of cesium beam frequency standard in atomic transition

The Ramsey spectral line of atomic microwave resonance obtained by this technique is shown in Fig. 6.26. The central part of the spectral line is very sharp and thus can be used as a frequency discrimination curve in a frequency standard.

The key techniques of spaceborne magnetic state-selected cesium beam clock include: the sealing and ultrahigh vacuum degree maintenance of cesium beam tube, the overflow prevention of cesium furnace and the collimation of cesium atom beam, the optical design of cesium atom beam, and the design of high-performance U-shaped microwave cavity and long-life electron multiplier.

The atomic state preparation based on magnetic state separation has the disadvantages of low cesium atom utilization and limited signal-to-noise ratio. Although the increase of cesium furnace temperature can enhance the intensity of the atomic beam and resonant signal, the working substance will be consumed too fast and the performance of the electron multiplier will drop sharply. Thus, the service life of the cesium clock will be shortened. The above problems can be effectively solved by using the laser-pumped cesium beam clock, which can use optical pumping, instead of magnetic state separation, to efficiently prepare the cesium atoms and leverage the fluorescence detection to avoid the restriction of electron multiplier on the service life of the entire clock.

# Chapter 7
# Generation and Assessment of Navigation Signal


Navigation signal quality has a great influence on the performance indicators of a navigation system including positioning, velocity measurement, timing accuracy, compatibility and interoperability, confidentiality, and anti-jamming capability. The monitoring and evaluation of satellite navigation signal quality have become an important part of the study of navigation satellite system construction.

The generation of navigation signal is complicated, as it involves a series of processes: baseband spread spectrum generation, modulation, amplification, and emission. The nonideal characteristics of all the devices in the process, including atomic clock, modulator, digital to analog converter, frequency converter, filter, high-power amplifier, multiplexer, and antenna, will lead to signal distortion. Besides, the combination of multiple signals through the same amplifier increases the nonlinear distortions of signal amplitude and phase. All the distortions caused by nonideal characteristics will lead to the deterioration of navigation signal acquisition and tracking performance on the user side and the demodulation of navigation information.

The assessment of navigation signal quality is different from continuous monitoring. Signal sampling and offline processing are carried out to acquire the evaluation information from a signal with high C/N, which can be obtained in the wired satellite test before launch or with a high-gain dish antenna in the orbit assessment.

This chapter will study the broadcasting link technology in depth, mainly focusing on the technology and design adopted by BDS. This chapter also analyzes the factors that influence the quality of satellite navigation signals, and discusses the parameters, methods, and algorithms related to the evaluation of navigation signal quality. As far as the BDS is concerned, the standard offline assessment targets 11 parameters, which will be discussed below in detail.





J. Xie et al., *Satellite Navigation Systems and Technologies*, Space Science and Technologies, https://doi.org/10.1007/978-981-15-4863-5_7

## 7.1 Navigation Signal Broadcasting Technology

The user reception performance of a navigation signal is mainly related to the signal design and channel transmission characteristics. However, the channel characteristics are usually nonideal and have an effect on signal performance. Therefore, the satellite payload designers should minimize this effect or make it easy to correct. The design of the broadcasted signal should also consider the channel characteristics and their effect to ensure smooth signal transmission and avoid excessive distortion. In summary, the navigation signal designer and the payload designer should communicate with each other so that the signal can reach the user with less performance loss.

A comprehensive ground test should be made on the navigation satellite before its launch. All the parameters, including signal frequency, output power, output spectrum, modulation signal characteristics, channel delay characteristics, and other parameters, are tested to ensure that the satellite navigation signal meets the system requirements. After being launched, the satellite will go through the on-orbit test and experiment, and network access confirmation to ensure that the satellite status and navigation signal can meet the system requirements after the signal transmission.

The BDS navigation system transmits signals through the broadcasting antenna according to the free space propagation model shown in Fig. 7.1. The broadcasting frequency bands of main BDS systems are basically the same as those of GPS due to the requirements of compatibility and interoperability. In terms of received signal power, considering that the power of BDS-3 public signals and authorized signals is equivalent to that of GPS signals, the power budget of the BDS-3 MEO-satellite broadcasting link will be roughly as shown in Table 7.1.

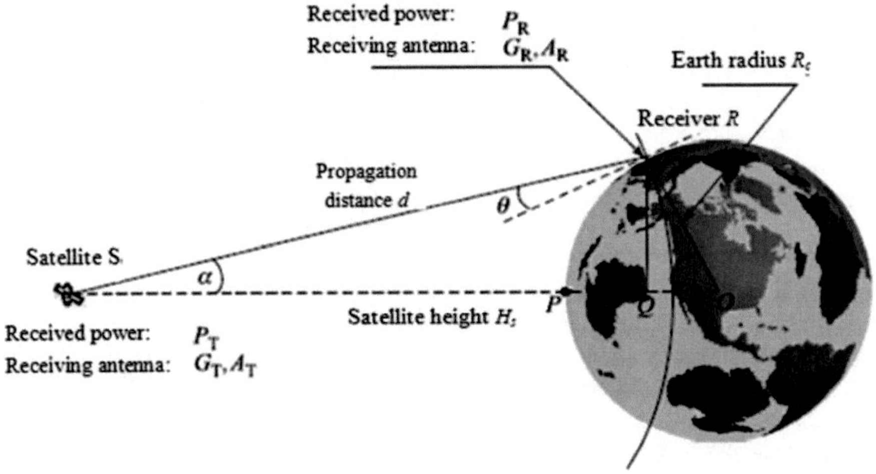


**Fig. 7.1** Free space propagation model of broadcasted BeiDou satellite signal

**Table 7.1** EIRP budget of signal broadcasting link in $B1$ frequency band (Satellite Altitude: 21,528 km)

| Item | Value |
|---|---|
| Satellite transmission EIRP/dBW | 34.3 |
| Center frequency/MHz | 1.575.42 |
| Spatial loss/dB | −184.9 |
| Atmospheric loss/dB | −0.5 |
| Polarization loss/dB | −1.0 |
| User antenna gain/dBic | 0 |
| Minimum ground received power for the entire frequency band/dBW | −152.1 |
| Minimum ground received power/dBW | −155.1 |

Satellite transmission power is the basic constraint on the signal-generating payload. Under this premise, the design of such payload products can be carried out.

### *7.1.1 Signal-Generating Payload*

Generally, the navigation signal generation can be implemented by Intermediate Frequency (IF) scheme to eliminate the effect of serious spectrum leakage when there is an imbalance between $I$ and $Q$. However, in the IF scheme, the undesired factors in the digital quantization, digital-to-analog conversion, up-conversion, filtering, and waveform power amplification will affect the signal performance.

The IF unit combines multiple signals in the same frequency band, and then the combined signal passes through the same RF channel to complete the mixing, filtering, and amplification. This scheme can ensure the delay stability between signals, and reduce the complexity of the signal-generating payload. The general signal generation process is shown in Fig. 7.2. The baseband or IF signal is first generated by a digital generation unit. Then the IF signal is converted into RF $L$ frequency band by an up-converter or modulator, and subsequently sent to the broadcasting antenna, and finally broadcasted to the ground through a filter, an amplifier, and an output multiplexer.

In the current BDS-3 system, the signal-generating payload is divided into two parts: navigation task subsystem and signal broadcasting subsystem. The navigation task subsystem completes the tasks such as digital generation and time management, and digital-to-analog conversion. The signal broadcasting subsystem completes the frequency conversion, filtering, amplification, combining, and signal transmission.

1. **Navigation task subsystem**

The navigation task subsystem is mainly to generate the navigation data in three frequency bands from the upload message, measurement, and remote control

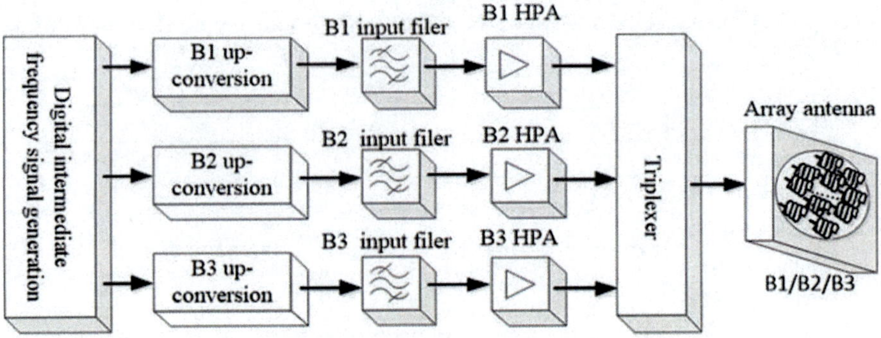


**Fig. 7.2** Block diagram of signal-generating payload

message, and to digitally modulate the navigation messages in the above three frequency bands to generate three IF signals.

The navigation task processing unit is to generate the navigation messages in three frequency bands, and after encoding, to generate the IF navigation signals through digital modulation. In the digital modulation, a constant envelope and a pre-distortion design are adopted to reduce the effect of nonlinear AM-AM and AM-PM on channel distortion.

Figure 7.3 shows the block diagram of the signal generation achieved by navigation task processing unit $B1$. The unit first generates the $I$ and $Q$ signal components in the $B1$ frequency band, and then filters out the out-of-band signals through the same digital low-pass filter to avoid spectral aliasing when changing to intermediate frequency. The filtered $I$ and $Q$ components are multiplied by sinusoidal digital carrier and cosine digital carrier, respectively, and subtracted with each other to obtain the IF signal. Then the signal is converted into an analog IF signal by pre-distortion and DAC. In the signal generation process, it is also necessary to design the anti-space

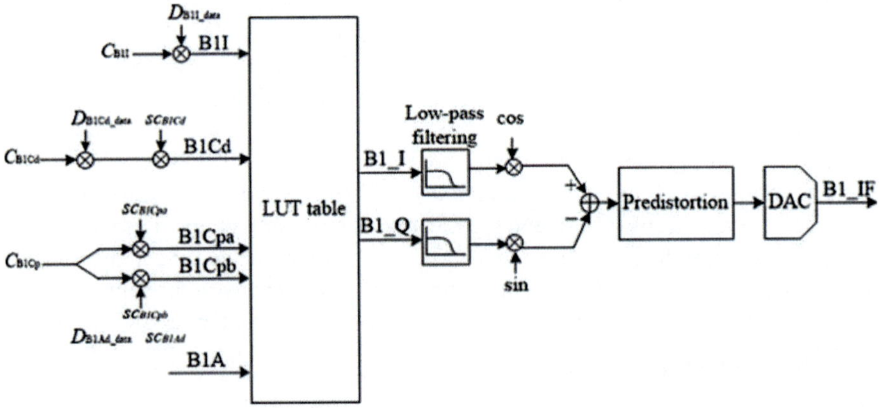


**Fig. 7.3** Block diagram of signal generation of navigation task processing unit $B1$

SEE solutions, such as three-mode redundancy, timing refresh, fault detection, and repair.

2. **Navigation signal broadcasting subsystem**

The navigation signal broadcasting subsystem is mainly to receive the IF signal, implement the up-conversion, filtering, and amplification, and provide triple-output to the antenna subsystem. While satisfying the in-band signal quality requirements, this subsystem also needs to suppress adjacent band interference.

(1) Antenna

The power level of a navigation signal directly affects the performance of the signal received by the user. In order to ensure that the power levels transmitted by different satellites and received by the user at any place and at any time are similar and stable, and that the user receiver operates within the appropriate dynamic reception range, the satellite system has adopted a "saddle" antenna design (Fig. 7.4).

When designing a navigation antenna of "saddle" pattern, attention should be paid to the gain difference between the subsatellite point and the Earth's edge. In order to ensure the user's receiving performance, the beam width in the real project is often slightly wider than its theoretical design value.

In order to obtain the required beams, an array antenna is usually used as a navigation satellite antenna. The antenna specifications of Galileo satellites are shown in Table 7.2.

The Galileo satellite antenna uses a planar array of 1.4 m diameter, and six triangular sub-arrays composed of 45 stacked patch elements.

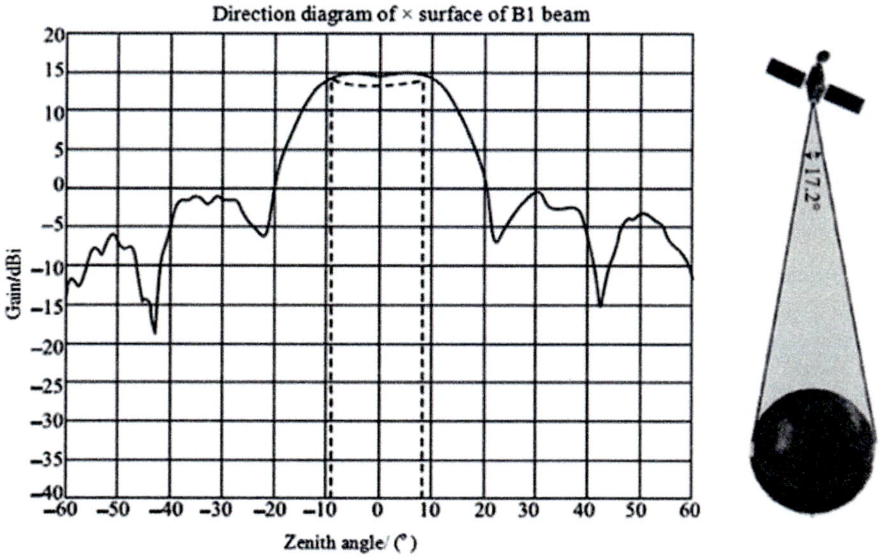


**Fig. 7.4** Design of navigation satellite antenna pattern

**Table 7.2** Specifications of Galileo Satellite Antennas

| | | |
|---|---|---|
| Frequency/MHz | Lower frequency | 1.145–1.237,1.259–1.299 |
| | Higher frequency | 1.555–1.595 |
| Coverage/(°) | | (Sight axis direction) 12.67 |
| Group delay in line of sight/mm | | (Error is better than) 30 |
| Minimum coverage gain/dBi | Lower frequency | 15.35 |
| | Higher frequency | 14.85 |
| Polarization mode | | Right circular polarization |
| Maximum axial ratio/dB | | 1.2 |
| Return loss/dB | | >20 |
| Mutual coupling between adjacent array elements/dB | | <−30 |

(2) Triplexer

The triplexer is mainly to filter and combine the three frequency band signals before they enter the antenna, in order to meet the out-of-band radiation requirements. A triplexer consists of three channel filters and a common pole, as well as a detection port and an output port. The signal enters the output triplexer from the Traveling Wave Tube Amplifier (TWTA), and then is transmitted to the antenna for radiation.

For a single GNSS system and the user on a certain frequency, the loss of the received signal component will decrease with the increase of signal bandwidth. This is beneficial to the phase characteristics and delay stability of the signal. Therefore, the amplitude-frequency response in the frequency band of the navigation signal is required to be flat to reduce the insertion loss. With the increase of signal bandwidth, the probability of interfering with other GNSS systems and other adjacent service frequencies will become greater.

In the application of a satellite navigation signal, an appropriate filter must be used to protect the quality of this signal in the frequency band without affecting the signals of other systems and users. In the design of a filter, it is required that the signal frequency band be as "wide" as possible, the amplitude of in-band response be as "flat" as possible, the phase be as "linear" as possible, and the out-of-band rejection be as "steep" as possible.

In order to achieve good filter characteristics (wide, flat, linear, and steep), a suitable filter design is required. For a higher order mechanical cavity filter, good in-band and out-of-band characteristics can be achieved more easily, but its volume and weight tend to be larger. The eighth-order filter structure is shown in Fig. 7.5

The frequency response of the eighth-order filter is shown in Fig. 7.6.

(3) Power amplification

The power amplifier can be a TWTA or Solid Status Power Amplifier (SSPA). The L frequency band TWTA (LTWTA) consists of a TWT Electronic Power Conditioner (EPC), a Linearized Channel AMPlifier (LCAMP), a TWT, and a high-power isolator.

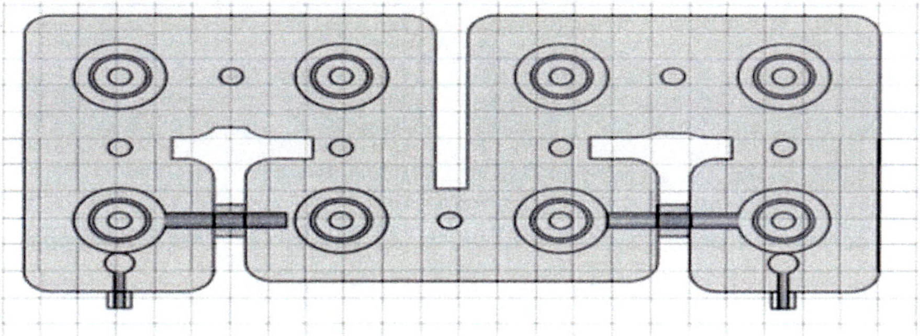

**Fig. 7.5** Structure of eighth-order filter

**Fig. 7.6** Frequency response of eighth-order filter

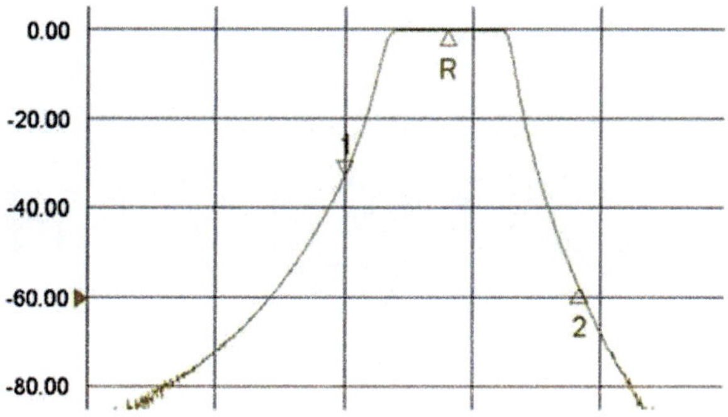


EPC and LCAMP are collectively called a multi-function component. The multi-function component is used to power the TWT to the required excitation level. It is integrated with TWT into LTWTA to amplify a low-power signal to suitable high power and realize the correction of TWT linearity at the same time. The LTWTA has two working modes: Fixed Gain Mode and Automatic Level Control (FGM/ALC). In addition, it has the functions of remote telemetry and overload protection. The schematic diagram and outside view of LTWTA are shown in Fig. 7.7 and Fig. 7.8, respectively.

The EPC converts the primary DC voltage provided by the satellite system into the voltage required by TWT and LCAMP. The LCAMP consists of a channel amplifier and a linearizer. It is powered by the EPC to achieve the TWT linearity correction and system gain control. Due to the antenna gain requirement and satellite power limitation, the TWTA transmitting the navigation signal generally operates near the saturation point to fully utilize the efficiency of the power amplifier, as shown in Fig. 7.9.

In order to achieve stable output and adapt to the possible level changes of the input navigation signal, the power amplifier uses an ALC to ensure the stability of the output signal. Therefore, in the amplifier design, it is necessary to pay attention to the

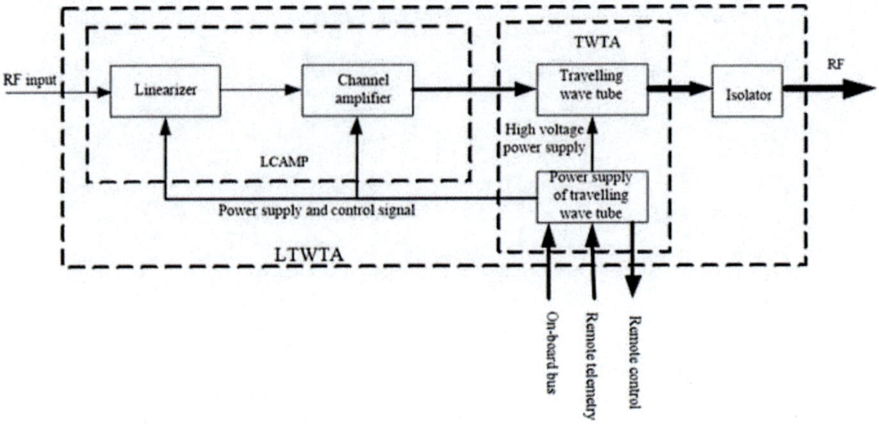


**Fig. 7.7** Block diagram of Traveling Wave Tube Amplifier

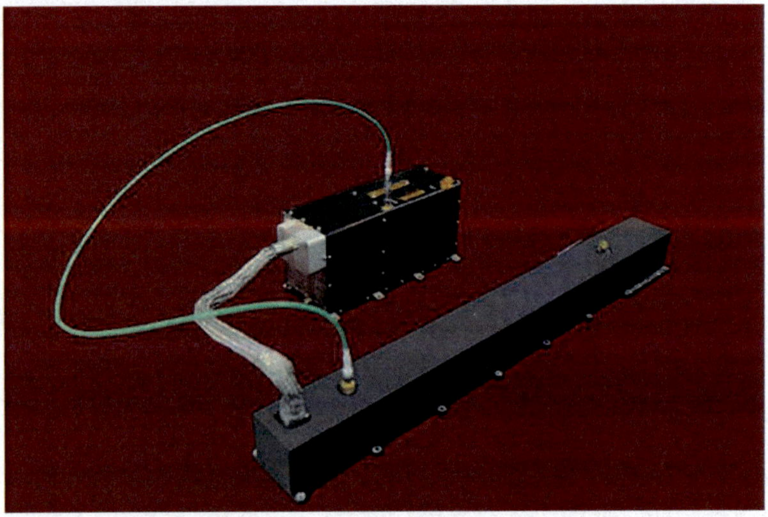

**Fig. 7.8** Outside view of traveling wave tube amplifier

characteristics of the power amplifier operating near the saturation point, including nonlinearity, amplitude-frequency response, and phase-frequency response.

When transmitting multiple signal components, the technique of constant envelope is used to eliminate the influence of saturation point nonlinearity, thereby improving efficiency and reducing losses. The multiplexing of multiple signal components in the same frequency band and the filtering (band limit) and distortion of the channel device will cause the correlation loss of the navigation signal.

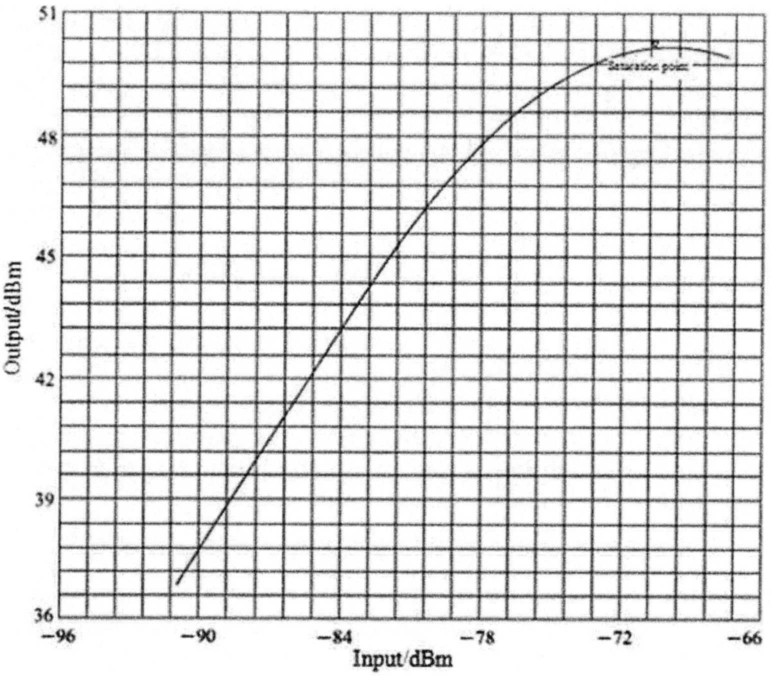


**Fig. 7.9** Input and output characteristics of TWT

(4) Up-conversion and pre-filtering

The frequency converter in the navigation signal broadcasting subsystem consists of a local oscillator link and a frequency conversion unit. The local oscillator link converts the reference frequency into local oscillator frequency, and the frequency conversion unit achieves the frequency shift of an intermediate frequency signal. To ensure effective power amplification and suppress the local oscillator and mirror image, an input filter should be added behind the converter, while producing as small influence on the constant envelope characteristic of the signal itself as possible.

The frequency converter needs to accurately convert three frequency bands into one component, and shall be precisely designed to avoid the leakage of local oscillator frequency into the signal band. The frequency characteristics of a signal include signal bandwidth, in-band frequency characteristics, and out-of-band frequency characteristics.

### *7.1.2 Nonideal Characteristics of Broadcasting Channel*

The nonideal characteristics of the channel include linear distortion and nonlinear distortion. The linear distortion is mainly caused by filter, up-converter, and triplexer, while the nonlinear distortion is introduced by the power amplifier. Different channel delays, filtering effect, up-converter non-ideality, and other factors may all cause the distortion of the navigation signal.

The modulation performance of a navigation signal is mainly measured by the amplitude deviation of each signal component, carrier phase deviation, carrier and pseudocode coherence, and pseudocode phase consistency. For a QPSK-modulated navigation signal, the amplitude deviation, carrier phase deviation, carrier and pseudocode coherence, and other indicators can be measured through the amplitude imbalance, orthogonality, and phase errors of the I/Q components.

The main error sources of the satellite are the modulator implementation and the quality of each reference clock signal. Therefore, it is necessary to adopt a unified reference clock signal to achieve strict phase synchronization and small noise through frequency synthesis. The user of the satellite navigation system can measure local transmission time by receiving the pseudocode or carrier phase of the recovered navigation signal. The signal transmission delay between the satellite and the user can also be accurately measured to achieve positioning.

The delay of the satellite system has a large impact on the user's navigation performance. In the process of signal generation, the signal transmission delay is related to the channel path of the signal itself and the delay stability of the product itself. The product delay stability is related to the product operation temperature (for example, the temperature of filter and amplifier) and especially the time frequency.

In order to control the delay of satellite navigation signal, the components and raw materials with stable temperature characteristics or linearity should be selected to realize the compensation for and self-calibration of delay deviation. To ensure the stability of navigation signal delay, the transmission path of the signal channel should be optimized, and the absolute value of product (or equipment) delay controlled. In addition, the front- and rear edge characteristics of the clock signal and pseudocode signal should be designed and guaranteed. Different frequency bands should be designed on the basis of the closed-loop delay check function to monitor and control the delay of the device.

1. **Linear distortion**

Main linear distortions include band limit, amplitude oscillation, amplitude slope, phase oscillation, and quadratic and cubic phases (equivalent to linear and quadratic group delay variations). The distortion is described by frequency response $H(f)$, and the spectrum of the output signal is given by

$$S_{\text{out}}(f) = H(f) \cdot S_{\text{in}}(f) \tag{7.1}$$

where $S_{\text{in}}(f)$ is the power spectral density of the ideal signal, and $S_{\text{out}}(f)$ is the power spectral density after the $H(f)$ distortion.

2. **Nonlinear distortion**

The nonlinear distortion is mainly caused by a high-power amplifier. The models for describing a high-power amplifier fall into two categories: memoryless models and memory models. Memoryless models include Saleh model, RAPP model, and lookup table model. Memory models include Weiner model, Hammerstein model, and memory polynomial model.

(1) Saleh model

When the signal bandwidth is narrower than the operating bandwidth of the high-power amplifier, a memoryless model, for example, the Saleh model or RAPP model in common use, can be used to describe nonlinear distortion, The feature of the memoryless model is that the amplitude and phase offset of the output signal at the current time are only related to the amplitude of the input signal envelope at the current time. The amplitude/amplitude (AM-AM) curve and amplitude/phase (AM-PM) curve are used to represent the nonlinear distortion characteristics. If the input signal envelope is $x(n) = r \cdot e^{j\theta}$, the AM-AM characteristic function is $A(r)$, and the AM-PM characteristic is $\Phi(r)$, then the corresponding output signal will be

$$y(n) = A(r) \cdot e^{j(\theta + \Phi(\gamma))} \tag{7.7}$$

The Saleh model is the result of in-depth analysis and research on the nonlinear distortion characteristic of TWTA. The AM-AM characteristic function and AM-PM characteristic function can be used separately, as shown below [35].

$$A(r) = \frac{\alpha_a r}{1 + \beta_a r^2} \tag{7.8}$$

$$\Phi(r) = \frac{\alpha_\phi r^2}{1 + \beta_\phi r^2} \tag{7.9}$$

The input/output characteristics of the amplifier can be fully characterized by four parameters, namely $\alpha_a$, $\beta_a$, $a_\phi$, and $\beta_\phi$.

The effect of the power amplifier model on the time domain waveform of BPSK (1) signal is shown in Fig. 7.10. The bandwidth of this signal is 4.092 MHz. It can be seen that the memoryless amplifier has a smaller influence on the symmetry of signal waveform. The results also show that this amplifier can hardly affect the zero-crossing of the discrimination function, but does affect the correlation loss and zero-crossing slope distortion.

(2) RAPP model

In general, the nonlinear distortion of SSPA is described by the RAPP model, while the AM-AM and AM-PM characteristics are expressed as [34]

$$\begin{cases} F_A(\rho) = \frac{\alpha_A r}{\left[1 + (r/A_{SS})^{2P}\right]^{1/2P}} \\ F_P(\rho) = \alpha_D \left(\frac{r}{1}\right)^2 \end{cases} \tag{7.10}$$

where $\alpha_A$ is the linear region gain of the amplifier; $A_{SI}$ is the input saturation voltage of the amplifier; and $P$ is the smoothing factor, which determines the smoothness

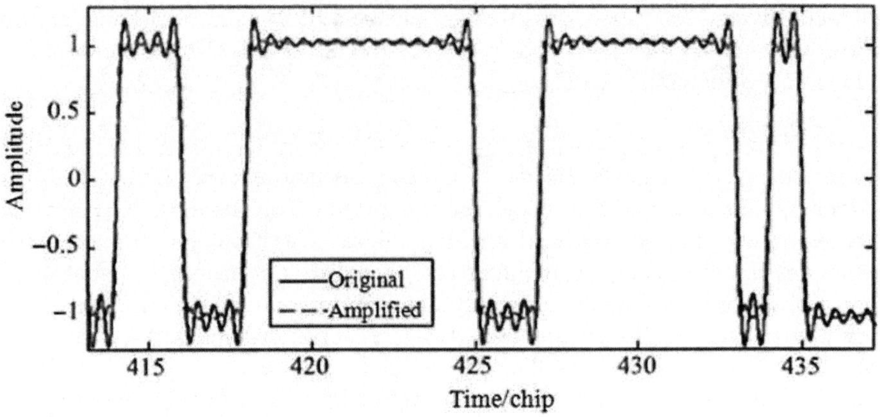


**Fig. 7.10** Effect of power amplifier on signal waveform

of the transition from the linear region to the saturation region. Generally, the value of $P$ is in the range of [2, 3]. A larger $P$ indicates that the input-output relationship before reaching the saturation point will be more linear. $\alpha_P$ is usually 0, that is, the AM-PM conversion effect of the RAPP model can be ignored.

(3) Weiner model and Hammerstein model

The memory effect of the power amplifier occurs when the bandwidth of the input signal is larger than the operating bandwidth of the amplifier. As the bandwidth of the input signal increases, this effect will gradually appear. The so-called memory effect means that the amplifier output is not only related to the current moment input, but also related to the past moment input. Therefore, the memory model of the power amplifier is usually more complicated than the memoryless model. The memory models in common use are Weiner model, Hammerstein model, Wiener-Hammerstein model, memory polynomial model, and cross memory polynomial model, which are all the simplified versions of the Volterra model.

Weiner model is similar to the Hammerstein model, a series of Linear Time-Invariant (LTI) system and memoryless nonlinear system [29]. The difference between them is that the order of the series in the Weiner model is the reverse of that in the Hammerstein model. In the Weiner model, the LTI system is in the front followed by the memoryless nonlinear system. On the contrary, in the Hammerstein model, the memoryless nonlinear system is in the front followed by the LTI system.

(4) Memory polynomial model

Another memory model commonly used for the power amplifier is the memory polynomial model. It can be regarded as a special parallel Weiner model or the equivalent form of a parallel Hammerstein model. Its input/output relational expression is [30]

$$y(n) = \sum_{\substack{k=1 \\ \text{kodd}}}^{N} \sum_{q=0}^{M-1} c_{kq}|x(n-q)|^{k-1}x(n-q) \tag{7.11}$$

### 7.1.3 *Pre-distortion Technique*

As mentioned earlier, there are linear and nonlinear distortions in the payload channel that can result in the signal quality deterioration. In order to ensure the quality of the navigation signal, the current BDS satellite payload has adopted some digital IF pre-distortion methods, including linear pre-distortion compensation and nonlinear pre-distortion.

1. **Linear pre-distortion compensation**

Linear distortion refers to the group delay and amplitude distortion of the channel that are mainly caused by filter, triplexer, and antenna, especially near the cutoff frequency. It should be noted that the group delay varies greatly. In the BeiDou satellite payload, an equalizer is added before the IF signal enters the DA, and an FIR filter structure is adopted to compensate for the linear distortion of the channel, so the influence of linear distortion on the signal can be effectively reduced.

The effect of linear channel distortion (especially group delay) on the signal can be seen from the signal constellation diagram. Take the Dual QPSK signal as an example. The filter characteristics are shown in Fig. 7.11. Based on the filter characteristics, Fig. 7.12 shows the constellation diagram of the band-limited Dual QPSK (10) signal in three cases: without distortion, with distortion, and with distortion and pre-distortion. It is obvious that, with the presence of linear distortion, the signal quality deteriorates seriously, and the phase point of the constellation is blurred. After the pre-distortion is added, the constellation diagram becomes clear again. Of course, in the actual channel, there is also the effect of nonlinear distortion, which may weaken the improvement of the FIR pre-distorter to some extent.

The biggest advantage of the FIR structure is stability and reliability, independent of channel variation. Moreover, it can be adjusted by on-orbit reconfiguration.

2. **Nonlinear distortion compensation**

The basic principle of nonlinear distortion compensation is to pass the baseband signal through a nonlinear pre-distorter whose nonlinearity is complementary to that of the power amplifier, so that the system as a whole exhibits the linear characteristic.

(1) Compensation technique of memoryless model

In the memoryless model of a power amplifier, the output at the current time depends only on the input at the current time, so the pre-distortion can be implemented by means of a memoryless lookup table.

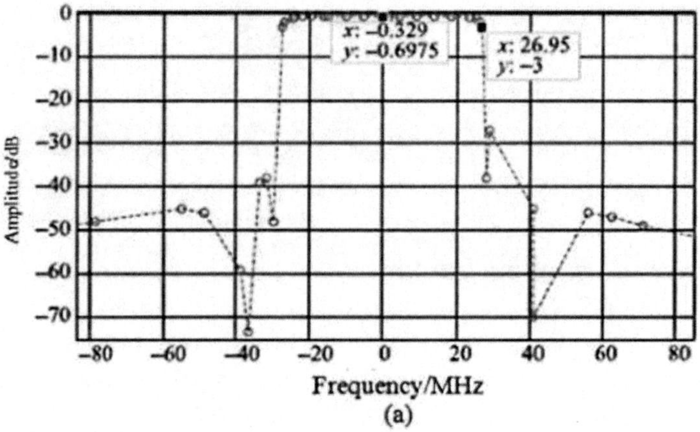


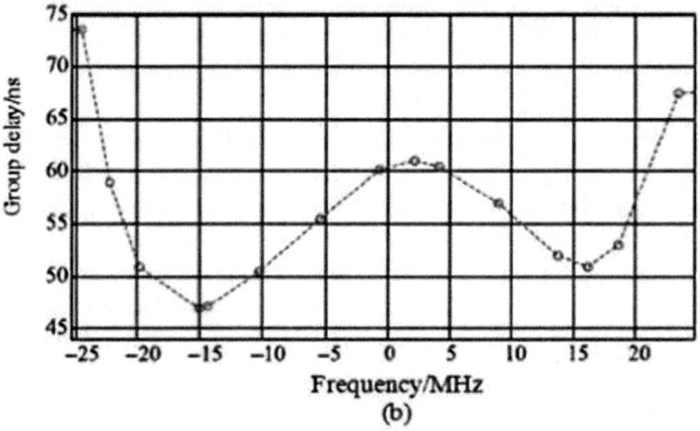


**Fig. 7.11** Filter Characteristics, **a** Amplitude-frequency characteristicS, **b** Group delay characteristics

The signal is pre-distorted by the pre-distorter. After passing through the high-power amplifier, the output signal satisfies a linear relationship with respect to the input signal. The memoryless lookup table can be implemented in a fixed or adaptive way. The fixed lookup table is simpler to implement, but may yield a degraded effect when the characteristics of the power amplifier change. The adaptive lookup table is an online learning method that can adapt the contents of the lookup table to the characteristics of the high-power amplifier so as to achieve better real-time performance.

(2) Compensation technique of memory polynomial model

For a memory model, the pre-distortion method based on memory polynomial is feasible. The parameters of the polynomial model are estimated at first, and then sent to the pre-distorter for pre-distortion operation. The main disadvantage of this method is that it can't track the time-varying parameters and is not suitable for online identification of pre-distorter parameters. In addition, cascading the lookup table to

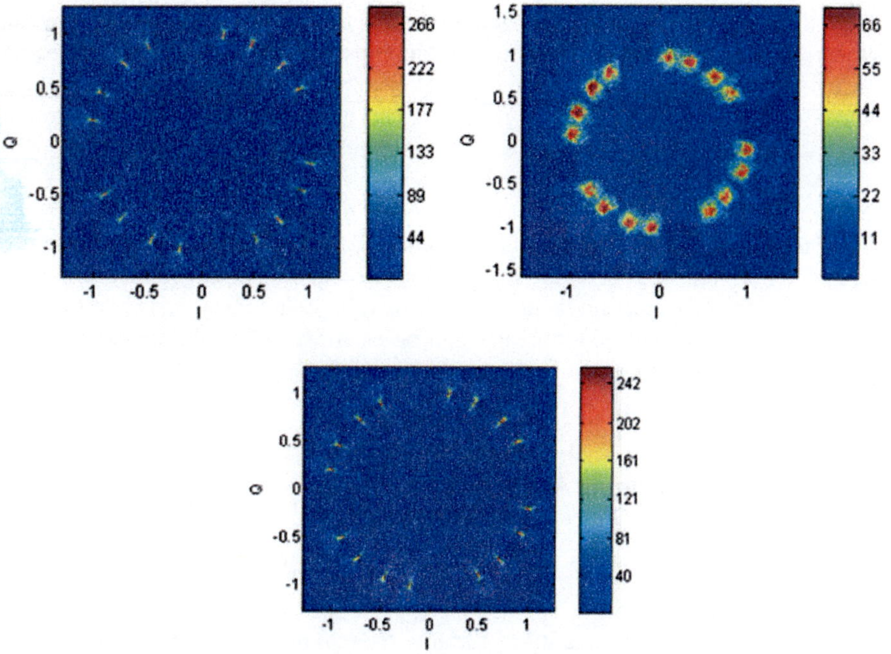


**Fig. 7.12** Dual QPSK signal constellation, **a** Without distortion; **b** With distortion; **c** With distortion and pre-distortion

the FIR filter is also an effective solution. The former method is widely used for the pre-distortion of memoryless amplifiers, while the latter method for the cancelation of memory effects. The latter solution is very simple and easy to implement in real-time systems.

### *7.1.4 Influence of the Antenna on Navigation Signal*

The influence of spaceborne transmission antenna on the navigation signal is mainly reflected in the following aspects. First, the amplitude-frequency characteristics and group delay characteristics of the antenna cause linear signal distortion, which can be compensated for in the abovementioned way. Second, the mismatch between non-ideality ports will cause the reflected signal to be fed back into the transmitted signal, thereby applying a certain influence on the navigation signal. Finally, the structure of the array antenna and the unsatisfactory design of each array channel may also cause the deterioration of signal quality.

The measurement of the distance between the satellite and the ground user is to measure the distance between the phase centers of the spaceborne and user-based navigation signal antennas. The phase center of the navigation signal refers to the center of the radiation source equivalent to the electromagnetic wave radiated by the antenna. On the spherical surface corresponding to the antenna service coverage, the phase fluctuation of the signal component at each point is the smallest. In practical use, the coincidence of the satellite mass center with the phase center of navigation signal antenna will benefit the user's positioning resolution.

Due to the constraints on the onboard system design and off-board product layout, the barycenter of the navigation satellite is often inconsistent with the coordinates of the phase center of navigation signal antenna. To solve this problem, the satellite mass center should be kept stable in implementation, and should be measured and calibrated before the satellite launch. In this way, the relationship between satellite mass center and antenna phase center can be established and the impact of their noncoincidence can be eliminated by modeling on the user side.

## 7.2 Navigation Signal Assessment

### *7.2.1 Influence Factors of Navigation Signal Quality*

The factors that affect the quality of satellite navigation signals originate from all the devices used in navigation signal generation. The possible influences of those devices are analyzed, and their effect on navigation signal quality is summarized in Fig. 7.13.

(1) Digital IF modulation

The digital Intermediate Frequency (IF) modulation of spread spectrum baseband signal can be carried out by many methods such as traditional QPSK, the constant envelope modulation of multiple signals in the same frequency band. The constant envelope modulation will introduce a spurious signal and intermodulation component, which will deteriorate the orthogonality of the navigation signal and change the power ratio between signal components. Especially, the intermodulation component will reduce the useful power efficiency. Moreover, constant envelope modulation can lead to the nonlinear distortion of amplitude and phase interaction.

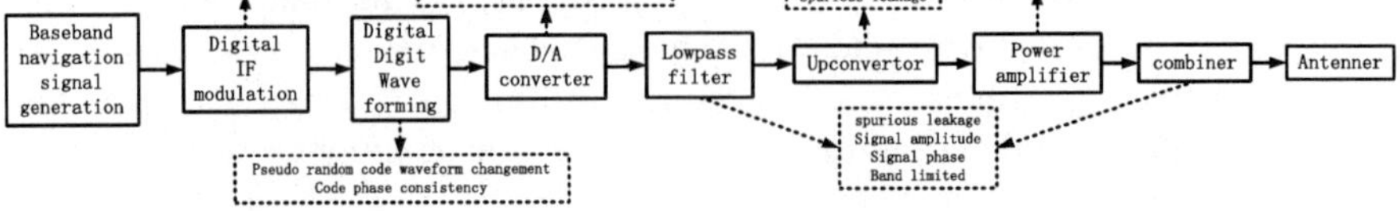


Fig. 7.13 Diagram of navigation signal generation and quality

(2) Digital waveform shaping

After numerical IF carrier modulation, the signal is processed by a digital waveform shaping device, which will change the amplitude and phase of the navigation signal, as well as the pseudorandom code waveform. As a result, the phase consistency between pseudorandom codes is affected.

(3) D/A converter

D/A converter is to convert a digital signal into an analog signal. Affected by the phase noise of the D/A-convertor oscillator, the coherence of the pseudocode of navigation signal carrier will deteriorate. The bit quantization characteristics can also increase the in-band spurs and change the pseudorandom code waveform, thus affecting the phase consistency of each navigation signal branch.

(4) Low-pass filter

Low-pass filter is to filter out the high-frequency components of the signal coming from the D/A convertor. The amplitude and phase distortion introduced by the passband amplitude-frequency and phase-frequency characteristics of the low-pass filter will affect the amplitude and phase of the navigation signal. The passband characteristics of the low-pass filter dominate the carrier leakage of the navigation signal and produce the band-limiting effect.

(5) Up-convertor

The up-convertor is to convert an IF navigation signal into an RF (Radio Frequency) signal. The phase noise produced by the local oscillator of the up-convertor deteriorates the carrier pseudocode coherence of the navigation signal. The internal passband filter of the frequency converter also has the same effect as the low-pass filter.

(6) Power amplifier

Power amplifier is to amplify the navigation signal. Undesired AM/AM (amplitude/amplitude) and AM/PM (amplitude/phase) transmission characteristics of the power amplifier introduce the amplitude and phase distortion of the amplified navigation signal. The difference in multiples of the amplifiers for each signal leads to the deterioration of in-band spurs, and increases the leakage of intermodulation components and out-of-band spurs.

(7) Combiner

As one of the nonideal characteristics of navigation signal generation devices, the transmission function of the signal channel introduces various forms of distortion.

### 7.2.2 *Process of Navigation Signal Assessment*

The assessment system of the satellite navigation signal is to measure the signal property of the navigation satellite at the stage of ground assembly work or to evaluate the navigation signal of the on-orbit satellite.

(1) Assessment system of on-orbit navigation satellite

The composition of the assessment system of the on-orbit navigation satellite is shown in Fig. 7.14. The high-gain receiving antenna is accurately aligned with the navigation satellite to be assessed. The downlink navigation signal is received when the elevation angle of the receiving antenna is appropriate. After the band-pass filtering, the signal-collecting device samples the satellite navigation signal and stores the digital navigation signal. Finally, the sample data of the stored navigation signal is used for the computer-aided offline processing and assessment of navigation signals.

(2) Navigation satellite assessment system in ground test

The composition of the navigation satellite assessment system in the ground test is shown in Fig. 7.15. After the navigation signal passes through the high-power attenuator, the signal-collecting device samples the satellite navigation signal and

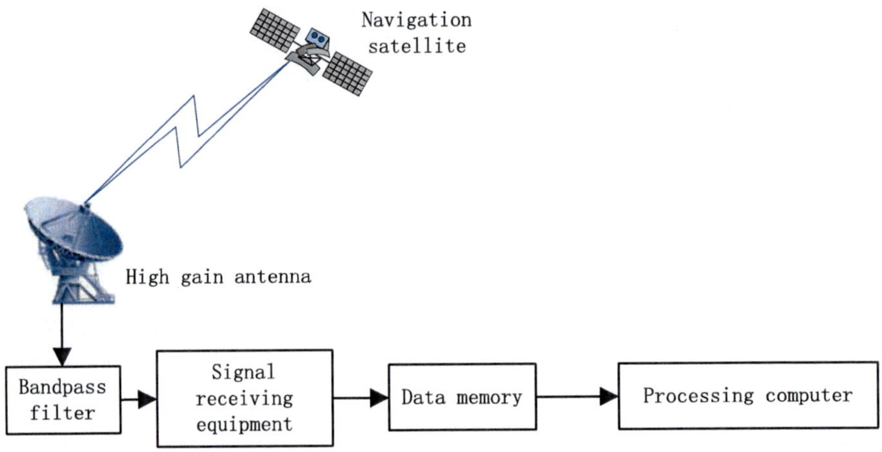


**Fig. 7.14** Navigation signal assessment system of on-orbit satellite

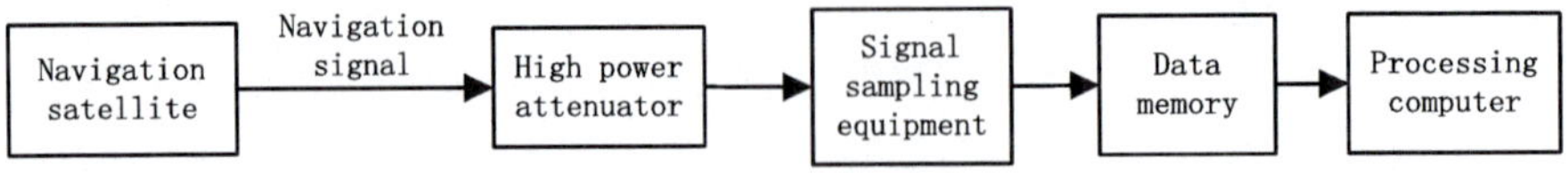


**Fig. 7.15** Navigation signal assessment system in ground test

stores the digital navigation signal. Finally, the sample data of the stored navigation signal is used for assessing the navigation signal with the aid of a computer.

The power level of the signal should be limited to the demanded input level. The sampling frequency of the signal-sampling device should meet the requirement of signal assessment accuracy. The sampling frequency is generally greater than 500 MHz to ensure that the navigation signal spectrum has no overlap. The length of navigation signal data should exceed the minimum length required by the assessment. The noise of the signal acquisition equipment itself shall not affect the navigation signal quality assessment. During the signal sampling, the number of quantized bits should be no less than 8 bits. The minimum loading period of the sampled navigation signal data is one integration period.

### 7.2.3 *Preprocessing of Evaluated Signal*

The sample data of the signal will be processed by a software receiver. The first processing step is to get the coherent integration data of $I$ and $Q$ components, including early, late, and prompt codes, through acquisition and tracking. This step is also called preprocessing. The block diagram of navigation signal preprocessing is shown in Fig. 7.16.

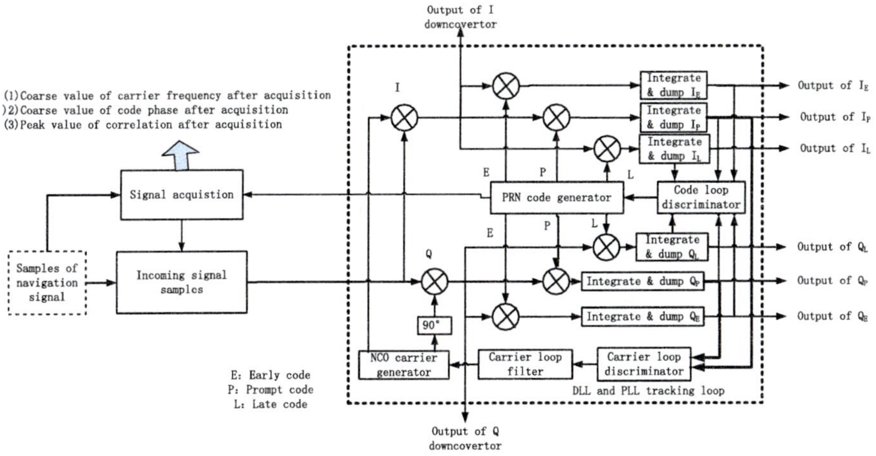


**Fig. 7.16** Block diagram of navigation signal preprocessing

## 7.3 Parameters and Methods of Navigation Signal Assessment

The assessment of navigation signal quality is to comprehensively evaluate the influence of baseband signal generation, modulation, frequency conversion, high-power amplification, and filtering on the quality indicators of the generated navigation signal, including time domain waveform, frequency domain characteristics, and modulation error (corresponding to correlation, stability, and consistency, respectively). As far as the BDS is concerned, there are 10 parameters to be measured. Furthermore, these assessment parameters and methods have been regarded as one of the important BDS standards.

### 7.3.1 *Power Spectrum Distortion*

Since the navigation signal contains both the spread spectrum code signal and the modulated information, the spread spectrum code is multiplied by the time domain of the information symbol and is equivalent to the convolution of the spectral line in the frequency domain. This results in the uncertainty in the power spectrum envelope. The effect on signal performance is reflected in the degradation of effective signal power and the distortion of the correlation function, which finally results in the deterioration of the ranging performance.

The power spectrum distortion between the actual navigation signal and the theoretical navigation signal is assessed at each frequency point under the same signal power. The composite power spectrum distortion is calculated according to the following formula:

$$\mathrm{H}_d(f) = \frac{S_r(f)/\int_{f_l}^{f_h} S_r(f)S_r^*(f)\mathrm{d}f}{S_0(f)/\int_{f_l}^{f_h} S_0(f)S_0^*(f)\mathrm{d}f} \tag{7.17}$$

where $\mathrm{H}_d(f)$ represents the amplitude ratio between actual navigation signal and theoretical navigation signal at each frequency point, $S_r(f)$ is the power spectrum of actual navigation signal, and $S_0(f)$ is the power spectrum of theoretical navigation signal. If the spectrum is not distorted and the transmission channel is ideal, then $\mathrm{H}_d(f) = 1$.

In order to eliminate the influence of information symbol on the shape of the power spectrum envelope of the navigation signal, the data of each spectrum analysis should be limited to one information symbol. According to the previous analysis, the actual deviation of the navigation signal from the theoretical power spectrum is evaluated in line with the principle shown in Fig. 7.17.

The assessment process is as follows:

(1) Band-pass filtering

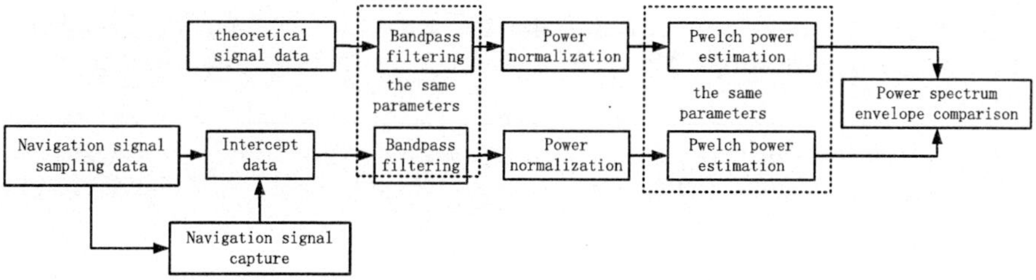


**Fig. 7.17** Estimation of power spectrum distortion

The starting-point number of the pseudorandom code obtained by preprocessing is taken as the starting point, and the navigation signal data sampled in 1 s is read. The read data is subject to band-pass filtering. The time-band filtered data is continuously divided into the data segments with equal length, and each data segment is represented as

$$x_i(n) = x(n + iD),\ n = 0, 1, \ldots, M - 1,\ i = 0, 1, \ldots, L - 1 \tag{7.18}$$

where $x_i(n)$ represents the $i$th sampled data of the $n + 1$th data segment; $x(n + iD)$ represents the $n+iD+1$th sample value of the navigation signal data sampled in 1 s; $L$ is the total number of data segments of the navigation signal data sampled in 1 s; and $M$ is the number of samples included in each data segment. The recommended values of $L$, $M$, and $D$ are $L = 200$, $M = 1e - 3 * f_s$, and $D = M/2$.

(2) Power spectrum estimation

The power spectrum estimation after the windowing of each data segment is

$$\tilde{P}^i_{xx}(f) = \frac{1}{\sum\limits_{n=0}^{M-1} w^{2(n)}} \left| \sum_{n=0}^{M-1} x_i(n) w(n) \exp(-j2\pi f n) \right|^2 \tag{7.19}$$

where $\tilde{P}^i_{xx}(f)$ is the power spectrum estimate of the $i$th data segment; $f$ is the frequency point of the spectrum estimation; and $w(n)$ is a window function. It is recommended to use the Hamming window function.

The Welch spectrum estimation of the actual navigation signal is

$$P^w_{xx}(f) = \frac{1}{L} \sum_{n=0}^{L-1} \tilde{P}^i_{xx}(f) \tag{7.20}$$

where $P^w_{xx}(f)$ is the Welch spectrum estimate of the actual navigation signal.

Simulate the theoretical navigation signal and obtain its Welch spectrum estimate.

(3) Spectrum comparison

Compare the theoretical spectrum value with the actual spectrum value to obtain the power spectrum distortion between the actual navigation signal and the theoretical navigation signal.

$$P^{w}(f) = 10\log\left(\frac{P_{xx}^{w}(f)}{P_{xx}^{\prime w}(f)}\right) \quad (7.21)$$

where $p$ represents the power spectral distortion between the actual navigation signal (in dB) and the theoretical navigation signal, and $P_{xx}^{\prime w}(f)$ represents the Welch spectrum estimate of the theoretical signal.

Perform the first-order straight line fitting of the power spectrum difference between the actual navigation signal and the theoretical navigation signal in the specified frequency band. Calculate the relative deviation value after the straight line fitting, with center frequency point as the monotonously changing center absolute value. Its maximum value is the evaluated value of the frequency power spectrum distortion obtained through the navigation signal synthesis.

### 7.3.2 *Inter-frequency Signal Phase Inconsistency*

Inter-frequency signal phase inconsistency refers to the degree of pseudorandom code synchronization of navigation signals in different carrier frequencies. It is used to estimate the relative delay between ranging codes in different frequency bands during modulation and transmission. The assessment of the inter-frequency phase inconsistency can be realized by selecting a signal component for each frequency band as a reference, and measuring the pseudorandom code phase difference between the signal component in the other frequency band and the reference signal component, as given by the following equation.

$$\Delta \mathrm{D}_{ij} = \mathrm{D}_i - \mathrm{D}_j,\ i = 1, \ldots, N \quad (7.22)$$

where $\mathrm{D}_j$ represents the phase delay value of the selected reference signal component of the $j$th signal; $\mathrm{D}_i$ represents the phase delay value of the measured navigation signal at the $i$th frequency; and $N$ is the total number of frequency bands.

The calculation process of inter-frequency signal phase inconsistency is shown in Fig. 7.18.

The assessment process is as follows:

(1) Signal phase difference measurement

Read a continuous segment of sampled data $\vec{D}_1 = \{d_i | i = 1, 2, \ldots, N_1\}$, where $N_1$ is the total number of sampled data, $N_1 = f_s \cdot C_{t1}$, where $C_{t1}$ represents the pseudorandom code period of the reference signal at the frequency 1. Take the starting-point number of the pseudorandom code as the reference point, and read the continuous segment of the sampled data $\vec{D}_2 = \{d_i | i = 1, 2, \ldots, N_2\}$ of the signal at

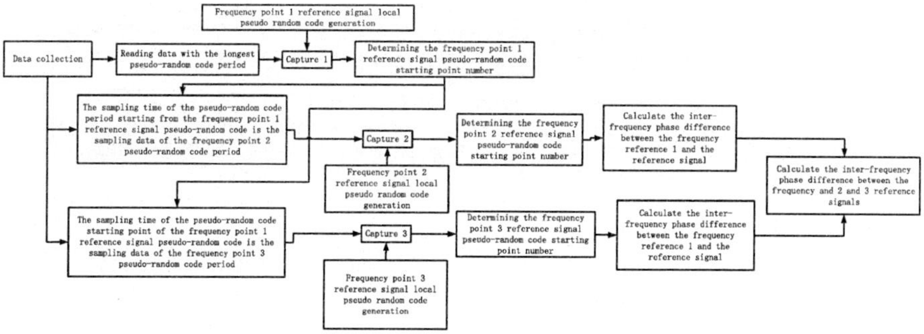


**Fig. 7.18** Calculation process diagram of inter-frequency signal phase inconsistency

the frequency 2. $N_2 = f_s \cdot C_{t2}$ ,where $C_{t2}$ is the pseudorandom code period of the reference signal at the frequency 2. Calculate the inter-frequency phase difference between frequency 1 and frequency 2.

$$\Delta T_{12} = \begin{cases} \frac{n_2}{f_s}, & n_2 \leq N_2 \\ \frac{n_2 - N_2}{f_s}, & n_2 > N_2 \end{cases} \tag{7.23}$$

where $\Delta T_{12}$ represents the delay time between the signal at the frequency 2 and the signal at the frequency 1; $n_2$ represents the starting-point number of the sampled data $\vec{D}_2$ of the reference signal at the frequency 2. Repeat the above process to calculate the difference $\Delta T_{13}$ between frequencies 1 and 3, and the difference $\Delta T_{23}$ between frequencies 2 and 3.

(2) Stability of phase inconsistency

The stability of the inter-frequency signal phase inconsistency can be calculated by fluctuation, mean, and variance. Assume that the phase inconsistency between the signal frequency $i$ and the reference signal $j$ can be obtained through continuous longtime measurement, and expressed as $\Delta D'_{ij}(k), k = 1, 2, \ldots, M$. The fluctuation of the inter-frequency signal phase difference is

$$W_{ij} = \max(\Delta' D_{ij}(k)) - \min(\Delta' D_{ij}(k)),\ k = 1, 2, \cdots, M \tag{7.24}$$

where max(·) refers to the maximum value and min(·) refers to the minimum value.

The mean and variance of the inter-frequency signal phase difference between the signal frequency $i$ and the reference signal $j$ are

$$E_{ij} = \frac{1}{M} \sum_{k=1}^{M} \Delta D'_{ij}(k) \tag{7.25}$$

$$\sigma = \sqrt{\frac{1}{M}\sum_{k=1}^{M}\left(\Delta D'_{ij}(k) - E_{ij}\right)^2} \tag{7.26}$$

### 7.3.3 Pilot and Data Signal Phase Inconsistency

The pilot and data signal phase inconsistency refers to the degree of pseudorandom code synchronization between the pilot component and the data component of the navigation signal in the same frequency band. Ideally, the pseudorandom codes of the components are completely aligned. But in practice, the pseudorandom codes may not be aligned after a series of signal transformations such as baseband generation, carrier modulation, power amplification, and antenna radiation. As far as the pseudocode phase consistency in the same frequency band is concerned, since the navigation signal components share the same carrier frequency and transmission channel, the influence of their transmission channel on the pseudorange is also basically the same. Therefore, the phase inconsistency measurement can be obtained by the following calculation formula:

$$\Delta P_{ij} = P_i - P_j,\ i = 1, \ldots, N \tag{7.27}$$

where $\Delta P_{ij}$ represents the phase difference between the $i$th and $j$th navigation signal components with reference to the $j$th navigation signal component, $P_i$ represents the output phase of the code discriminator of the $i$th signal component, $P_j$ represents the output phase of the code discriminator of the $j$th signal component, $N$ is the total number of the signal components in the same frequency band.

The calculation process of pilot and data signal phase inconsistency is shown in Fig. 7.19.

The assessment process is as follows:

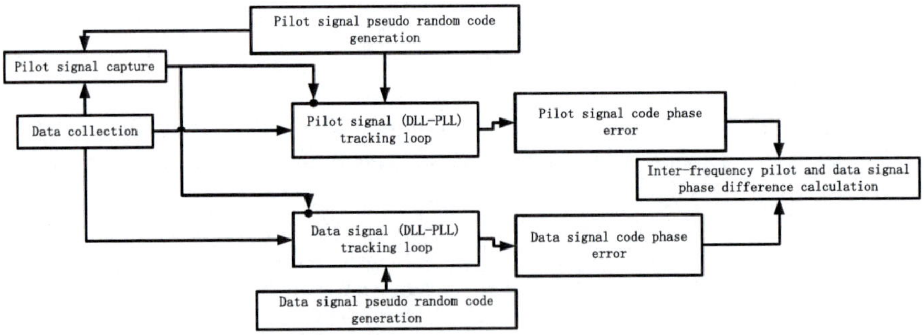


**Fig. 7.19** Calculation process diagram of pilot and data signal phase inconsistency

(1) Signal phase difference measurement

Calculate the difference between the tracking phase of the pilot signal and that of the data signal.

$$\tilde{D}_i = P_{pi} - P_{d(u \cdot i)},\ i = 1, 2, \ldots, m$$

where $\tilde{D}_i$ denotes the $i$th integral period of the pilot-signal tracking loop, $\vec{P}_p = \left[ P_{p1}\ P_{p2}\ \ldots\ P_{pm} \right]$ represents the tracking phase of the pilot signal at a certain time, $P_{d(u \cdot m)}$ denotes the tracking phase of the data signal in the $u \times m$ integral period, and $u$ is the multiplier of the pseudorandom code period.

(2) Deviation of phase difference

The difference between the tracking phase of the pilot signal and that of the data signal is

$$E_{pd} = \frac{1}{m}\sum_{i=1}^{m} \tilde{D}_i \tag{7.28}$$

$$\sigma_{pd} = \sqrt{\frac{1}{m}\sum_{i=1}^{m} \left(\tilde{D}_i - E_{pd}\right)^2} \tag{7.29}$$

where $E_{pd}$ represents the mean of the phase difference between pilot signal and data signal, and $\sigma_{pd}$ represents the standard deviation of the phase difference between pilot signal and data signal.

### *7.3.4 Deviation of Effective Power Rate of Signal Component*

Due to the introduction of intermodulation terms and the influence of noise, the power rate of each signal component in the same band may deviate from the design value. The deviation of the effective power rate of each signal component refers to the deviation from the designed power rate.

The calculation process of this deviation is shown in Fig. 7.20.

The assessment process is as follows:

(1) Correlation peak calculation

Considering that the periods of the components are inconsistent, the correlation peaks should be normalized. According to the ratio of the period of the evaluated signal component to the period of the 1st signal component, the correlation peaks captured by the respective signal components are

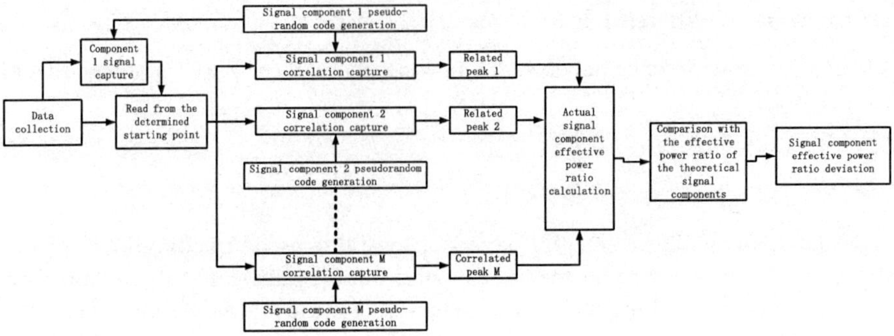


**Fig. 7.20** Calculation process diagram of the deviation of effective power rate of the signal component

$$\bar{A}_i = \left(\frac{T_{c1}}{T_{ci}}\right)^2 \cdot A_i \tag{7.30}$$

where $\bar{A}_i$ represents the correlation peak of the normalized $i$th signal component, and $T_{ci}$ represents the period of the $i$th signal component.

(2) Calculation of effective power rate

Calculate the proportion of the effective power of each signal component in the actual signal power:

$$A_r(i) = \frac{\bar{A}_i}{\sum_{j=1}^{M} \bar{A}_j} \tag{7.31}$$

where $A_r(i)$ represents the ratio of the effective signal power acquired by the $i$th signal component to the actual signal power, and $M$ is the total number of signal components to be evaluated in the frequency band.

(3) Calculation of the deviation of effective power rate

Calculate the deviation of the effective power rate of each signal component

$$A_d(i) = 10\log\left(\frac{A_r(i)}{A_t(i)}\right) \tag{7.32}$$

where $A_d(i)$ represents the deviation of the effective power rate of the $i$th signal component in dB, and $A_t(i)$ is the ideal power rate of the $i$th signal component.

## *7.3.5 Correlation Loss*

The correlation loss is defined as the attenuation of the effective power of the received satellite signal relative to the total power, which is the difference between the actually received signal power and the signal power in the ideal reception. The correlation loss reflects not only the signal power, but also the distortion of the navigation signal during transmission. The correlation loss is calculated according to the following equation:

$$\mathrm{CL[dB]} = \underset{\text{Ideal-input}}{P_{\text{CCF}}}\,[\text{dB}] - \underset{\text{Real-input}}{P_{\text{CCF}}}\,[\text{dB}] \tag{7.33}$$

where $\underset{\text{Ideal-input}}{P_{\text{CCF}}}$ represents the correlation power in the ideal reception, and $\underset{\text{Real-input}}{P_{\text{CCF}}}$ represents the correlation power of the actually received signal. The signal power $P_{\text{CCF}}$ is calculated according to the following equation:

$$P_{\text{CCF}}[dB] = \underset{\text{over all } \varepsilon}{\max} \left(20 \log_{10}(|\text{CCF}(\varepsilon)|)\right) \tag{7.34}$$

The calculation process of correlation loss is shown in Fig. 7.21.

The assessment process is as follows:

(1) Band-pass filtering

Take the starting-point number of the pseudorandom code obtained by preprocessing as the starting point, and read the filtered theoretical navigation signal data.

$$\vec{D}_t = \{d_{ti} | i = 1, 2, \ldots, N_t\} \tag{7.35}$$

where $d_{ti}$ is the $i$th filtered theoretical navigation signal data sample; $N_t$ is the total number of the samples to be evaluated, $N_t = f_s \cdot T_t$, where $T_t$ is the pseudorandom code period of the signal component to be evaluated.

The power of the filtered theoretical navigation signal data is normalized as

$$\vec{D}'_t = \left\{ \frac{d_{ti}}{K_t} \middle| i = 1, 2, \ldots, N_t \right\} \tag{7.36}$$

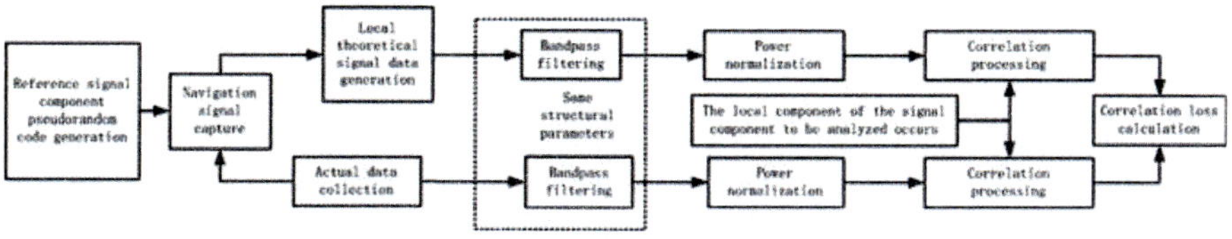


**Fig. 7.21** Calculation process diagram of correlation loss

where $\vec{D}_t$ is the theoretical navigation signal data obtained after power normalization; $K_t$ is the power normalization coefficient of theoretical navigation signal data, $K_t = \sqrt{\sum_{i=1}^{N_t} d_{ti}^2}$.

(2) Calculation of correlation loss

Correlate the actual filtered navigation signal data with the component to be evaluated, and then determine the peak of correlation amplitude.

The correlation loss is calculated by

$$C_L = 10\log\left(\frac{A_t^2}{A_r^2}\right) \tag{7.37}$$

where $C_L$ is the correlation loss of the signal component to be evaluated, $A_t$ is the correlation peak of the theoretical signal component to be evaluated, $A_r$ is the correlation peak of the actual signal component to be evaluated.

### 7.3.6 Zero-Crossing Deviation of S-Curve

*S*-curve is the phase-detection curve of the "early minus late" correlation value in the receiver code tracking loop, which varies with the phase-detection algorithm of the receiver. When the navigation signal is not distorted or the cross-correlation function after distortion is still symmetric about the *y*-axis, the code ring lock point will coincide with the peak point of the correlation curve, and the code tracking deviation will not be introduced. At the zero-crossing point of *S*-curve, the code tracking error is zero.

In reality, due to the influence of channel transmission distortion and the nonlinear effect of power amplifier, the *S*-curve is asymmetrical. When the code locking point deviates from the peak point of the correlation curve, the code tracking deviation will be generated to introduce the ranging error that will affect the positioning accuracy. The zero-crossing deviation of the phase-detection curve is calculated numerically by

$$\mathrm{R}_{sd} = \varepsilon \times c \tag{7.38}$$

where $\mathrm{R}_{sd}$ represents the ranging error caused by signal distortion, $\varepsilon$ is the zero-crossing deviation of *S*-curve, and $c$ is the propagation velocity of electromagnetic wave.

To quantify the effect caused by the offset of the zero-crossing of *S*-curve and make a reasonable compromise between parameter complexity and practical value, a noncoherent power discriminator over a wide "early minus late" spacing $\delta$ is considered. The zero-crossing of S-curve is determined by

**Fig. 7.22** Typical *S*-curve

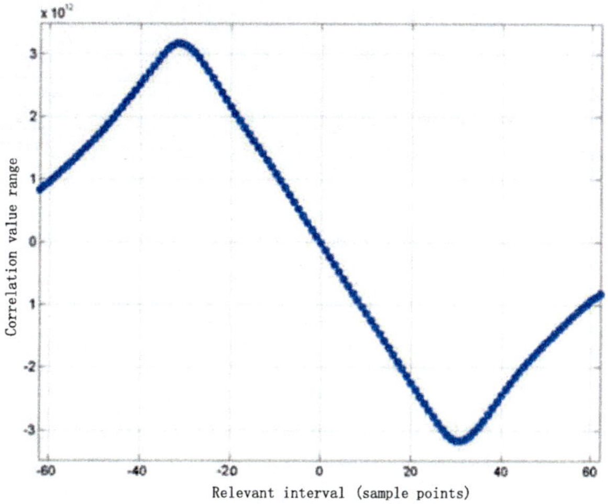


$$S\,\text{Curve}(\varepsilon,\delta) = \left|\text{CCF}\left(\varepsilon - \frac{\delta}{2}\right)\right|^2 - \left|\text{CCF}\left(\varepsilon + \frac{\delta}{2}\right)\right|^2 \tag{7.39}$$

where CCF($\varepsilon$) represents the cross-correlation function of the reference signal and the received signal.

For the typical noncoherent power discriminator, the *S*-curve is shown in Fig. 7.22.

At the locking point, the discriminator output satisfies the condition $S\,\text{Curve}(\varepsilon_b(\delta),\delta) = 0$. Therefore, the *S*-curve deviation is defined as the difference between the maximum offset and the minimum offset by considering all the $\delta$ values in the interval $[0, \delta_{\max}]$. The S-curve deviation is calculated by

$$S_b = \max_{\text{over - all - }\varepsilon}(\varepsilon_b(\delta)) - \min_{\text{over - all - }\varepsilon}(\varepsilon_b(\delta)) \tag{7.40}$$

For the BOC-modulated signal, there may be multiple zeros in the discrimination curve. In this case, the zero-crossing point nearest to the main peak of the correlation curve should be selected. When calculating the *S*-curve deviation, it is necessary to determine the range of the $\delta$ value with different modulation methods.

The calculation process of the zero-crossing deviation of *S*-curve is shown in Fig. 7.23.

The assessment process is as follows:

(1) *S*-curve calculation

Set the relevant parameters of the DLL-PLL tracking loop, and realize the continuous tracking of the *S*-curve during each integration period.

$$S_j(\varepsilon, e(i)) = \left|C_F\left(\varepsilon - \frac{e(i)}{2}\right)\right|^2 - \left|C_F\left(\varepsilon + \frac{e(i)}{2}\right)\right|^2 \tag{7.41}$$

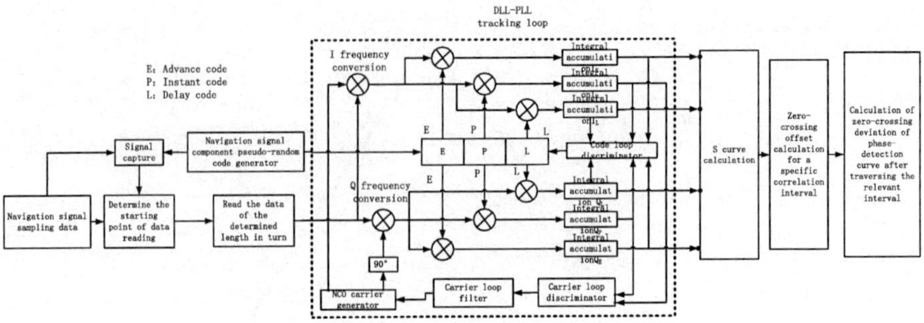


**Fig. 7.23** Calculation process diagram of zero-crossing deviation of *S*-curve

where $e(i)$ is the ith correlation interval, $S_j(\varepsilon, e(i))$ is the *S*-curve point in the jth integration period, $\varepsilon$ represents the delay of the reference signal relative to the received signal, $C_F()$ represents a cross-correlation function of the reference signal and the received signal.

(2) Calculation of zero-crossing deviation

Calculate the deviation of the zero-crossing point of *S*-curve:

$$S_j\left(\delta_j^i, e(i)\right) = 0 \; j = 1, \cdots, M \tag{7.42}$$

where $\delta_j^i$ represents the zero-crossing deviation of *S*-curve in the jth integration period. Calculate the mean value of the zero-crossing deviation of the *S*-curve:

$$\bar{\delta}_i = \frac{1}{M}\sum_{j=1}^{M} \delta_j^i \tag{7.43}$$

Calculate the zero-crossing deviation of *S*-curve:

$$\delta_t = \max\left(\left[\,\bar{\delta}_1 \; \bar{\delta}_2 \; \ldots \; \bar{\delta}_{S_N}\,\right]\right) - \min\left(\left[\,\bar{\delta}_1 \; \bar{\delta}_2 \; \ldots \; \bar{\delta}_{S_N}\,\right]\right) \tag{7.44}$$

## 7.3.7 *Zero-Crossing Slope Deviation of S-Curve*

The zero-crossing slope deviation of *S*-curve is defined as the *S*-curve slope difference between the real receiver discriminator and the ideal receiver discriminator. Under the white-noise condition, the tracking accuracy of the code tracking loop can be calculated by

$$\sigma_\varepsilon^2 = 2B_L R_n(0) T_p / K^2 \tag{7.45}$$

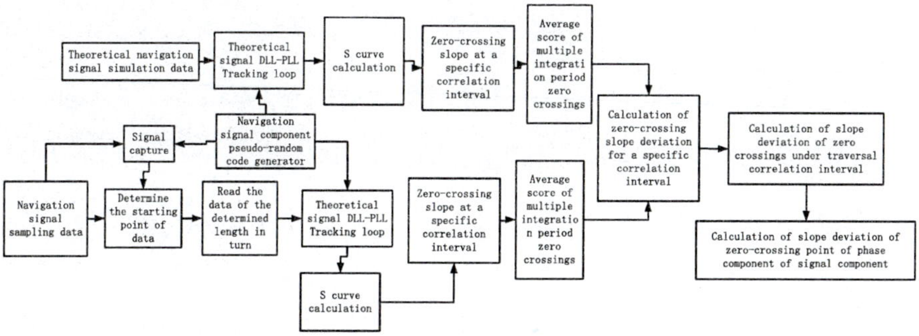


**Fig. 7.24** Calculation process diagram of zero-crossing slope deviation of *S*-curve

where $\sigma_\varepsilon^2$ is the code tracking error variance, $B_L$ is the bandwidth of code locked loop, $R_n(0)$ is the noise power of discriminator output, $T_p$ is the pre-detection integration time, and $K$ is the zero-crossing slope of *S*-curve.

The code tracking error variance is inversely proportional to the square of the zero-crossing slope of *S*-curve. The larger the zero-crossing slope of *S*-curve, the higher the code tracking accuracy.

The calculation process of the zero-crossing slope deviation of *S*-curve is shown in Fig. 7.24.

The assessment process is as follows:

(1) Calculation of the zero-crossing slope of *S*-curve

Set the relevant parameters of the DLL-PLL tracking loop, and realize the continuous tracking of the *S*-curve during each integration period. The zero-crossing slope in each integration period is given by

$$L_r(j) = \left.\frac{dS_j(\varepsilon, e(i))}{d\varepsilon}\right|_{S_j\left(\delta_j^i, e(i)\right)=0,\ j=1,\cdots,M,\ i=1,\ldots,S_N} \tag{7.46}$$

where $L_r(j)$ represents the zero-crossing slope of *S*-curve in the jth integration period. Average the slopes of multiple *S*-curves over the same correlation interval $e(i)$:

$$L_r^i = \frac{1}{M}\sum_{j=1}^{M} L_r(j),\ j = 1, \ldots, M \tag{7.47}$$

where $L_r^i$ represents the mean value of the zero-crossing slope of *S*-curve in the correlation interval $e(i)$.

(2) Calculation of zero-crossing slope deviation

Calculate the zero-crossing slope deviation of *S*-curve between the theoretical signal data and the actual signal data.

$$\Delta L_i = L_a^i - L_r^i,\ j = 1, \ldots, M \tag{7.48}$$

The zero-crossing slope deviation of $S$-curve is calculated by

$$\Delta L = \max\left( |\Delta L_1|\ |\Delta L_2|\ \ldots\ |\Delta L_{S_M}| \right) \tag{7.49}$$

where $\Delta L_i$ is the zero-crossing slope deviation of $S$-curve in the jth integration period.

### 7.3.8 *Distortion of Baseband Waveform*

The evaluation of baseband waveform distortion is to analyze the edge characteristics of the pseudorandom code chip of the recovered navigation signal. The effect of baseband distortion on signal performance is reflected in the distortion of the correlation function, which will cause the deterioration of the ranging performance. The distortion of the baseband waveform is evaluated by calculating the width difference between the actual code chip and the theoretical code chip, and its fluctuation.

The calculation process of baseband waveform distortion is shown in Fig. 7.25. The assessment process is as follows:

(1) Generation of baseband waveform

After the tracking becomes stable, read the navigation signal data sampled in the kth integration period as

$$\vec{I}_k = F_L\left( I_k(1)\ I_k(2)\ \ldots\ I_k(N_l) \right) \tag{7.50}$$

where $N_l$ is the sample number in one integration period, $\vec{I}_k$ is the waveform data obtained after low-pass filtering in the $k$th integration period, $F_L()$ denotes digital

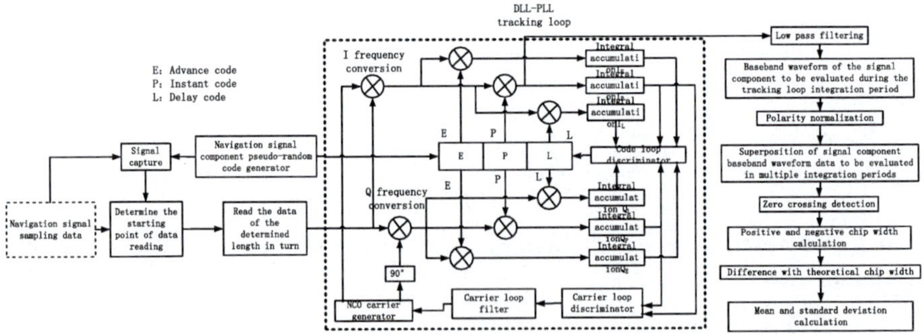


**Fig. 7.25** Calculation process diagram of baseband waveform distortion

low-pass filtering, and $I_k(i)$ is the $i$th sampled data. $\vec{I}_k$ should be processed through polarity normalization to obtain $\hat{I}_k$, which is the baseband waveform data in the $k$th integration period.

$$\hat{I}_k = \vec{I}_k \cdot \mathrm{sign}\Big(\max\Big(\vec{I}_k * \vec{C}\Big)\Big) \tag{7.51}$$

where $\vec{C}$ represents the local pseudorandom code sample of the signal to be evaluated.

(2) Distortion evaluation

After the polarity normalization, the multiple integration periods are accumulated and averaged to obtain the time domain waveform of the evaluated chip baseband. The rising edge and falling edge of the chip waveform around the zero-crossings point are processed by linear fitting in order to determine the positive and negative chip widths. The difference between the calculated width and theoretical width is used to represent the distortion of the baseband waveform.

### 7.3.9 Coherence Between Carrier and Pseudorandom Code

The non-coherence between the carrier and the pseudorandom code refers to the phase difference between the pseudorandom code and the carrier. For example, if the phase difference is φ rad when the code level is +1, then the phase difference should be π + φ rad when the code level is −1. In the actual signal generation process, the signal is affected by various nonlinear factors related to the channel and stability of the carrier. When the code level is +1 or −1, the actual phase difference will be φ + Δφ rad or π + φ+Δφ rad, respectively, where Δφ is the phase non-coherence caused by various nonlinear factors related to the channel and stability of the carrier.

The calculation process of the coherence between the carrier and the pseudorandom code is shown in Fig. 7.26.

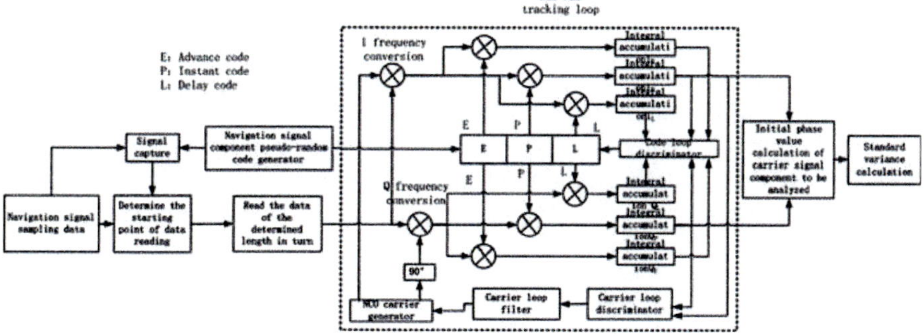


**Fig. 7.26** Calculation process diagram of the coherence between carrier and pseudorandom code

The assessment process is as follows:

(1) Calculation of initial carrier phase

After the navigation signal preprocessing is completed and the signal component tracking becomes stable, the initial carrier phase of the signal component to be evaluated will be

$$\theta(k) = \arctan\left(\frac{\sum_{i=1}^{N_k}\left[\cos(\alpha_k(i)) \cdot c_{k,i} \cdot S_{k,i}\right]}{\sum_{i=1}^{N_k}\left[\sin(\alpha_k(i)) \cdot c_{k,i} \cdot S_{k,i}\right]}\right) \tag{7.52}$$

where $\theta(k)$ is the initial carrier phase calculated from the $k$th integration period, $N_k$ is the number of actual navigation signal samples in the $k$th integration period, $\alpha_k(i)$ is the $i$th sample value of the local carrier phase in the $k$th integration period, $c_{k,i}$ is the $i$th sample value of the local code in the $k$th integration period, $S_{k,i}$ is the $i$th sample value of the actual navigation signal sample in the $k$th integration period.

(2) Coherence calculation

Read a certain number of initial carrier phase values, and calculate the variance of the carrier phase as shown below:

$$\sigma_t = \sqrt{\frac{1}{P}\sum_{i=1}^{P}\left(\theta(i) - \frac{1}{P}\sum_{j=1}^{P}\theta(j)\right)^2} \tag{7.53}$$

where $\sigma_t$ is the coherence between carrier and pseudorandom code, $P$ is the number of initial carrier phase values.

### 7.3.10 *Deviation of Signal Phase*

When multiple signals are multiplexed, some signals are modulated on the in-phase branch and the other signals are modulated on the orthogonal branch. Because the signals are affected by nonideal modulators and noise, the carrier phase of each signal may deviate from its theoretical phase value.

The navigation signals that are multiplexed with $N$ components can be represented as the vector summation of all useful signals and intermodulation components.

$$s(t) = \sum_{i=1}^{N} s_i(t)e^{j\theta_i} + s_m(t) \tag{7.54}$$

where $s_i(t)$ is the baseband signal of the $i$th navigation signal component, $\theta_i$ is the initial carrier phase of the $i$th navigation signal component, $N$ is the number of useful navigation signal components, and $s_m(t)$ is the intermodulation component introduced by constant envelope modulation.

The phase deviation of the signal component is defined as

$$\Delta\theta_{\text{ji}} = \left(\theta_j^t - \theta_i^t\right) - \left(\theta_j^r - \theta_i^r\right) \tag{7.55}$$

where $\Delta\theta_{\text{ji}}$ is the phase deviation between the $j$th and $i$th components of navigation signal, $\theta_j^t$ and $\theta_i^t$ are the initial carrier phases of the $j$th and $i$th components to be evaluated, respectively, $\theta_j^r$ and $\theta_i^r$ are the ideal initial carrier phases of the $j$th and $i$th components, respectively.

The calculation process of signal phase deviation is shown in Fig. 7.27.

The assessment process is as follows:

(1) Calculation of initial carrier phase deviation

During the integration period of all signal components, the deviation between the initial carrier phase values of the signal components is

$$\Delta\theta_{i,j}(k) = \theta_i(k) - \theta_j(k) \tag{7.56}$$

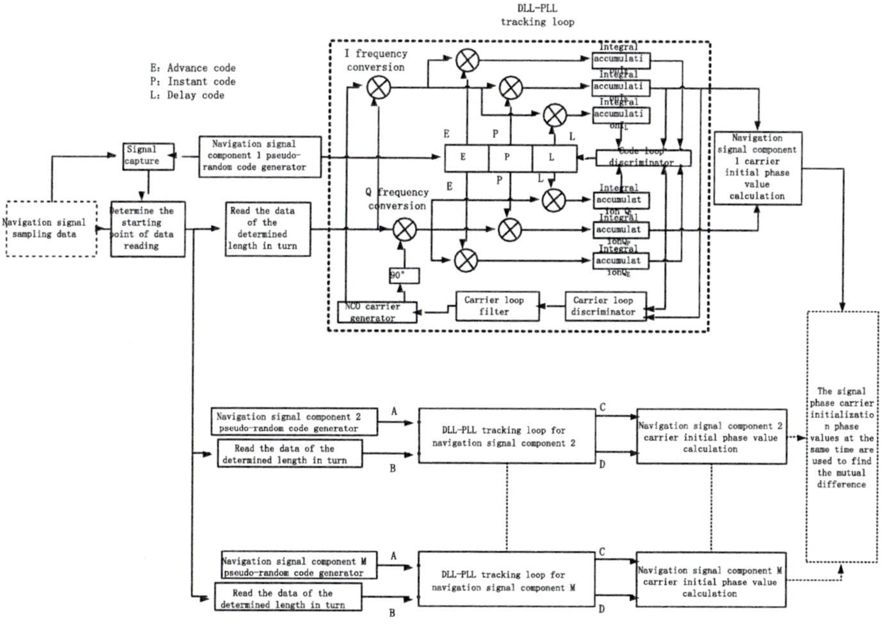


**Fig. 7.27** Calculation process diagram of signal phase deviation

where $\Delta\theta_{i,j}(k)$ is the initial carrier phase deviation between the $i$th and $j$th signal components in the $k$th integration period, $\theta_i(k)$ and $\theta_j(k)$ are, respectively, the initial carrier phases of the $i$th and $j$th components in the $k$th integration period.

(2) Calculation of signal phase deviation

The maximum value of the absolute value of the initial phase deviation is the phase deviation between the two signal components, as calculated below:

$$\hat{\theta}_{i,j} = \max\left(\left[\, \Delta\theta_{i,j}(1) \;\; \Delta\theta_{i,j}(2) \;\; \cdots \;\; \Delta\theta_{i,j}(N_\theta) \,\right]\right) \tag{7.57}$$

where $\hat{\theta}_{i,j}$ is the phase deviation between the $i$th and $j$th signal components, and $N_\theta$ is the total number of samples with the initial carrier phase deviation.

# Chapter 8
# Satellite Navigation Information Management

The satellite navigation system controls the satellite constellation in the space segment through the measurement of the ground segment, and regularly uploads the navigation information to satellites. Then the satellites process the system information, the information uploaded by the ground segment, and the information of the satellites themselves according to the design requirements, and configure such information into the navigation messages to be broadcasted to users, so as to complete the satellite control and status monitoring. The Global Navigation Satellite System (GNSS) has diverse information interfaces between satellite and ground station and between satellites, with various types of information from a lot of sources in complex path relationships. Numerous information exchanges happen among multiple satellite-ground systems, among multiple ground stations, among multiple satellites, and among multiple onboard equipment.
Satellite navigation information management technology is just to comprehensively design the transmission, processing, and storage of all kinds of information mentioned above, so as to ensure the security and reliability, accuracy (during transfer), and integrity of such information as well as the continuity, stability, and availability of navigation services. Besides, great concern is also shown to the time characteristics of information in the management process. The GNSS fully embodies the achievements of highly continuous information transmission and management.
This chapter firstly gives an overview of satellite navigation information, including information transmission channels and information classification, satellite information management, and system architecture. Then the strategy and evaluation technique of satellite information management are analyzed. Finally, the management technique of intra-satellite navigation messages is addressed.



J. Xie et al., *Satellite Navigation Systems and Technologies*, Space Science and Technologies, https://doi.org/10.1007/978-981-15-4863-5_8

## 8.1 Overview of Satellite Navigation Information

The satellite navigation system consists of a space segment, a ground segment, and a user segment. Among the system components, the satellite navigation information is transmitted through various satellite-ground communication uplinks and downlinks, through inter-satellite communication links in the space segment, and through ground communication links in the ground segment.

### *8.1.1 Information Contents in the Space Segment and Ground Segment*

The information to be transmitted and managed in the space segment and ground segment mainly includes satellite-ground operation control and satellite-ground TT&C information.

(1) Satellite-ground operation control information

Between space segment and ground segment, information is transmitted through satellite-ground uplinks to achieve the navigation message injection, satellite-ground range measurement, and operation management and control. Between space segment and ground segment/user segment, the transmitting of navigation signals, the information transmission, and the broadcasting and monitoring of navigation messages are completed through satellite-ground downlinks.

In the ground segment, the time and space coordinate data of the navigation system are established and maintained, the basic navigation, satellite-based augmentation, and other service operations are completed, and various service information and operation control information are injected into navigation satellites. In order to complete the operation service of navigation systems, the navigation signals downloaded by the space segment and the satellite-ground time synchronization data are collected in the ground segment to generate the satellite broadcast ephemeris and clock difference, which are then uploaded to space segment.

In the space segment, the broadcast ephemeris and clock difference uploaded by the ground segment are received, saved, and arranged into navigation messages to be broadcasted to the user segment through downlinks.

In the user segment, the ephemeris and clock difference downloaded by space segment are received and demodulated, with which the user's three-dimensional position and velocity, and time information in the spatial and temporal references of GNSS are calculated.

(2) Satellite-ground TT&C information

Between space segment and ground segment, information is transmitted through satellite-ground TT&C links to achieve the space segment telemetry and telecontrol, and implement the constellation configuration maintenance, long-term satellite management, satellite state monitoring, and other functions.

Through the satellite-ground TT&C information transmission, some activities are completed in the ground segment, such as the determination of navigation satellite orbit, the telemetry, reception and processing of navigation satellite engineering information, the generation, transmission and management of telecommands, the emergency treatment of faults, as well as the maintenance and management of navigation constellation and the operation management of inter-satellite links.

In the ground segment, the ranging signals and telemetry signals downloaded by the space segment are collected, the orbit measurement and determination during the launch, orbit injection, and long-term operation of satellites are performed, and the telemetry data for the satellite state monitoring are analyzed to ensure the normal operation of navigation constellation. In the space segment, the telecommands uploaded by the ground segment are received and executed to complete the management and control of satellites and the operation management of inter-satellite links.

### *8.1.2 Internal Information Contents of Satellites in the Space Segment*

Through a reasonable constellation configuration design for the GNSS, several navigation satellites with independent functions are arranged at the specified orbital positions, so as to provide the Positioning, Navigation, and Timing (PNT) services. All the satellites in the space segment are networked into a constellation through inter-satellite links to achieve inter-satellite high-precision ranging and data information transmission, thus completing autonomous navigation and one-station management and control. The satellite system itself is also capable of autonomous management and control, integrity information monitoring, and health information management.

Based on stable and reliable inter-satellite links in the GNSS, inter-satellite ranging and time calibration are conducted to realize multi-satellite measurement, increase the observations, and improve geometric observation structure in autonomous orbit determination. By using the inter-satellite measurement information, the satellites' orbital position and clock difference are autonomously calculated and corrected, thus achieving the satellite-ground joint precise orbit determination and improving the accuracy of orbit determination and time synchronization and the accuracy of positioning and services in the whole system.

In addition, on the basis of satellite-ground link communication, overseas satellites are monitored and injected with information with the help of the inter-satellite link information transmission based on the satellite information system, thus achieving their "one-station measurement and control".

In the space segment, the information on power supply and distribution, temperature, data and time, signal level, and equipment operation state are stored in each satellite. The satellites' on-orbit viability and reliability can be improved by the autonomous integrity monitoring, autonomous health management, and other control functions based on the satellite information system, thus reducing the management and control task in the ground segment.

### *8.1.3 Internal Information Contents in the Ground Segment*

The stations in the ground segment communicate with each other through the ground communication links, and support the information exchange between the space segment and the ground segment through various satellite-ground communication links.

The ground segment consists of the main control station, monitoring stations, time synchronization/injection stations, and TT&C center (and its TT&C stations). The main control station communicates with the monitoring stations to transmit the navigation signal monitoring data of the space segment, and with the time synchronization/injection stations to upload the navigation messages and operation control instructions of the space segment. The TT&C center communicates with its TT&C stations to receive and process the telemetry data of the space segment, generate and transmit the telecommands, and manage and control the satellites and inter-satellite links in the space segment.

The information in the ground segment is mainly in the form of continuous real-time data packets in a fixed format, including the observation information of ground segment (the pseudorange and phase data observed by the receiver), the onboard observation data, the ephemeris information broadcasted by satellites in real time, the meteorological data and the calculation result information generated by ground segment (satellite ephemeris, satellite clock difference, ionospheric delay parameter, difference integrity information, etc.).

### *8.1.4 Information Contents of Channel Interaction Support*

Based on the interstation, satellite-ground and inter-satellite communication links, and spaceborne information system, information channels are established between the communication links with relatively independent functions. Through the network protocols between system segments, the ground segment completes the satellite-ground information interaction. The space segment completes the identification, processing, distribution, and execution of all kinds of satellite-ground and inter-satellite information, achieves the information interaction and integration, supports the interaction management of multisource, multifrequency, and multi-channel satellite information, and improves the information transmission capacity and system reliability.

## 8.2 Transmission Channels and Classification of Satellite Information

### *8.2.1 Satellite Information Transmission Channel*

Generally, satellite navigation information is transmitted interactively through three types of channels, including satellite-ground operation control link, satellite-ground TT&C link, and inter-satellite link. The satellite information transmission channels are shown in Fig. 8.1.

The satellite-ground operation control links have functions such as ranging and data transmission, as well as navigation message injection, navigation service

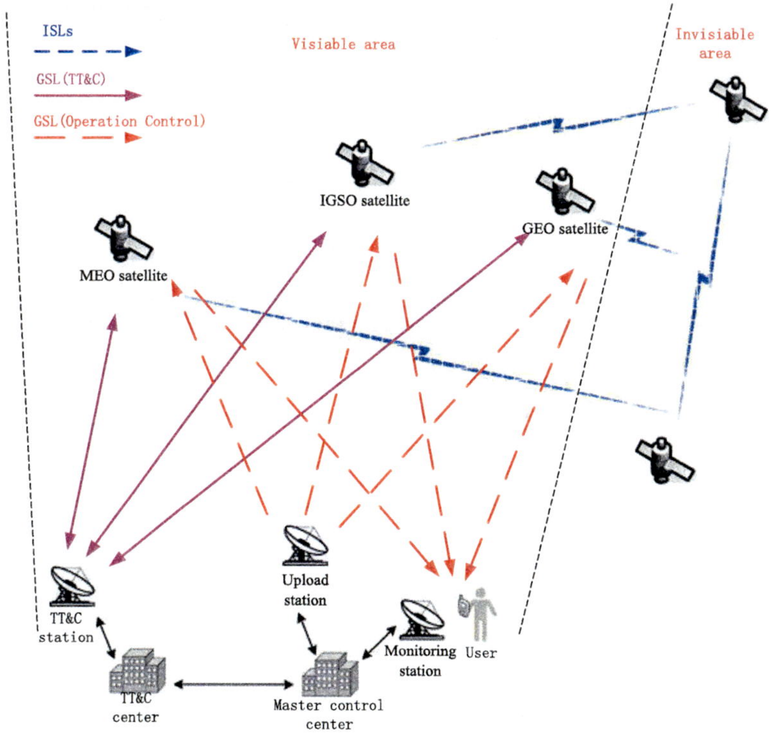


**Fig. 8.1** Schematic diagram of satellite information transmission channels

management and control, and navigation signal broadcasting. Through the information injection by ground operation control uplinks, key navigation data injected into navigation satellites, such as broadcast ephemeris, satellite clock difference, and satellite almanac, can be maintained and updated, and navigation services can be managed and controlled. The satellite downlink channels arrange the received navigation messages into signals, and then broadcast them to users through the navigation antennas. In addition, the ground stations realize the satellite-ground time synchronization through the information injection via uplinks and the pseudorange measurement and satellite-ground bidirectional comparison via downlinks.

The satellite-ground TT&C links provide the wireless transmission channels among satellites, ground TT&C stations and ships, as well as the functions such as ranging and velocity measurement, and satellite telecontrol and telemetry.

The TT&C uplinks are mainly used to upload the command information given by domestic visible satellites and overseas invisible satellites, and to implement the satellite control measures. The TT&C downlinks are mainly used to download the telemetry information of domestic visible satellites and overseas invisible satellites to monitor the satellite status.

The inter-satellite links assist the ground station in uploading the telecommands to the satellites invisible at the ground stations and in downloading the telemetry information from those satellites, so as to achieve inter-satellite ranging, time synchronization and precise orbit determination, autonomous navigation data transmission, integrity monitoring, and real-time data transmission.

### *8.2.2 Classification of Satellite Information*

Different satellite navigation information has different functions, purposes, and characteristics. Therefore, information classification is distinguished according to the needs for different information so as to design the targeted strategies of information transmission and processing.

1. Classification by Function and Purpose

According to the function and purpose, satellite navigation information can be classified as follows. Navigation service information is used to directly provide navigation services for users; and Satellite control and monitoring information is used to monitor and maintain the navigation constellation system and provide system support for navigation services. Through this classification, satellite system interfaces can be defined to support the interface design for large engineering systems, the satellite service planning, and the process control.

**Satellite service information** is used to provide navigation services for users, including navigation message information, navigation integrity information, location report information, etc.

**Satellite control information** is used for the control and maintenance of satellites and the constellation system, including spaceborne equipment telecommand information, constellation configuration information, operation control commands and parameters, software on-orbit reconstruction information, etc.

**Satellite monitoring information** is used to monitor the working and health conditions of satellites, including the engineering telemetry information of spaceborne equipment, exception message, control information response, attitude information, environmental monitoring and measurement information, the information downloaded from spaceborne memory, etc.

Table 8.1 shows the typical categories of satellite navigation information classified according to function and purpose.

2. Characteristics of Information Sources

Satellite navigation information can be classified into periodic information and random information according to the characteristics of the information source. This classification method can clarify the information generation characteristics, support the mathematical model description of the information process, and provide a basis for modeling and simulation demonstration.

**Periodic information** is the data generated by information sources at equal time intervals, such as spaceborne equipment temperature and other monitoring parameters. This information is generally collected at fixed time intervals and then transmitted to the ground to monitor the working status of spaceborne equipment. Such data can be modeled by using periodic functions.

**Random information** is the data generated by information sources at varying time intervals, such as control command response data, and fault data. Such data is randomly triggered by a specific event. According to different data amount and timeliness requirements, a random process can be modeled for description purposes.

The information source of random information generally contains several independent incremental processes, each of which will produce no more than one data packet in any very short time interval $\delta t$. Denote the number of generated data packets as $N(t)$. That is, when $\delta t \to 0$, if $O(\delta t)$ is an infinitesimal with higher order than $\delta t$, then the probability of generating more than one data packet will be

$$P\{N(t+\delta t)-N(t)>1\}=O(\delta t) \tag{8.1}$$

For the process in which the information source has the same data generation intensity at any time interval, the stationary independent incremental process, the data generation can be described by a homogeneous Poisson process. Denote the mean of $N(t)$ in unit time as intensity $\lambda$. If $\delta t \to 0$, then

$$\mathrm{P}\{N(t+\delta t)-N(t)=1\}=\lambda\cdot\delta t+O(\delta t) \tag{8.2}$$

For the process in which the information source has different data generation intensities at different time intervals, time-varying independent incremental process, the data generation can be described by a non-homogeneous Poisson process.

**Table 8.1** Information classification according to function and purpose

| No. | Category | Name | Function, property, and contents |
|---|---|---|---|
| 1 | Satellite service information | Navigation service information | Global basic navigation information, and integrity information |
| 2 | | Increased service information | Position report, message communication data, data link transmission data, etc. |
| 3 | | Extended service information | Various types of user data transmitted to inter-satellite links, and other extended data |
| 4 | Satellite control information | Engineering telecommand information | Various engineering telecommands, used for the startup and shutdown and control of spaceborne equipment |
| 5 | | Constellation configuration data | Constellation configuration table, routing table, timeslot table, etc. |
| 6 | | Operation control command information | All kinds of operation control commands for navigation service control |
| 7 | | Operation control parameter information | All kinds of operation control parameters, used for navigation service parameter configuration and autonomous operation parameter configuration |
| 8 | | Reconfiguration information | Various software and FPGA reconfiguration information |
| 9 | Satellite monitoring information | Engineering telemetry information | All kinds of engineering telemetry data, used for monitoring the working and health conditions of spaceborne equipment |
| 10 | | Exception message | Exception alarm data and fault diagnosis information, used for satellite fault analysis |
| 11 | | Control information response | The response information used to reflect the onboard reception and processing of control information |
| 12 | | Attitude information | Attitude angle, attitude angle velocity, and other attitude data |

(continued)

**Table 8.1** (continued)

| No. | Category | Name | Function, property, and contents |
|---|---|---|---|
| 13 | | Monitoring and measurement information for space environment | Variety of monitoring and measurement data on electromagnetics, plasma, electron, and proton |
| 14 | | Monitoring and measurement information for mechanical environment | Measurement data on vibration, impact, acceleration, and other mechanical factors |
| 15 | | Memory download information | Onboard computer programs, parameters, memory data, etc. |

These random processes are used to describe the statistical characteristics at the moment of data generation. In addition to the temporal characteristics of data generation, data size is also an important element of information source modeling. In general, a certain basic length can be specified so that the data size of each type of process can be expressed as a number relative to the basic length.

In the information source modeling, the processes shall be generally decomposed for description according to the characteristics of various kinds of information and then fused into a composite model as the basis of system design and simulation.

## 8.3 Satellite Information Management and System Architecture

### *8.3.1 Concepts and Definitions*

In the satellite design, navigation information management is usually implemented in the form of information flow design. The terms for information flow are defined as follows:

**Information flow** refers to the process in which the information with a specified flow direction and format travels from the source to the destination.

**Information flow channel** is the path through which the information travels from the source to the destination.

**Information flow node** refers to the equipment, components, and modules for information processing in the information flow channel.

**Information flow network** is a network structure composed of multiple information flow channels and nodes.

The design process of satellite information flow is generally as follows: first, the information type, quantity, channel requirements, and processing function/performance requirements of information flow in the satellite system are determined, and then, the information flow is designed on this basis under comprehensive consideration of various factors such as reliability, safety, efficiency, and timing sequence to meet the requirements such as the information system architecture, information management strategy, information flow function, and performance evaluation.

### 8.3.2 Technology Elements of Information Management

1. Information Management Technology Element—Reliability

For reliability, one or more of the following measures is (are) taken to ensure normal or degraded operation of the information flow network in case of failure.

(1) Multicenter distributed information flow network

This information flow network is based on the distributed information system architecture, and has several information management centers with centralized management and control function. In case of the failure of an information management center, it supports the interactive replacement of the core information processing function to complete the basic management of satellite navigation information.

(2) Redundant backup of information flow channels

The information flows through key spaceborne information channels as well as the information channels between satellites and ground segment systems and between constellation satellites. In the process of transmission from the source to the destination, the communication equipment backup and multi-channel backup are optional modes for redundant backup.

(3) Redundant backup for information flow node

For all kinds of equipment that process information in the information flow channel, the corresponding isolation/redundancy measures are taken. For the common parts and modules among equipment (such as power supply and information cross-linking), it is necessary to ensure that there is no single-point failure or common cause failure.

(4) Redundant backup for the interface at each node

Redundancy can be achieved by switching between the main and backup interfaces. Hot backup, cold backup, and other redundant ways are optional.

(5) Redundant capacity of information flow channel

Redundant information capacity and transmission rate are provided to avoid the local deadlock of the information flow network and the blockage of the information flow channel.

(6) Software reliability design

The satellite software is designed with redundancy, fault tolerance and robustness, and a certain degree of fault detection and processing. The system resources occupied by software design and implementation have a certain redundancy. The telemetry information in the software is reasonably set to accurately reflect the software operation state, so as to provide guarantees for on-orbit use, testing,and maintenance.

2. Information Management Technology Element—Security

For security, the following measures are taken to prevent the local damage of the information flow network that affects satellite security.

(1) Information source identification

Fully identify and analyze the information sources that affect the security of the information system, and take the targeted design measures to reduce risks.

(2) Security of important information sources

Strengthen the storage and processing protection of important information sources, save important status data in a distributed redundant way, and support its on-orbit injection.

(3) Security of information control commands

Carry out multilevel authentication for key control instructions, and take measures to prevent abnormal execution of instructions.

(4) Information transmission security

Minimize the information transmission errors, and ensure the correctness and legality of information sources and contents. When similar or associated information from multiple channels is input into an information flow node, an appropriate screening strategy should be taken to avoid information competition.

3. Information Management Technology Element—Efficiency

For efficiency, the following measures are taken to improve the information management efficiency of the system while ensuring reliability and security.

(1) Trade-off between demand and cost

Systematically and comprehensively analyze and evaluate the demand for reliability, safety and other elements, and the cost of their engineering design implementation, and reasonably allocate and dispatch the information management resources.

(2) Streamlined system configuration

On the premise of sufficiency and user-friendly operation, reduce the quantity and scale of equipment hardware and software as much as possible, simplify the information connection relationships at all levels, weaken the coupling degree between information layers till the system decoupling, and reduce the design iterations.

(3) Layering design

Reduce the complexity of the information flow network, weaken the coupling degree between layers in the system architecture to ensure that the changes in a single layer won't affect other layers, and enhance the flexibility in the expansion of information flow network.

(4) Ensuring the product maturity

Minimize the risk and probability of the occurrence of unexpected equipment problems in orbit by using standard or finalized or proven products.

4. Information Management Technology Element—Time Sequence

For timing sequence, the following measures are taken to ensure that the information transfer relationship between nodes in the information flow network is matched and the processing logic is correct.

(1) Control the information timeliness. The time length of information generation, transmission, and processing shall be made clear and controlled.
(2) Specify the sequence of response events. The sequence and interval of response events at each information flow node shall meet the requirements of the design logic.
(3) Comprehensively consider the time start, time synchronization accuracy, time calibration method, time information transmission delay, and other factors for each piece of information and each event.

### *8.3.3 Information System Architecture*

Generally, the architecture of a navigation satellite information system is with distributed network topology and centralized function control. A hierarchical processing structure including system level, subsystem level, equipment level, and component level is established to reduce the complexity of information flow relationship and effectively improve the function, performance, and reliability of spaceborne information systems.

In the distributed information flow network, the primary information processing node is generally the system-level computer that completes the centralized management and control of satellite functions. The slave nodes are generally the intelligent processing units with a single function. They provide local information support for the master node to achieve centralized control.

To ensure the robustness of information system architecture and reduce the risk of on-orbit system-level failure, the slave nodes can also support the upgrading and expansion functions through the targeted design. In the case of the non-recoverable failure of the master node, it can be replaced by a slave node. In this case, the distinction between primary and slave nodes lies only in system management convention.

The backbone structure of the information flow network in the satellite information system is usually of bus type. The data bus serves as the information exchange center to transfer various onboard information flows.

The data bus can be in a single level or multilevels. The single-level mode is a mode with a single master node and multiple slave nodes, as shown in Fig. 8.2.

The multilevel mode is a mode with multiple master nodes and multiple slave nodes, as shown in Fig. 8.3.

1. System Architecture Topology

The topology of the information system can be abstractly modeled by a graph, characterized by network characteristic parameters, and evaluated by index.

The topological graph is formed by two elements, namely vertex and edge. For the information system topology, the vertex corresponds to the information node and the edge corresponds to the connection between nodes.

A graph $G$ is defined as a set pair $(V, E)$, denoted as $G = (V, E)$, where $V$ is a set whose elements are called vertices and $E$ is a subset of different elements of $V$, whose elements are called edges.

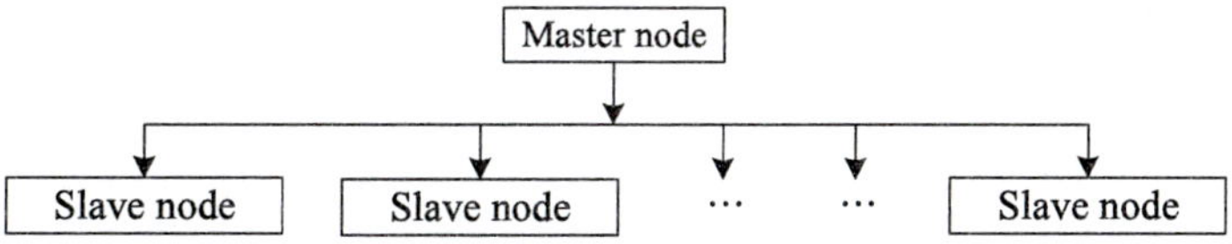


**Fig. 8.2** Single-level mode of data bus

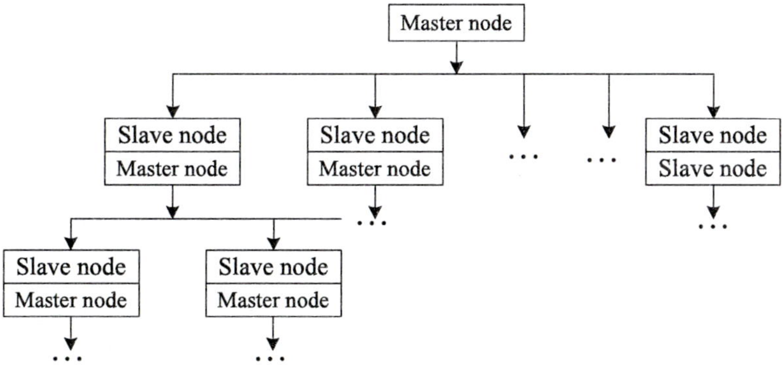


**Fig. 8.3** Multilayer mode of data bus

The graph composed of point set $V(G)$ and edge set $E(G)$ constitutes the information flow network. Each edge ei in the edge set $E(G)$ corresponds to a pair of points $(u, v)$ in the $V(G)$.

Information flow network is usually a directed network, that is, any edge $(u, v)$ in the network isn't identical with $(v, u)$.

Network features are divided into static features and dynamic features.

Static features include degree distribution, average distance, clustering coefficient, and so on. Dynamic features include survivability, reparability, congestion, and other features.

2. Protocol System of Information Flow Network

The protocol system of the information flow network is generally designed in the principle of layering. The protocols are described and implemented at different abstraction layers to reduce the degree of coupling between layers. Mature or proven protocols shall be used as much as possible.

The protocol of the application layer shall be designed according to the specific task. The protocols of transmission layer, network layer, data link layer, and physical layer shall be designed or implemented in accordance with the relevant standards. For example, the protocol for each layer of 1553B bus shall meet the technical requirements, the requirement for information flow in the bus, and the adopted electrical and information formats, as specified in the standard. For the core service data types with high information transmission reliability (such as telecommand data and operation control injection data), a reliable transmission mechanism shall be designed to provide reliable data transmission services that are defined in a standardized way.

An integrated mechanism of intra-satellite, inter-satellite and satellite-ground information transmission shall be established to support the changing and upgrading of protocols and the flexible information transmission in upper-level applications. The service and interface of each layer are standardized to shield the influences brought by the changes of underlying hardware interface and protocol. To this end, the change of one layer will not affect other layers.

Through the integrated design of the satellite-ground and the inter-satellite information flow protocol systems, the protocol format for each satellite-ground and inter-satellite transmission layer is designed according to the convention. By comprehensively considering the information transmission capacity, compatibility, and interoperability as well as technology evolution perspectiveness, an appropriate framework system can be developed and the execution hierarchy of the integrated design can be reasonably determined to support seamless information transmission between the TT&C and operation control systems in the ground segment and the constellation in the space segment, and provide a unified operation control interface for ground users. As a result, the information format conversion by ground users is simplified, and the ground operation and control become easier and more efficient.

A protocol system of information flow network based on the Space Link Protocol of CCSDS (Consultative Committee for Space Data Systems) is shown in Fig. 8.4 [5, 6].

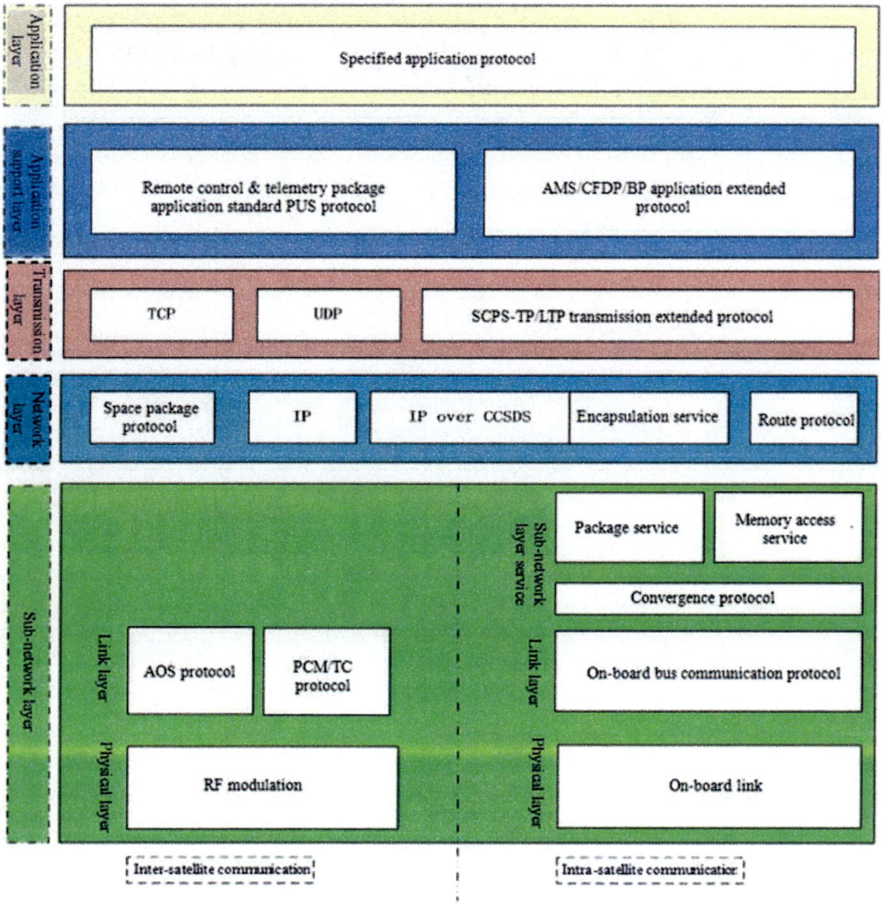


**Fig. 8.4** Protocol system of the information flow network

In the figure, AMS is Asynchronous Message Service, CFDP is CCSDS File Delivery Protocol, BP is Bundle Protocol, TCP is Transmission Control Protocol, UDP is User Datagram Protocol, LTP is Licklider Transmission Protocol, and IP is Internet Protocol.

3. Satellite Information System Architecture Based on 1553B Bus

A spaceborne information system architecture based on 1553B bus is composed of information flow nodes such as various communication link devices and information processing devices, as shown in Fig. 8.5.

In this system architecture, the information flow design involves the key devices in each spaceborne subsystem. Among them, the link communication devices include inter-satellite link communication devices and satellite-ground TT&C communication devices as well as navigation service communication devices. In addition to

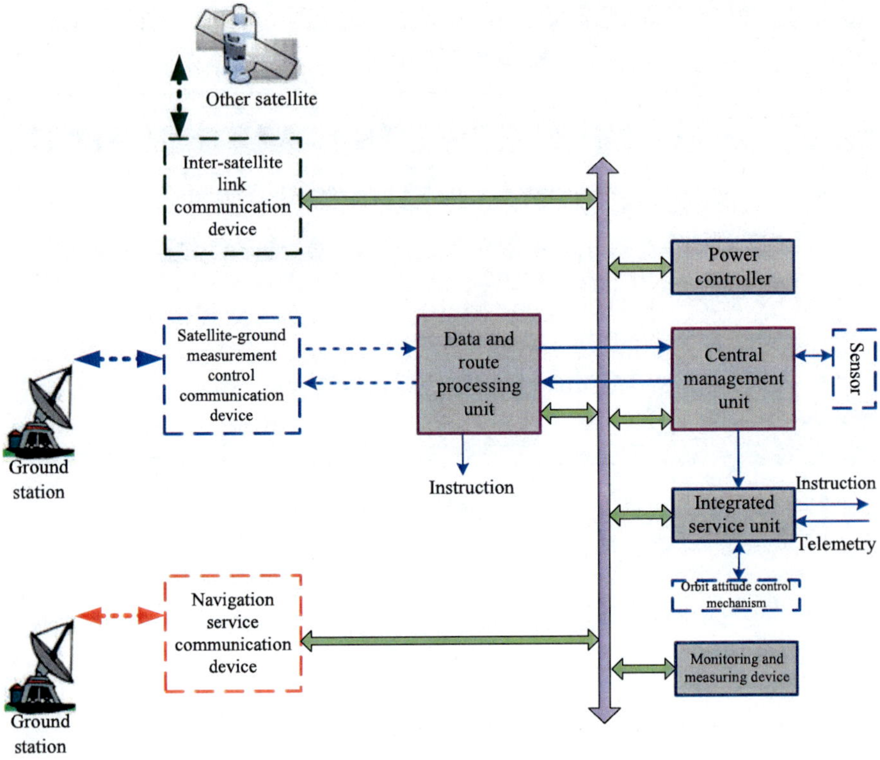


**Fig. 8.5** Spaceborne information system architecture based on 1553B bus

power controller, monitoring and measuring devices, and other platform support facilities, the information processing devices include central management unit, data processing and routing units, and integrated service unit. All these equipment jointly constitute the spaceborne information system.

With the management unit of the spaceborne integrated electronic center as the core and the 1553B bus as the trunk of the data bus network supplemented by various point-to-point data transmission interfaces, the main information nodes and related components are linked together into whole spaceborne information system architecture to complete the integrated information management and satellite operation scheduling.

In extreme cases, if both primary and backup computer modules in the central management unit are in trouble, the central management unit can be used as the Remote Terminal (RT) of bus 1553B after being equipped with the built-in fault detection module. In addition, the data processing and routing units are started as the bus controller (BC) of bus 1553B, replacing the central management unit to complete the centralized management and control of various satellite functions.

In this spaceborne information system architecture, the information flow design mainly includes control information flow design and data flow design. The control information flow includes TT&C uplink flow, operation control uplink flow, and self-control flow. The data flow includes TT&C downlink flow and operation control downlink flow.

(a) Control information flow design

(1) TT&C uplink flow. The TT&C uplink flow is mainly to transmit the instructions of this satellite and forward the instructions of other satellites. The satellite applies hierarchical management to the telecontrol upload processing. For the subsystem equipment connected to the central management unit, data processing and routing units, and integrated service unit, the instructions are sent directly by data processing and routing units, or from the central management unit through the bus to the integrated service unit, and then to subsystem equipment. For the intelligent terminal connected to the data bus, the instruction data is sent to the terminal for decoding and execution. For the telecommand data to be sent to other satellites, the data processing and routing units will, according to the result of routing selection, send the telecommand data to the inter-satellite link communication device for channel transmission.

The TT&C upload instruction data is under multilevel authentication. The information injected into the data frame is authenticated and checked for validity and integrity based on hardware and software. In addition, each integrated service unit checks the validity of TT&C upload instruction data to be distributed by itself.

The onboard software can store continuously received TT&C upload instructions through the buffer storage, and execute them in sequence either in real time or delayed time.

(2) Operation control uplink flow. In the mode of ground operation control and navigation, the navigation payload establishes the satellite time system, receives the uplink data for ranging, and generates and transmits the navigation signals based on the ground operation control information.

The navigation communication device receives the signals uploaded by the ground station, recovers the injected information, and carries out precise ranging based on the satellite's local time, and then sends the injected information and ranging value via the bus to the information processing and routing devices.

For the operation control data to be sent to other satellites, the navigation communication device sends the telecommand data to the data and routing processing unit, which, according to the result of routing selection, sends the operation control information to inter-satellite link communication device for channel transmission.

(3) Self-control flow. The satellite has the functions of time management, important data management, bus management, attitude and orbit control management, autonomous energy management, and autonomous thermal control management, which can improve the on-orbit survivability and reliability of the satellite, as well as its autonomous management ability.

**Time control information flow** is used for time synchronization and maintenance. Its purpose is to achieve high synchronicity between ground time reference and spaceborne atomic frequency reference in the whole navigation system, and finally to establish and maintain a high-precision time system for satellite navigation.

The onboard time systems mainly include satellite platform time system, inter-satellite link time system, and navigation service payload time system. They can provide the time information necessary for the functioning of spaceborne subsystems.

After the navigation payload is powered ON and stabilized, the satellite introduces the reference frequency and time of high-stability clock signals generated by the time-frequency system of navigation payload to provide a unified reference for all the onboard time systems. With the high-stability clock signal emitted by payload as the reference, the satellite platform equipment and payload equipment achieve and autonomously maintain the time synchronization till the unified time, namely navigation service time. The ground operation control system can set and adjust the navigation service time by means of initial synchronization, time phase adjustment, and reference frequency modulation and phase modulation. In addition, the ground TT&C system can set and adjust the satellite platform time by means of timing and calibration.

Information flow for the control of important data: For the important data that influences the operation stability and security of the whole satellite, it shall be stored at multiple onboard information flow nodes by means of distributed redundant storage based on the bus network, where the data can be protected and updated.

In case of abnormality, important data shall be requested for through the network and recovered from storage in time to ensure the continuity of the software operation process and maintain the running state of the system. By the selection of important data, the software operation continuity, bus communication bandwidth, storage capacity, overall demand, and other factors shall be considered.

**Information flow for bus control**: through autonomous bus management, the data communication of 1553B bus can be controlled, and the data interaction between the core computer and each lower computer can be completed. By using dual redundancy, autonomous switching between working bus and redundant bus as well as automatic management of bus fault can be achieved.

When using the bus for communication, the 1553B bus controller checks the register status and the working mode of the bus, for example, whether the transmission is timed out, whether the handshake fails, whether the message format is correct, and whether the BC state is set. If the settings and inspection results change or become abnormal, the interface chip of the bus shall be reinitialized. The BC is responsible for monitoring the 1553B bus in dual redundancy. When the communication of a terminal node is found abnormal, it will be automatically switched to another bus. Moreover, the switching between the main and backup interfaces of a stand-alone device shall be considered to improve the reliability of data transmission.

**Information flow for attitude and orbit control**: through the information interaction between the central management unit or integrated service unit and measurement sensor or actuator, the satellite controls the actuator operation, overcomes the influence of the satellite itself and of the disturbance torque in space, and completes the attitude control and orbit control and management at various mission stages from satellite-rocket separation to on-orbit operation and finally to the end of satellite life. The design of the satellite system has taken full consideration of the capabilities of component backup and recombination. Therefore, in case of control system failure, the faulty component can be autonomously switched to the backup. In the case of single-component failure, the subsystem can maintain its normal attitude through the component backup or cross recombination without affecting the normal completion of the required tasks.

**Information flow for energy management and control**: the management unit of the onboard integrated electronic center interacts with the power controller and equalization manager to receive the telemetry information. By using the telemetry information and the telecommands of the power subsystem, it completes the management of the onboard energy operation mode and battery pack. In the longtime light exposure, the equalization management and half-charge state layup of the battery pack can be completed. By using the shunt regulation circuit, the switch shunt of the solar cell circuit is realized to stabilize the busbar voltage within the specified range. By using the charge regulation circuit, the onboard battery pack is charged in the way of trickle charge. The charging/discharging management of the battery pack is achieved in the Earth shadow season. By using the discharge regulation circuit, the discharge regulation of the battery pack is realized to stabilize the busbar voltage within the specified range.

**Information flow for thermal control**: the central management unit and the integrated service unit jointly complete the automatic temperature control of onboard instruments and equipment as well as the automatic diagnosis and treatment of faults in the thermal control circuit. Under normal circumstances, they can complete the following functions: automatic temperature control of the general heating circuit, automatic temperature control of the atomic-clock heating circuit, autonomous switching between primary and backup heating circuits, autonomous switching between battery temperature thresholds, and heating circuit management in multiple temperature control modes. Under abnormal or faulty conditions, they can diagnose and handle the faults in general heating circuit, battery heating circuit, atomic-clock heating circuit, and the heating circuit in multiple temperature control modes.

(b) Data flow design

(1) TT&C downlink flow

The TT&C downlink flow mainly transmits the telemetry information generated by this satellite and that transferred by other satellites to this satellite. The satellite dynamically schedules the transmission of telemetry data through a virtual channel and multiplexing according to different operation states or telecommands.

For current satellite time and attitude and other important telemetry information that needs to be downloaded quickly, the specific scheme of rapid transmission shall be designed to ensure the required downlink updating frequency and time delay. The time-delay telemetry data is sampled, stored, and compressed. The storage time covers the continuous invisibility periods of the ground TT&C station.

The telemetry parameter information, including the telemetry data generated by this satellite and that transferred by other satellites to this satellite, is organized by the management unit of the onboard integrated electronic center, and then collected and transmitted by the data processing and routing units through the TT&C downlink channel.

The acquisition of telemetry parameters is under hierarchical management. For the intelligent terminal connected to the bus, the telemetry data is acquired through the bus. For scattered data sources, different types of telemetry data scattered in different areas are collected by the integrated service unit.

The telemetry parameters are generally measured through continuous multiple acquisitions and filtering to reduce the influence of noise on parameter accuracy.

(2) Operation control downlink flow

The navigation service payload receives the broadcast ephemeris and clock difference uploaded by the ground segment, and then saves and arranges them into navigation messages and signals to be broadcasted to the user segment through downlinks.

In addition, the on-orbit reconfigurability of onboard software is an important function of the whole satellite. To deal with possible on-orbit satellite failures, restore the software defects or meet new requirements, and give full play to the role of software, the online modification programs or data channels provided by onboard information flow network shall be utilized to ensure that key communication devices and information processing devices have fully software functions and the capabilities of modularization and on-orbit data injection.

The on-orbit reconfiguration of onboard software is generally initiated by the ground segment according to the on-orbit working condition and system requirements of the satellite. The injection program is activated by the ground command. During the on-orbit software reconfiguration, the satellite shall be able to work normally and download the telemetry data related to on-orbit reconfiguration to the ground for observation of the reconfiguration process state. The satellite shall be able to verify the correctness of injected data through the check code, and support the memory download, so as to facilitate the comparison and supplement of injection programs on the ground.

4. Spaceborne Information System Architecture Based on Ethernet Switch

A spaceborne information system architecture based on an Ethernet switch is shown in Fig. 8.6.

With Ethernet as the intra-satellite data bus, the information system is centered on the switch. All the spaceborne devices are connected to the corresponding switch ports. If the number of terminals accessing the network is changed, the switch ports can be expanded or increased.

In the distributed system architecture based on a switched network, information is exchanged, distributed, and shared among multiple node equipment with computing power through the network. This architecture is with an open end, flexible configuration, and strong adaptability. Each computing node is equal in status. The tasks of the

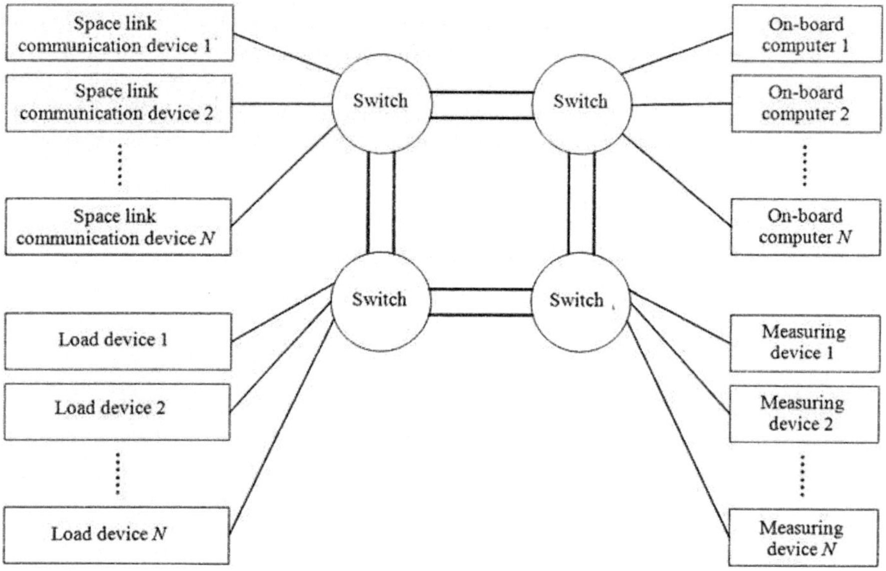


**Fig. 8.6** Spaceborne Information System Architecture based on Ethernet switch

information system can be distributed to all the modular nodes. So the system-level dynamic management and reconfiguration of resources can be achieved more easily, and the task reliability and maintainability are higher.

The architecture based on Ethernet switch can effectively improve the function, performance, and reliability of the spaceborne information system. This is mainly reflected in the following aspects:

(1) **Collaborative task management among multiple computing units.** Through the task management mechanism, the system's application tasks are distributed to different computing nodes. These tasks can be independently developed, and then integrated into the same distributed platform through system integration. The application tasks running at different computing nodes access the shared resources in the information system through the onboard network, thus providing a basis for further improving the onboard information processing capacity and autonomous management capacity.
(2) **Dynamic migration and reconfiguration of tasks**. Through the task status management mechanism, the status of application tasks at a computing node can be monitored, and the interaction between nodes can help determine the running status of global application tasks. After identifying the failure in a task, the task configuration service and external-interface redirection service are invoked through the task changing policy to migrate the task, so as to ensure the effective isolation of the faulty computing node.
(3) **System health management**. The minimum redundancy granularity of the modular information system can reach the module level. By monitoring the

self-check status and health information of each functional module, the health management mechanism can evaluate the health status of each module and assign the tasks according to priority. As the key stand-alone unit in the information system, the switch works in the mode of hot backup.

## 8.4 Strategy and Evaluation Technique of Satellite Information Management

### *8.4.1 Information Flow*

To ensure the matching of information transfer relations and the correctness of processing logic among various nodes in the information flow network, the information flow timing is generally designed in accordance with the following principles:

(1) The sequence of response events of information flow at various nodes shall meet the requirement of the design logic.
(2) The timing relationship shall adapt to the change of communication interface performance in the whole life cycle of a satellite.
(3) In the information flow relationship, the information flow shall be subject to the matching test and verification when the interface at any node is changed.
(4) The timing relationship shall be simplified provided that the functional and performance requirements are met.
(5) For the core service data types that require high information transmission reliability (such as telecommand data and operation control injection data), a reliable integrated mechanism of inter-satellite and satellite-ground transmissions that is defined in a standardized way shall be designed and provided.

In particular, the information flows for satellite time shall be specifically designed. For example, the timing relationships between information flows, such as the time synchronization between satellite and ground or between spaceborne equipment, shall comprehensively consider such factors as time start, time synchronization accuracy, time calibration mode, and time information transmission delay.

The property of time delay is defined as follows. The total delay $T$ is calculated by the following equation:

$$T = t_s + t_t + t_p + t_q \tag{8.3}$$

where *ts* (sending delay) is the time required by the device to send a data frame, that is, the time from the sending of the first bit of the data frame to the completion of the transmission of the last bit. *tt* (transmission or propagation delay) is the time for a signal to travel a certain distance in a channel. The signal propagation distance inside the satellite is short, so the propagation time delay is negligible. Where a long distance is required for the signal propagation between satellites or between satellite and ground, tt shall be reckoned. *tp* (processing delay) is the time for a device to process data after receiving it. *tq* (queuing delay) is the time when the input or output data is queuing in the device for processing.

## 8.4.2 Information Usability Management

Information usability is the ease of use by the ground system or user at various stages of satellite navigation, such as TT&C/operation control command preparation, uplink injection, satellite status monitoring and control, routine maintenance, and task planning.

Information usability management is an important aspect of information management technology in the satellite navigation, generally including autonomy management, convenience management, exception handling, and so on.

1. Autonomous Management

Satellite autonomy is the most direct manifestation of information usability. The autonomous management of navigation satellite is oriented to satellite platform information and navigation service information. The autonomous management of satellite platform aims to improve the survivability of satellites on orbit and ensure their normal operation. The autonomous operation of navigation service aims to free the satellite from ground support and provide the service capacity within a certain range of positioning accuracy.

(1) Design of autonomous management of the satellite platform

Based on the information processing platform, the onboard energy, heat, and time as well as the satellite attitude and orbit can be autonomously managed and controlled to ensure normal smooth satellite operation and reduce the long-term management efforts of ground TT&C system. With the loss of ground uplink signal, the satellite has a certain autonomous management capacity, and each subsystem of the satellite platform can work normally within a certain time frame.

(2) Design of autonomous operation of navigation service

In the absence of longtime support from ground stations, the satellite continuously corrects the long-term forecast ephemeris and clock parameters uploaded by ground stations through inter-satellite two-way ranging, data exchange, and spaceborne processing, autonomously generates the navigation messages, and maintains basic constellation configuration. The satellite, without the ground support, can satisfy the ranging accuracy requirement of an NSS user within a certain time frame.

2. Convenience Management

Convenience is the centralized reflection of the difficulty of implementation by the ground system or user. Convenience management is mainly oriented to the convenience of satellite TT&C, operation control, and other control processes, with the aim of improving the convenience and efficiency of satellite control.

(1) Constellation-wide manipulation and management

Based on inter-satellite communication links, the whole constellation can be manipulated and managed. The indirect control and monitoring of overseas invisible satellites, the generation of ground commands, the analysis of telemetry data, and other operations can be carried out in a way same as the operations of domestic satellites with the help of domestic visible satellites. As a result, the manipulation of overseas satellites and the user experience can be improved.

(2) Telecommand usability

The upload rate of telecommand information is improved to ensure the upload efficiency of a large amount of ground control data. The command-carrying capacity is improved to relieve the pressure and reduce the difficulty and workload of ground operation at the time-critical ground control phases, such as the powered phase before the satellite injection. When the telecommands and operation control data are uploaded to overseas satellites through inter-satellite links, the services of confirmation and automatic repeat are provided to effectively ensure the reliable management of overseas satellites by ground TT&C and operation control systems.

(3) Telemetry parameter usability

The download rate of telemetry parameters is improved to provide a solid foundation for the download of constellation-wide satellite telemetry data at a high bit rate through domestic visible satellites. The design of telemetry parameters shall take comprehensiveness, real time, continuity, autonomy, and flexibility into consideration, ensure the updating frequency of key state parameters and the intuitiveness and readability of telemetry parameters, help monitor the state of overseas invisible satellites, improve the monitoring efficiency of ground TT&C station, provide strong support for ground troubleshooting, and comprehensively strengthen the usability of telemetry data.

3. Exception Handling Management

The exception handling management of a satellite is oriented to autonomous fault management and on-orbit satellite reconfiguration. The former aims to improve the satellite's autonomous fault handling capability on orbit, while the latter aims to provide key equipment functions with reconfiguration flexibility.

(1) Autonomous fault management of satellite

With the function of autonomous fault detection, isolation, and recovery, the satellite can classify the fault information by severity, and acquire the health information based on the onboard information network. It can also autonomously manage and recover foreseeable faults in attitude and orbit control, thermal control, energy, information transmission channel and key stand-alone information processing unit, and complete autonomous operations such as automatic power-OFF, reset, and mode conversion. Autonomous fault detection, isolation, and recovery can reduce the difficulty and workload of satellite troubleshooting by the ground, increase the troubleshooting speed, and improve the system's robustness.

(2) Usability of on-orbit reconfiguration

The on-orbit reconfiguration can deal with possible on-orbit satellite failures, restore the software defects, or meet new requirements. By using the online modification programs or data channels provided by the onboard information flow network, the faults unable to recover autonomously can be corrected online through full-function modular software and the correctness can be verified.

### *8.4.3 Integrated Information Management*

The reliability and security of the satellite system can be improved through the sharing of information in a spaceborne information flow network, namely the application of integrated information management, the integrated function, and performance. For example, information is transmitted interactively in multiple information flow channels, and key performance data is shared among spaceborne equipment.

To ensure the reliable and effective transmission of navigation information, the inter-satellite and satellite-ground information transmission shall be reasonably designed in accordance with the configuration of multiple types of information channels in the satellite system to support the information interaction among satellite-ground TT&C and operation control channels and inter-satellite link channels and to realize mutual information backup and transmission. Based on the integrated inter-satellite and satellite-ground information transmission systems, the control channels for ground users are expanded and the control information interfaces are unified to provide a strong guarantee for reliable control by ground users.

A design of multi-channel interactive transmission in the NSS is shown in Fig. 8.7.

To reduce the coupling between systems and the complexity of each system, the strategies of design elements of multi-channel interactive transmission generally include the following:

(1) Each satellite-ground information transmission channel shall first meet its own general information requirements, and try to maintain the original system and data format. Generally, a protocol system integrating the data link layer and network layer shall be established.
(2) The hardware interfaces of information processing devices in various channels shall be connected, and their resource configuration shall meet the requirement of mutual backup and transmission.

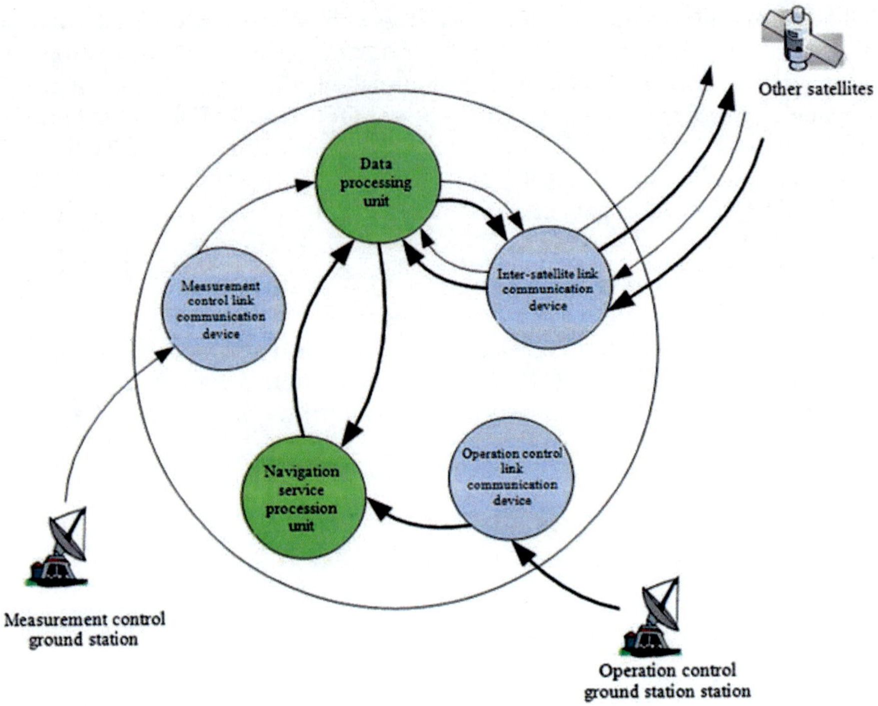


**Fig. 8.7** Example of multi-channel interactive transmission

(3) The software design of information processing devices in various channels shall support the flow of information among multiple channels, and comprehensively consider the influence on general functions and resource support.

## 8.4.4 *Functions and Performance of Information Flow*

After the evaluation of functions and performance, the design results of satellite information flow are compared with design requirements and iterated to finally optimize information flow design. The evaluation of functions and performance of information flow mainly include the determination of information flow evaluation indicators, the simulation analysis of information flow, and the testing and demonstration of information flow.

1. Evaluation Indicators of Information Flow

The evaluation indicators of information flow mainly include completeness, accuracy, and real time.

(1) Completeness refers to the difference in type, quantity, and parameter range between the objects reflected in the information flow and the real objects.

Here, the objects generally refer to various information flow nodes; the parameters include the characteristics of various information types; and the parameter ranges refer to the number and value intervals of parameters.

(2) Accuracy refers to the difference between the parameter value of an object in the information flow and its actual value, which can be represented by deviation.
(3) Real time refers to the length of time for information flow generation, transmission, and processing.

From the perspective of the information flow network, the technical indicators for network administrators and users to evaluate the performance of information flow network include throughput, capacity, utilization rate, response time, delay, delay variation, etc.

2. Simulation Analysis of Information Flow

The evaluation indicators of information flow are calculated through analysis, simulation, testing (including experiments), and other methods, with which the information flow design is verified. Through the establishment of mathematic and semi-physical models, the information flow scheme is simulated, tested, and analyzed by using software and sample data, so as to provide a basis for information flow design and improvement.

In the simulation of information flow, the functions of all kinds of information flow channels and the actual capacity of the information flow network are generally analyzed to obtain the simulation evaluation results of functional and performance indicators of the information system. It can also cover the untestable items in the information flow design or the items with high verification cost, and examine the situation of multipath information and multi-event concurrency.

The main contents of information flow simulation are as follows:

(1) Whether the information channels outside the satellite are smooth.
(2) Whether the information transmission channels among backbones, branches, and nodes are smooth.
(3) Whether the information flow is still smooth after the failure of a key node.
(4) Whether the information flow relationships in the operation of the information flow network meet the design expectation.
(5) Whether the information flow relationships with high complexity or frequency involving multiple subsystems are smooth.
(6) The transmission capacity and utilization rate of each information flow channel.
(7) Whether the information processing capacity of each node meets the requirement.
(8) Whether multiple information flows are competing for transmission and processing.
(9) Whether the information flow concurrency will cause network congestion.

3. Testing and Demonstration of Information Flow

The main purpose of information flow testing and demonstration is to check the information transmission process, and detect and solve the problems found in its operation and testing, so as to achieve the satellite function correctness, the coordination between systems, and the acceptance of performance parameters, and to ensure that the operation of the information flow network meets the requirements of satellite design and application.

The testing and demonstration of information flow are usually carried out level by level from equipment level to subsystem level and system level, and are included in various tests at these levels. The main purposes of information flow testing include

(1) To check the correctness, rationality, and matching of information transmission design.
(2) To check the correctness and matching of the information interfaces between nodes.
(3) To verify the correctness and matching of internal and external information interfaces.

## 8.5 Satellite Message Management Technology

### *8.5.1 Classification of Satellite Message Information*

The navigation messages mainly include ① the satellite time, ephemeris, clock difference, and basic integrity information necessary for positioning and navigation services; ② the information supporting the positioning and navigation, such as ionospheric model parameters, network-wide almanac, the parameters of time synchronization with other systems, and EOP parameters; ③ and the differential enhancement bias for further improving the positioning and navigation accuracy.

Different types of navigation messages have different functions and real-time requirements for satellite processing. Among them, the real-time requirement for differential enhancement information is the highest, followed by the basic integrity information, ephemeris, and clock difference information. The real-time requirement for supplementary navigation information, such as network-wide almanac, the parameters of time synchronization with other systems, and EOP parameters, is the lowest.

The navigation messages are autonomously generated by a satellite, or injected by the ground operation control system. Most of the information is injected by a ground operation control system. In normal operation and control mode, little information is autonomously generated by satellite, mainly including satellite time information, satellite health information, and the encrypted identification of navigation messages.

To upload and broadcast the navigation messages, the mapping relationship shall be established. The processing process of satellite navigation messages is to unpack the uplink messages according to the agreed mapping relationship and timing sequence, and then repack, frame, and broadcast the downlink messages according to their broadcast format and timing sequence.

In addition to navigation message information (including ephemeris, almanac, satellite clock, etc.), the uplink information also includes the information on satellite control and management (including control parameters, control instructions, etc.). After analysis, the satellite control information is used to control onboard equipment or condition as needed. The corresponding telemetry data is sent to the ground after the control becomes effective. During processing, the navigation messages are stored in order by group number or time to facilitate the downstream framing.

### *8.5.2 Message Reception and Processing*

The uploaded navigation messages are usually defined by information type. Different navigation messages are defined as different information types. For example, the uplink messages such as ephemeris, satellite clock, basic integrity, ionospheric model correction parameters, almanac, EOP parameters, and the time synchronization with other systems are defined as different information types in terms of the format. When being received by satellite from the ground operation control system, the uplink information is firstly checked for CRC by the hardware in the link layer. The wrong information found will be discarded, while the correct information found will be processed by the information processing unit.

The onboard information processing unit deals with the uplink navigation messages mainly in three ways: check and verification, decryption, and parsing.

(1) The check and verification of an uplink message is to compare the calculated check value of uplink data with the preset check value in the message. If they are the same, the uplink data is judged to be correct; otherwise, it is judged to be incorrect.
(2) Decryption is to authenticate and decrypt the uplink data and obtain the original message.
(3) Uplink message parsing is to parse the information type after the verification and decryption while judging the correctness of uplink information according to the length rationality, the rationality of internal key parameters, and other criteria. Then the messages are processed and stored for the convenience of use according to the information types, applicable time frame, and different broadcast branches.

### *8.5.3 Message Format*

Downlink information formatting is mainly to organize and arrange various downlink messages through framing and encryption according to the satellite-user interface documents formulated by the satellite navigation system and the protocol between the ground operation control system and satellite system.

Message framing is to determine the framing format and frame the downlink messages in each channel by using the stored uplink messages according to the current time reference of the satellite navigation system (such as the GPST of GPS system and the BDT of BDS system). A message frame usually consists of a frame header and an information bit. The frame head mainly contains synchronous head, cycle count, second count, subframe number, and other information.

It is worth noting that the framing timeslot of messages in each channel should be reasonably arranged to meet the requirement of real-time information broadcasting. With high real-time performance, the fast-changing differential information shall be framed in the time frame closest to the point of signal broadcasting.

Message encryption is to generate the onboard secret key information, which is combined with the original downlink message through aspecific operation to complete the message encryption.

The arrangement of "B-CNAV-1" and "B-CNAV-2" navigation messages in the BDS system will be presented as follows.

1. Arrangement of "B-CNAV-1" Message

The "B-CNAV-1" navigation message is broadcasted in the B1C signal, and its data is modulated on the B1C data component. The length of each message frame is 1800 sign bits, the symbol rate is 100 s/s, and the broadcasting period is 18 s. The basic frame structure is defined in Fig. 8.8.

The B-CNAV-1 is a combination of data frames and data blocks so that the navigation messages can be broadcasted with great flexibility. The subframe 2 broadcasts the fixed messages in a fixed period. The subframe 3 broadcasts and pages the variable information. Different pages are broadcasted in a random sequence as needed by the user. The subframe 3 can define up to 63 page types. Currently, four valid page types have been defined, namely types 1, 2, 3, and 4.

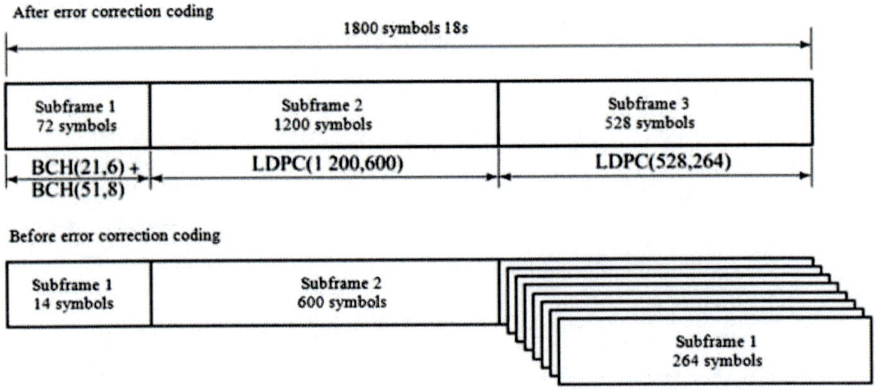


**Fig. 8.8** Basic frame structure of B-CNAV-1

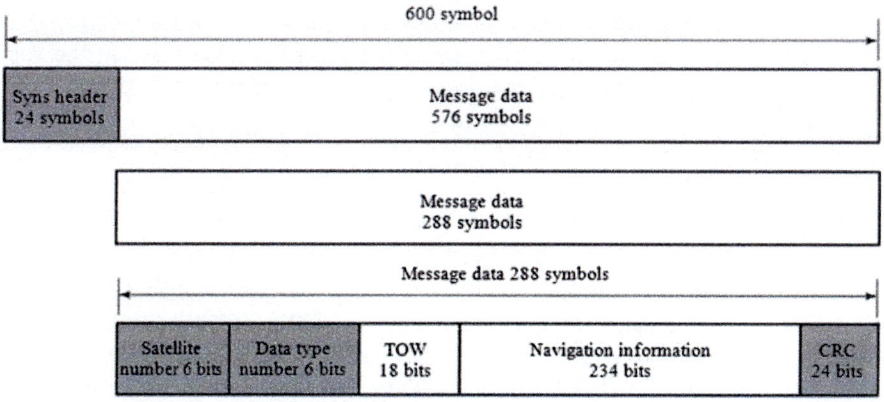


**Fig. 8.9** Basic frame structure of B-CNAV-2

2. Arrangement of "B-CNAV-2" Message

The "B–CNAV-2" navigation message, containing basic navigation information and global basic integrity information, is broadcasted in the B2a signal, and its data is modulated on the B2a data component. The length of each message frame is 600 sign bits, the symbol rate is 200 s/s, and the broadcasting period is 3 s. The basic frame structure is defined in Fig. 8.9.

For the "B-CNAV-2" navigation message, a maximum of 63 page types can be defined. Currently, 8 valid page types have been defined, namely types 10, 11, 303, 1, 32, 33, 34, and 40. The broadcasting sequence of these information types can be dynamically adjusted, but the types 10 and 11 remain sequential. Whenever receiving a "B-CNAV-2" message, the user shall identify its type (MesType).

The formatting of navigation messages can provide the message broadcasting with great flexibility and real-time performance according to the user's need so that the user can receive useful information in the shortest time.

The design characteristics of messages in China's BDS system are

(1) The messages are formatted into the combination of data frames and data blocks so that there will be no strict correspondence between frames and system time and the randomness of broadcast messages can be enhanced. The design of data blocks can enhance the expandability of message contents. If new parameters are needed, new data blocks can be added.
(2) The settings of page types are increased. The parameter format corresponding to each page type is fixed. The message broadcasting based on page type is still quite flexible. The satellite can broadcast the message type that the user applies for, thus enhancing the real-time performance of navigation messages.
(3) The inter-satellite cross-broadcasting strategy can be implemented. The navigation system is a vast network of satellites. The broadcasting of the same navigation messages by all the satellites is not only a waste of resources, but also an obstacle for improving the Time To First Fix (TTFF). Instead, the satellites can

be grouped. Different groups of satellites will broadcast different messages. In this way, this strategy can greatly improve the TTFF and the real-time performance of navigation messages while ensuring that the user receives signals from different groups of satellites.

### *8.5.4 Main Points of Satellite Message Management*

The satellite navigation system has many types of navigation messages, whose injection frequency and real-time requirements are different. The satellite adopts the processing strategies based on the characteristics of messages. For the differential enhancement information with the highest real-time requirement, the timing sequence of ground upload and satellite broadcasting shall be strictly restricted, and the ms-level real-time performance shall be guaranteed by both the satellite and the ground operation control system. For basic navigation parameters such as ephemeris and clock difference, a sufficient storage space shall be reserved by considering both message updating frequency and ground uplink strategy.

In the system design, the priority levels of message processing shall be established according to the real-time requirement of message information. The higher the real-time requirement of the message is, the higher will be the processing priority given. For the messages with a higher real-time requirement, more attention shall be paid to their real-time processing. For the messages with a lower real-time requirement, more attention shall be paid to their integrity. Efforts shall be made to ensure that the unprocessed messages will not be lost. If necessary, a buffer space shall be opened to store these messages before processing.

Compared with navigation messages, the satellite control information needs more attention to the correctness and reliability of its execution. The necessary redundancy-check measures shall be taken for information such as two-out-of-three, which means that a control instruction will be sent three times in succession with the accuracy of at least 2/3 before being executed by the satellite.

The data storage for messages requires very high reliability, so necessary detection and error correction shall be designed to prevent spatial single event upset so as to ensure the data correctness. A large data storage capacity must be reserved for messages. The message storage area shall be designed in a linear, double-buffer or multi-buffer area according to the usage strategy of downlink messages to ensure the convenience and rapidness of message use and avoid the sending-receiving conflicts in the storage area.

Usually, the messages are arranged in one timing sequence before being broadcasted. The timing sequence for the arrangement shall be determined according to the real-time requirements of various types of information. The real-time performance of fast-changing differential information is better than that of slow-changing differential information and other information, so the timing sequence shall be carefully designed to give priority to the arrangement of fast-changing differential messages.

# Chapter 9
# Autonomous Operation Technology of Navigation Satellites

A Navigation Satellite System (NSS) consists of several satellites in space. The TT&C and service management of satellites relying solely on ground TT&C and operation control will bear a heavy load of data transmission and processing, and will be limited by the continuous visibility of satellites to ground stations. Excessive reliance on the TT&C and operation management of ground stations will make the safety and serviceability of NSS in wartime a worry. Therefore, improving the autonomous management function and level of navigation constellation and satellites is of great significance to relieve the operation management pressure of ground stations and enhance the stability and safety of the whole system.

The autonomous operation technology of navigation satellites mainly includes the following three aspects: autonomous navigation technique, autonomous health management technique, and autonomous integrity monitoring technique.

(1) The autonomous navigation technique of navigation satellite is, without the longtime management and support of ground operation control and TT&C system, to ensure continuous, stable and reliable navigation services in the on-orbit satellite operation period through the autonomous operation management of constellation system, by means of inter-satellite measurement and data exchange as well as autonomous data processing technique.

(2) The autonomous health management technique of a navigation satellite is to ensure its healthy trouble-free operation during its autonomous operation, that is, to percept external environment and detect various state parameters of the satellite independent of ground stations, and then take this information as a basis for independent judgment and decision-making, and perform the corresponding operations in order to realize normal task management, fault isolation, and recovery and ensure the safe and stable operation of the satellite on orbit.



J. Xie et al., *Satellite Navigation Systems and Technologies*, Space Science and Technologies, https://doi.org/10.1007/978-981-15-4863-5_9

(3) The autonomous integrity monitoring technique of a navigation satellite is to monitor the autonomous integrity of service information during the autonomous operation of the navigation satellite, to ensure a timely and effective alarm to the user when the navigation signals transmitted by the satellite cannot be used for the expected positioning operation, so as to prevent the user from being misled by abnormal navigation signals.

During the autonomous operation of a navigation satellite, the basic navigation service is guaranteed by autonomous navigation technique, the healthy trouble-free satellite operation by autonomous health management technique, and the reliability, and correctness of navigation signals by autonomous integrity monitoring technique. Therefore, this chapter will mainly address the autonomous navigation technique, autonomous health management technique, and autonomous integrity monitoring technique.

## 9.1 Autonomous Navigation Technique

Navigation Satellite System (NSS) is a positioning, velocity measurement, and timing system with the satellite as a time-space reference source. Therefore, the determination of satellite orbit and clock difference as well as the broadcasting of relevant information is the key to maintain the NSS operation.

At present, the information on orbit and clock difference is generated and transmitted in the following mode. The ground monitoring stations collect the satellite observation data and send it to the main control station, which completes the orbit determination and regularly uplinks the ephemeris parameters to the satellite. The disadvantage of this mode is that the main control station in the ground operation control system bears a heavy workload and a large operation risk. If the ground operation control system does not work normally or is damaged, the infrastructure in the whole navigation system will be unable to function normally and provide normal Positioning, Navigation, and Timing (PNT) services.

### *9.1.1 Introduction of Autonomous Navigation*

Satellite autonomous navigation is the process where, in the absence of longtime support from the ground system, the constellation satellite continuously corrects the longtime forecast ephemeris and clock parameters previously uploaded by ground stations through inter-satellite two-way ranging, data exchange, and spaceborne processor filtering, autonomously generates the navigation messages, and maintains basic constellation configuration, in order to meet the user's need for navigation and positioning. This process is the core to ensure that the navigation satellite can continue to provide navigation services during its autonomous operation period.

In addition, autonomous navigation can achieve the backup of ground operation control and TT&C functions, improve the survivability of the satellite navigation system in a complex environment, and enhance its operational reliability.

1. **Autonomous Navigation of Non-navigational Satellites**

The positioning and timing achieved by the satellites other than navigation satellites (including satellites working in the form of networking, such as the communication and remote sensing constellation satellites) using autonomous navigation techniques are only used to realize their own functions rather than provide the PNT services for other satellites. The autonomous navigation of these satellites is of great significance to improve satellite performance and expand the space application.

Currently, most of China's satellites still base their navigation on land- and ship-based TT&C. The satellite navigation and TT&C techniques based on GNSS are being actively explored. The satellite TT&C system based on land and ship will face severe challenges in the future. In China, there are few ground stations for civil satellites. With the increase of satellites, huge navigation pressure will overwhelm the limited number of ground stations. To achieve high orbit determination accuracy in the future, a large number of widely distributed ground stations must be established at the cost of expensive manpower and material resources. Moreover, the navigation relying on ground stations will greatly increase the operation cost of satellites. In case of military applications, once the ground TT&C system is attacked, the whole space-based system would be paralyzed and would be unable to maintain normal information acquisition and distribution, and the space advantage will be completely lost.

In recent years, with the continuous progress of microsatellite technology, microsatellites play an increasingly important role in the fields of communication, remote sensing, and navigation. For microsatellites, high-precision orbit determination is the premise to ensure their normal working. With the increas of small satellites in low and medium Earth orbits, traditional equipment resources for ground TT&C and orbit determination are becoming more and more unfit for actual needs. It is urgent to realize the autonomous navigation of satellites.

First, autonomous navigation technology can reduce the workload on ground equipment and the cost of ground support needed to keep satellites working. Second, due to the timely determination of satellite position, the utilization rate of satellite observation data can be improved. For military satellites, autonomous navigation can reduce their dependence on ground stations and improve their survivability. Even if ground stations are forced to halt the tracking and measurement for a period of time, the continuity of flight missions can still be maintained. Therefore, the practical significance of this technology is especially obvious.

Especially since the 1980s, with the increasing development of satellite application requirements, more and more flight missions cannot be completed by a single satellite. However, a constellation often consists of many satellites and requires a certain degree of cooperation among satellites. Therefore, the member satellites shall have not only a certain degree of absolute navigation accuracy, but also a certain degree of relative navigation accuracy. The method of single satellite autonomous navigation is not the best choice.

At present, the method most frequently studied is to achieve autonomous constellation navigation through inter-satellite relative measurement. This method is a very promising navigation means in military applications. Its significance is that the constellation can still work when the GNSS satellites cannot be used or when the navigation satellites in this constellation are disturbed or attacked. Inter-satellite relative measurement information can achieve autonomous constellation operation independent of ground stations or navigation satellites to the greatest extent.

1. **Autonomous Navigation of Navigation Constellation**

As a spatial and time datum, the navigation constellation provides the PNT services for many users, including communication satellites and remote sensing satellites. To this end, navigation satellites are of great significance to realize autonomous navigation.

Currently, the navigation ephemeris is updated on the basis of the ground operation control system through several steps, such as the data collection by monitoring stations, the processing by ground operation control system, and the ephemeris uplink. The ephemeris uplink usually requires the satellite visible to the ground uplink station. The autonomous navigation technology provides a way of rapidly updating the navigation ephemeris, each satellite autonomously updates its ephemeris within a much shorter period of 5–15 min. The shortening of the ephemeris updating cycle means a shorter clock-difference prediction period and higher prediction accuracy. Autonomous navigation can reduce the age of ephemeris data and improve the accuracy of navigation ephemeris. Therefore, autonomous navigation technology is an effective way to improve the accuracy of ephemeris broadcasting.

At present, the satellite integrity information is acquired by ground stations, and then sent to the main control station, where the information is finally arranged into navigation ephemeris and periodically uploaded to the satellite. An independent data source and data processing mode of the autonomous navigation system are different from that of the ground operation control system. The generated navigation ephemeris is used to check the accuracy of the ephemeris uploaded by the ground operation control system, and then the satellite integrity information is timely sent out. Therefore, the autonomous navigation technology provides a means to check and verify the broadcasted ephemeris online quickly, thus reducing the data transmission steps and improving the real-time integrity monitoring capability.

Due to the great role of GNSS in military and civilian uses, the GNSS may become the primary target of adversaries in wartime. In the current GNSS operation, the generation and uplink of navigation ephemeris are completed independently by the main control station. The destruction of the main control station will paralyze the satellite navigation system, so the main control station is the weakest link in the navigation system and the most likely target for a preemptive strike. The autonomous navigation provides a mode of autonomous ephemeris updating to guarantee that the satellite navigation system can still provide the navigation and positioning services even if the ground operation control system is destroyed.

In addition, autonomous navigation technology can also cope with major natural disasters. When the ground operation control system is destroyed by a serious natural disaster, the satellite navigation system dependent on it will lose the navigation and positioning ability. As a result, key infrastructures in the transportation, civil aviation, electric power, communication, and other fields cannot function normally, thus affecting the normal operation of the national economy and the implementation of rescue and relief tasks.

### *9.1.2 Principle of Autonomous Navigation*

Since autonomous navigation is of great significance to improve the satellite performance and expand the space application, all the countries have been attaching great importance to the research and on-orbit demonstration of autonomous navigation technology for more than 50 years. The commonly used autonomous navigation techniques include the autonomous navigation technique based on external reference datum and the autonomous navigation technique based on relative observation.

1. **Autonomous Navigation Technique Based on External Reference Datum**

The autonomous navigation technique based on external reference datum is a navigation method that relies on external reference sources (such as celestial bodies) to determine the satellite's position. Here, it generally refers to the autonomous navigation method based on the satellite's measurement of the pointing direction to the reference objects (such as stars, the sun, the moon and the Earth), the measurement of the Earth's geomagnetic field intensity, and other information. Based on the adopted different sensors and measurement information, different autonomous navigation systems or methods have been developed.

Although the astronomical navigation based on a single sensor has relatively poor accuracy, its theoretical and technical research has received wide attention from many countries such as the United States, Russia, Germany, France, Japan, and China due to its great development potential and non-negligible practical value in the military.

(1) Autonomous orbit determination technique based on star sensor measurement

It is impossible to accomplish the task of autonomous orbit determination by using the existing star sensor alone, because the change of spacecraft orbit is not enough to change the pointing direction of the line of sight from spacecraft to star in the inertial space. Therefore, some countries are now committed to developing a new type of star sensor. This sensor can not only observe stars, but also detect and track nonstellar objects (asteroids, comets, planets, man-made satellites, etc.), and determine both the satellite attitude and orbit. The autonomous vision system developed by the Technical

University of Denmark is just a research project based on the above principle and has been tested on the satellite TEAMSAT. Although this new star sensor failed the autonomous orbit determination, it did achieve the goal of nonstellar target detection. This technology is still under continuous development.

Combining with other sensors (for example, magnetometer), a star sensor can achieve autonomous orbit determination. The star sensor can provide an accurate attitude of the spacecraft in inertial space, and effectively estimate and correct the error of the Earth geomagnetic field model.

The autonomous navigation based on starlight refraction/dispersion is one of the most interesting researches at present. This technique determines the satellite orbit by using a star sensor to measure the refraction of starlight passing through the atmosphere at the Earth's edge and indirectly sensing the horizon. When the atmospheric refraction model is relatively accurate, this technique can achieve higher navigation accuracy. From the mid-1970s to the mid-1980s, it was studied by the United States, France, and other countries.

The American research on the autonomous navigation technique based on starlight refraction can be traced back to the 1960s. In the implementation of the Apollo program, the autonomous navigation techniques based on celestial occultation, starlight refraction in the atmosphere, and starlight attenuation when passing through the atmosphere were studied. In 1975, Draper Laboratory, funded by the US Office of Naval Research and the DARPA, studied and demonstrated the starlight refraction technique. In 1979, the Office of Naval Research, Draper Laboratory, and the University of Maryland jointly proposed a sensor design for measuring the starlight refraction, namely bicolor refractometer. The test results showed that this sensor could achieve high measurement accuracy. To study the optical properties of atmospheric refraction, the laboratory Draper analyzed the error of the starlight refraction technique and simulated the whole navigation process on LEO, GPS orbits, and geosynchronous orbit, respectively. The results showed that the navigation accuracy of this technique was up to 50–200 m.

In the early 1980s, France also conducted research on the autonomous navigation technique based on starlight refraction. This research was organized by the CNES with the participation of Matra and SAGEM. Its measurement system consists of a gyro package and two strapdown CCD sensors. The random drift of the gyro is 0.01°/h, and the measuring accuracy of the star sensor is 10″. With low Earth orbit measurement satellite SPOT as the background, this research deeply analyzed and simulated the navigation principle, measurement scheme, natural environment constraints on system observation, error distribution, system performance optimization, and other aspects. The navigation accuracy of this system is 300 m. It is a backup system in the orbital flight segment of the space shuttle Hermes, and has been further tested on the satellite SPOT-3 flying on orbit.

The integrated navigation based on a star sensor has also been studied in China. The researchers proposed to use the starlight angle measured by a star sensor and a horizon sensor as the main observation, and the indirectly measured geocentric distance as supplementary observation (calculated from the geocentric direction vector directly measured by the horizon sensor). When the accuracy of the horizon sensor is 0.002° and the adoption period is 15 s, the mode of "single starlight angle + geocentric direction vector + geocentric distance" is used for navigation with the position error of 0.2 km and the velocity error of 0.32 m/s. Chinese researchers achieved the autonomous sea satellite navigation based on onboard radar altimeter and star sensor, studied the importance of sea level model, and put forward a simple algorithm for preprocessing the altimeter data and an orbit determination algorithm with Kalman filtering. This navigation can achieve the accuracy up to 100 m, but requires an accurate sea level model.

(2) Autonomous orbit determination technique based on the sun-Earth-moon optical measurement

The optical features of the sun, the moon, and Earth are outstanding, and have been widely used in space measurement. In the early 1990s, the US Air Force Phillips Laboratory first proposed the autonomous navigation technique based on sun-Earth-moon measurement and entrusted Microcosm to develop it (known as MANS). MANS uses a dedicated autonomous navigation sensor to observe the directions of the sun, the moon, and Earth's center, that is, to determine the spacecraft's orbit and triaxial attitude according to the sun-Earth-moon azimuth information. This system is the main component of the TAOS autonomous orbital-survival flight test. The STEP platform launched by the Taurus rocket of the US Air Force carries the MANS autonomous navigation system. With two autonomous navigation sensors as basic measurement components, this system completes the data processing and extended Kalman filtering by using an onboard computer from Honeywell. The performance of the MANS system is analyzed by a ground simulator. Under normal operating conditions, the MANS autonomous navigation system is expected to achieve the following accuracy after 60 s of orbital maneuver:

LEO: 500 m ($3\sigma$), 0.08″ ($3\sigma$) (single sensor)
LEO: 100 m ($3\sigma$), 0.03″ ($3\sigma$) (double sensor).

The theory and algorithm of autonomous orbit determination based on the sun-earth-moon optical measurement information had been studied in china. On the solar and lunar visible arcs, the nonlinear least square method or Kalman filtering algorithm is used to solve the position vector of satellite in the inertial coordinate system. On the solar and lunar invisible arcs, the method of satellite orbit prediction is adopted.

(3) Autonomous navigation technique based on X-ray pulsar

In 1967, the first radio pulsar was discovered by Dr. Bell at the University of Cambridge in England. So far, nearly 3,000 pulsars have been discovered and cataloged, including more than 140 pulsars with good periodic X-ray radiation performance.

Basically, the pulsar navigation research can be divided into three stages: the discovery and preliminary research of pulsar (1967–1974), the exploration of pulsar navigation theory (1974–2003), and the formation and experimental verification of X-ray pulsar navigation theory (from 2004 up to now). In 1974, NASA proposed the interstellar navigation concept based on radio pulse signals. In 1981, the US Institute of Communication Systems put forward the spacecraft navigation conception based on pulsar X-ray sources. In 1999, the US Air Force launched the Advanced Research and Global Observation Satellite and conducted nine space science experiments, including using X-ray sources for satellite navigation. This experiment was based on the occultation method, but still belonged to the category of traditional astronomical navigation. The observation data accumulated in the experiment triggered the later rapid development of the pulsar navigation technique.

In early 2004, Dr. Schock from the University of Maryland deeply studied the basic physical characteristics of pulsars from the perspective of a modern satellite navigation system, and proposed the autonomous spacecraft navigation theory based on X-ray pulsar. Subsequently, the US Department of Defense launched the pulsar navigation research program, known as XNAV. This program was implemented in three stages: feasibility demonstration, key technologies development and ground experiment verification, and space flight testing. So far, the NASA Goddard Space Flight Center has built a ground experiment verification system named "Goddard X-ray navigation testbed", where the software and hardware for pulsar navigation are now verified before being tested in the International Space Station (ISS). In addition, the ESA, Russia, Germany, the UK, Japan, India, Australia, and other countries or international organizations have also conducted the research and experimental verification for the pulsar navigation technique.

Since 2005, China has been developing the micro-channel plate X-ray pulsar detector and focusing on X-ray pulsar silicon drift detector. On November 10, 2016, China successfully launched the "Pulsar Test satellite 01", conducting several on-orbit experiments such as Crab pulsar observation. The 100 mas pulsar angular position measurement based on X-ray was achieved for the first time in China to demonstrate the X-ray photon measurement and optical focusing techniques. The acquired pulsar observation data has been published on the BDS website to promote the development of pulsar navigation technology.

Although the concept of pulsar navigation was put forward more than 30 years ago, the autonomous navigation using X-ray pulsar has only been studied since 2004, based on the idea of time ranging in modern satellite navigation. Therefore, the autonomous navigation based on X-ray pulsar is an exploratory and prospective new concept (or method) with an important strategic research value and a broad engineering application prospect. It is expected that in the next 5–10 years, the accuracy of orbit determination using X-ray pulsar will be 100 m, and the accuracy of time synchronization will be 10 ns. Such high accuracy can ensure the long-time high-precision autonomous operation of navigation constellation. Moreover, this navigation technique will be expanded to other application fields.

2. **Autonomous Navigation Technique Based on Relative Observation**

A constellation often consists of many satellites, so the information on the inter-satellite relative motion will be ignored by simply using the method of the single satellite autonomous positioning. It has become a research hotspot to make full use of the inter-satellite relative motion law and study the relative autonomous navigation using inter-satellite relative measurement.

Some countries have done fruitful research on autonomous navigation using inter-satellite relative measurement information, and proposed many navigation methods. Those methods include the determination of relative position and attitude by using a miniature laser scanning rangefinder developed by the Japan Aerospace Exploration Agency; Ananda's research on GPS autonomous navigation; and Menn's research on GPS autonomous navigation based on satellite ranging.

The US GPS Block IIR satellite is a specific application of the autonomous navigation based on inter-satellite relative measurement, with the abilities of inter-satellite ranging and on-orbit navigation data processing. Each satellite regularly measures the distances to other satellites and transmits the measurement results to other satellites in the constellation through inter-satellite communication links. Each satellite uses an onboard processor to calculate the corrected values of navigation parameters, so as to improve its navigation accuracy and enhance its autonomous survivability. In the 1980s and early 1990s, the autonomous navigation of the satellite Block IIR was systematically studied, and a series of overall plans were proposed. In the late 1990s, a series of issues were deeply studied, including the establishment of an uplink simulation system (including data generation, autonomous navigation system algorithm, and system performance evaluation), the processing of clock difference and pole motion, and other issues.

The Block IIR satellites can autonomously operate for 180 days to guarantee the navigation and positioning services. On the 180th day, the User Range Error (URE) can still reach $\pm$ 7.4 m, which is only 1/1300 of the URE of Block IIA satellite (the URE of Block IIA on the 180th day is 10,000 m). Therefore, even if the ground monitoring system is temporarily destroyed, high-precision GPS satellite navigation and positioning can still be maintained to build up the anti-destruction ability of GPS system. This is extremely beneficial in wartime.

As far as the autonomous navigation based on the inter-satellite relative measurement is concerned, many scholars have put forward many distinctive solutions. In the research on the HEO satellite navigation, LeMay proposed the method of autonomous orbit determination based on inter-satellite ranging and angle measurement. According to this method, if the state of one satellite is known, the orbit of the other satellite can be determined by using the relative range and the angle in the inertial coordinate system. Mark proposed a batch filtering method based on the measurement of the inter-satellite relative position vector and also used it for the autonomous determination of two satellite orbits. The system provides a high-precision autonomous orbit determination method independent of either the signals of GPS constellation or ground stations or any prior orbit information. By using a series of relative position vectors in the inertial reference system and the orbital dynamics model of two satellites, the state vectors of six orbital parameters of each satellite as well as their drag coefficient can be estimated. Through the variance analysis, the position accuracy obtained under the right configuration can be up to 1 m.

The Galileo navigation system also considers the navigation technique based on inter-satellite observation. In this system, the on-orbit autonomous navigation is realized through inter-satellite measurement. Due to the limitation of onboard computing ability, a dynamic model that is too complex cannot be adopted. Therefore, the inter-satellite measurement and the satellite-ground measurement shall be integrated to update the ephemeris so as to improve the accuracy of satellite ephemeris and clock. Through the ground simulation test, the URE can reach the decimeter level. Therefore, the relative constellation positioning based on inter-satellite measurement can not only realize the self-survival of the constellation in wartime, but also improve the prediction accuracy of ephemeris and clock difference in peacetime, so as to improve the service performance of the whole navigation system.

### *9.1.3 Design of Autonomous Navigation*

Depending on how the inter-satellite measurement data is processed, the algorithms of spaceborne autonomous orbit determination can be divided into two categories: centralized data processing and distributed data processing. According to the way of onboard orbit prediction, the algorithms of autonomous orbit determination can be classified into the algorithm based on onboard orbit integrator and the algorithm based on uplink longtime ephemeris prediction.

The onboard carrier of autonomous navigation function is autonomous navigation software, whose core is autonomous orbit determination and time synchronization. Here, a solution for distributed autonomous orbit determination and time synchronization based on the uplink longtime forecast ephemeris parameters will be presented. It is mainly characterized by the storage of uplink longtime (such as 90-day) ephemeris and clock predictions at the onboard data processing unit. The combination of these forecast parameters and state transition matrix can achieve the prediction of onboard orbital and clock parameters. Significantly different from the solution based on the onboard integrator, the onboard data processing unit in this solution does not need the built-in numerical orbit integrator. This design can minimize the consumption of expensive computing resources in the onboard processor.

The distributed autonomous orbit determination and time synchronization system based on the uplink longtime forecast parameters, as shown in Fig. 9.1, includes the self-satellite system, other satellite systems, and the ground operation control system.

The inter-satellite measurement and data transmission relationships exist between the self-satellite system and other satellite systems. Inter-satellite measurement is the pseudorange measurement based on the transmitting and receiving time of the ranging signal. The measurement information includes ionospheric error, phase center error, multipath effect at the signal reception, signal transmitter delay, signal receiver delay, relativistic delay, signal transmitter clock error, signal receiver clock error, and so on. The data transmission information related to autonomous orbit determination and

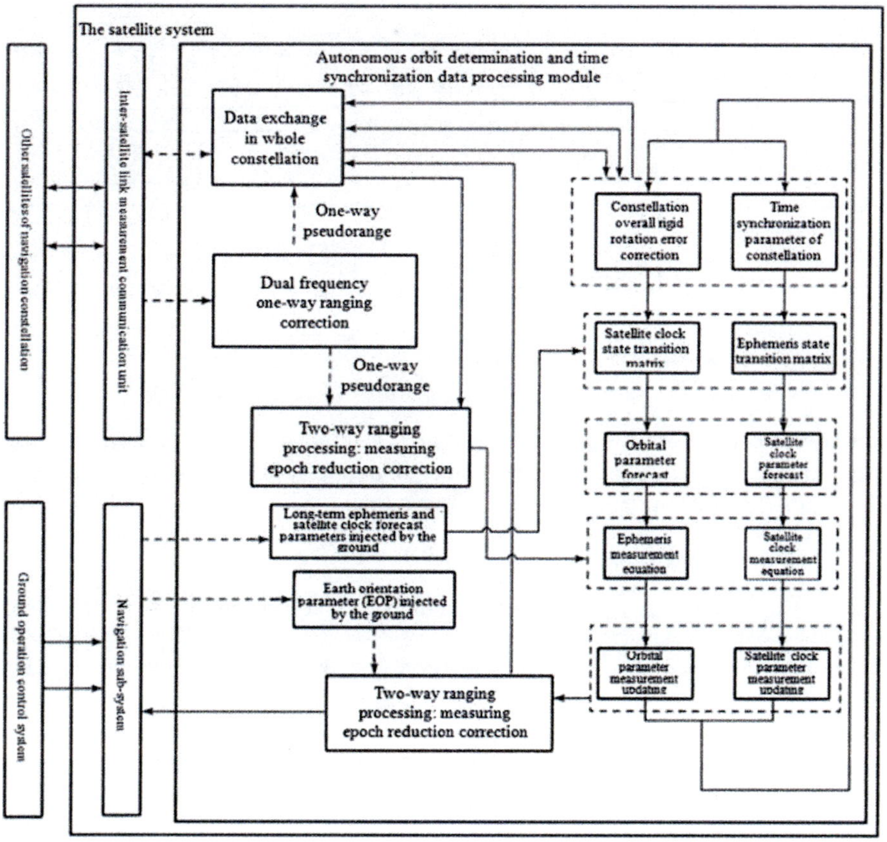


**Fig. 9.1** Distributed autonomous orbit determination system based on the uplink longtime forecast parameters

time synchronization includes autonomous navigation message and integrity information, clock difference correction, orbital inclination correction, orbital ascending node correction, and dual-frequency one-way pseudorange measurement.

There is a one-way data uplink between the ground operation control system and the satellite system. The uplink data types include the forecast parameters for ionosphere, troposphere, Earth polar motion, UT1-UTC deviation, and uplink longtime ephemeris and clock difference.

The functional block diagram for the distributed autonomous orbit determination based on the uplink longtime forecast ephemeris is shown in Fig. 9.2.

It can be seen from the functional block diagram that the onboard autonomous orbit prediction is achieved by using the ephemeris state transition matrix and uplink longtime forecast ephemeris parameters, and autonomous orbit-filtering correction is obtained by combining the ephemeris measurement equation and inter-satellite two-way pseudorange reduction with the onboard autonomous orbit prediction. The

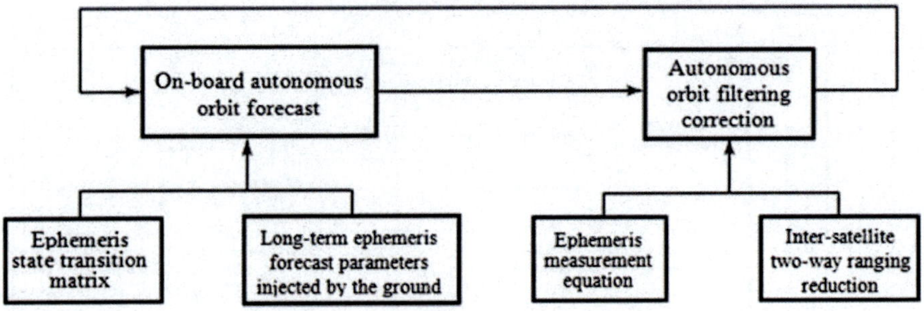


**Fig. 9.2** Functional block diagram for the autonomous orbit determination algorithm based on the uplink longtime forecast ephemeris parameters

above two processes are synchronized with inter-satellite ranging to give the result of autonomous orbit determination in real time, including autonomous orbit prediction information and inter-satellite ranging information.

1. **Method of Satellite Autonomous Orbit Prediction**

   (1) Theoretical basis of onboard autonomous orbit prediction

The orbital state variable at $t_{k-1}$, namely $\widehat{\boldsymbol{X}}_{k-1}^{+}$, the ground longtime forecast ephemeris parameter at $t_{k-1}$, namely $\boldsymbol{X}_{k-1}^{-}$, and the ground longtime forecast ephemeris parameter at $t_k$, namely $\boldsymbol{X}_k^{*}$, are known. Denote the difference between the orbital state variable at $t_{k-1}$ and the ground longtime forecast ephemeris parameter at $t_{k-1}$ as $X_{k-1}$. That is,

$$x_{k-1} = \hat{X}_{k-1}^{+} - X_{k-1}^{*} \tag{9.1}$$

Then the difference between the orbital state variable at $t_k$ and the ground longtime forecast ephemeris parameter at $t_k$ can be calculated by the following equation:

$$x_k = \Phi(t_{k-1}, t_k)x_{k-1} \tag{9.2}$$

where $\Phi(t_{k-1}, t_k)$ is the state transition matrix from $t_{k-1}$ to $t_k$. So the orbit prediction at $t_k$, namely $\hat{\boldsymbol{X}}_k$, is determined as

$$\hat{X}_k^{-} = X_k^{*} + x_k \tag{9.3}$$

Equations (9.1)–(9.3) constitute the autonomous prediction process of onboard orbital state variable. Figure 9.3 shows the autonomous orbit prediction based on the uplink longtime forecast ephemeris parameters.

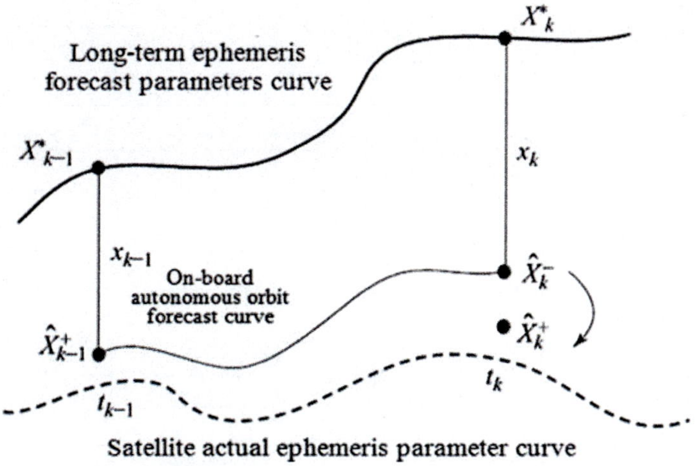


**Fig. 9.3** Autonomous orbit prediction based on the uplink longtime forecast ephemeris parameters

The process of deriving $\widehat{\boldsymbol{X}}_k^-$ from $\widehat{\boldsymbol{X}}_{k-1}^+$, as shown in Fig. 9.3, is just the onboard autonomous orbit prediction process. The process of deriving $\widehat{\boldsymbol{X}}_i^+$ from $\widehat{\boldsymbol{X}}_k^-$ is a one-step measurement updating process of onboard autonomous orbit filtering correction, which is not described here.

It can be seen that two types of data, namely ground longtime forecast ephemeris parameter $X^*$ and state transition matrix $\Phi$, are used in the onboard autonomous orbit prediction process. This process also involves the selection of the type of state variable $X$ in the autonomous orbit determination algorithm.

(2) Calculation process of onboard autonomous orbit prediction

Figure 9.4 shows the calculation process of autonomous orbit prediction based on the uplink longtime messages. The input data includes the ECEF coordinates of the satellite at $t_{k-1}$, denoted as $(R_{k-1}, \dot{\boldsymbol{R}}_{k-1})$; and the prediction time length $\Delta t = t_k - \mathrm{t}_{k-1}$, where $\mathrm{t}_{k-1}$ is the current time and $t_k$ is the target prediction time. The content of longtime navigation message G is as follows:

$$\mathbf{G} = \left(a^{1/2}, e, i_0, \Omega_0, \omega, M_0, \dot{\Omega}, \dot{i}, \Delta n, C_{\mathrm{uc}}, C_{\mathrm{us}}, C_{\mathrm{ic}}, C_{\mathrm{is}}, C_{\mathrm{rc}}, C_{\mathrm{rs}}\right) \tag{9.4}$$

where $a$ is the semi-major axis of orbit; $e$ is orbital eccentricity; $i_o$ is the orbit inclination angle at the reference epoch time toe; $\Omega_0$ is the longitude of ascending node at toe; ω is the argument of perigee; m∘is the mean anomaly at toe; $\dot{\Omega}$ is the rate of right ascension of the ascending node; $\dot{i}$ is the rate of orbit inclination angle; $\Delta n$ is mean motion difference from computed value; $C_{\mathrm{uc}}$ is amplitude of cosine harmonic correction term to the argument of latitude; $C_{\mathrm{us}}$ is amplitude of sine harmonic correction term to the argument of latitude; $C_{\mathrm{ic}}$ is amplitude of cosine harmonic correction term to the angle of inclination; $C_{\mathrm{is}}$ is amplitude of sine harmonic

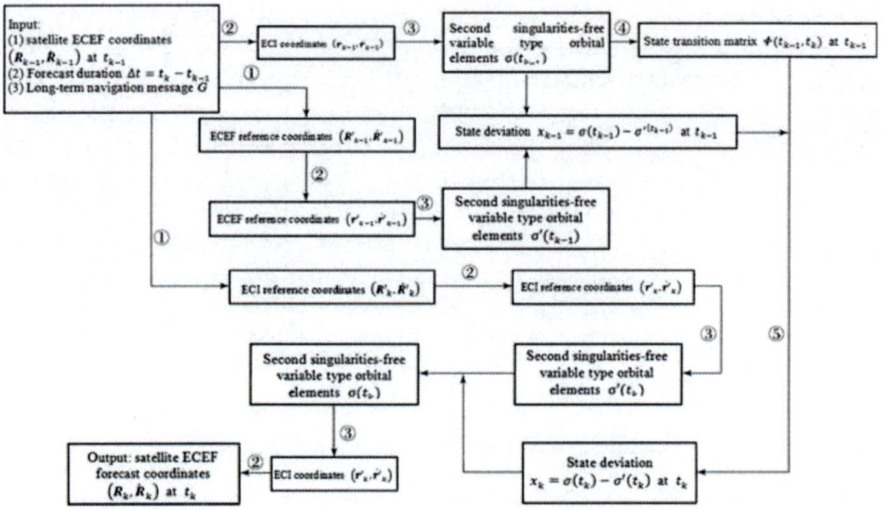


**Fig. 9.4** Calculation process of autonomous orbit prediction based on the uplink longtime messages

correction term to the angle of inclination; $C_{\text{rc}}$ is amplitude of cosine harmonic correction term to the orbit radius; $C_{\text{rs}}$ is amplitude of sine harmonic correction term to the orbit radius.

The output data is the forecast result, namely the ECEF coordinates of the satellite at $t_{\text{k}}$, denoted as $(R_{\text{k}}, \dot{\boldsymbol{R}}_k)$.

As can be seen from the above calculation flowchart, the autonomous orbit prediction based on the uplink longtime messages mainly involves the following five steps:

(1) Calculate the ECEF (Earth-centered, Earth-fixed) coordinates at the specified time according to the navigation message G, $(R_t, \dot{\boldsymbol{R}}_k)$.
(2) Convert the ECEF coordinates $(R, \dot{\boldsymbol{R}})$ to or from the Earth-Centered Initial (ECI) coordinates $(r, \dot{r})$.
(3) Convert the ECI coordinates $(r, \dot{r})$ to the Type-II non-singular variable orbital element, namely $\sigma$.
(4) Calculate the state transition matrix $\Phi\ (t_{k-1}, t_k)$ from $t_{k-1}$ to $t_k$ according to $\sigma$.
(5) Calculate the state deviation at $t_k$, namely $x_k$, according to the state deviation at $t_{k-1}$, namely $x_{k-1}$, and state transition matrix $\Phi\ (t_{k-1}, t_k)$.

The serial numbers of the above five calculation steps correspond to the callouts in Fig. 9.4. Among them, the calculation method of step (1) is the ephemeris algorithm for users, and that of step (5) is shown in Eqs. (9.1)–(9.3). The calculation methods of the other 3 steps will be presented below.

(1) Conversion between ECEF and ECI coordinates

The ECEF coordinate system here is the coordinate system where the navigation message is located. The conversion of the satellite state variable in space from ECEF

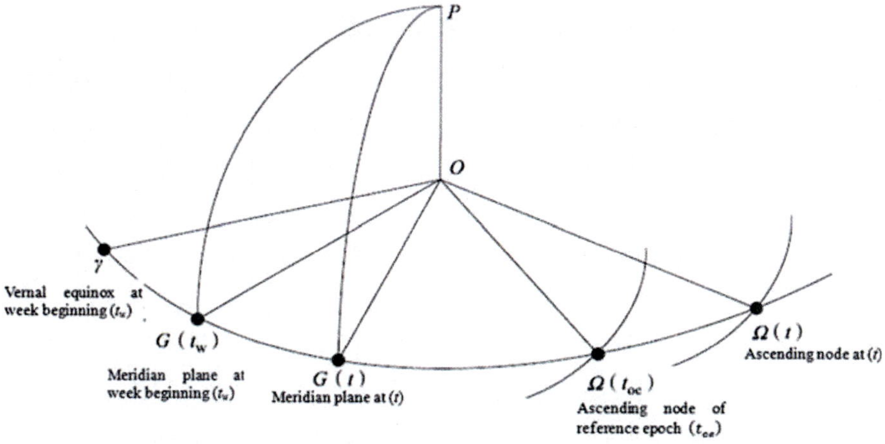


**Fig. 9.5** Coordinate transformation diagram

to ECI is intended to calculate the state transition matrix of the satellite, mainly because the dynamic equation of the satellite is based on the inertial coordinate system. Figure 9.5 shows the coordinate system transformation diagram, where $O$ is the Earth's center, $G(t)$ is the Greenwich meridian plane at the current time, y is the vernal equinox at the beginning of a week ($t_W$), $G(t_W)$ is the Greenwich meridian plane at $t_W$, Ω (toe) is the orbital ascending node at the reference epoch time toe of navigation message, and $\Omega(t)$ is the orbital ascending node at the current time.

The difference between Ω(toe) and $\Omega(t)$ is the result of satellite orbit change and is also a reflection of satellite orbit dynamics law. The difference between $G(t_W)$ and $G(t)$ is the result of Earth's rotation, and is the reflection of Earth rotation angle in the period from $t_W$ to $t$. The difference between $G(t_W)$ and $\gamma$ corresponds to the Greenwich Sidereal Time (GST) at $t_W$. $t_W$ is the time corresponding to zero seconds within a week, and changes once a week.

The starting measurement point of ECEF coordinate system is $G(t)$, while that of ECI system can be $G(t_W)$ or $\gamma$. If the vernal equinox at $t_W$ is chosen as the starting measurement point of the ECI system, the GST at $t_W$ shall be calculated. For simplicity, $G(t_W)$ is chosen as the starting measurement point in this book. Thus, the conversion relationship between the ECEF coordinates of $t_k$, namely ($R_k$, $\dot{\boldsymbol{R}}_Z$), and its ECI coordinates ($r_k$, $\dot{r}_Z$) can be expressed by the following equation:

$$\boldsymbol{r} = \boldsymbol{R}_Z(-\theta)\boldsymbol{R} \tag{9.5}$$

$$\dot{\boldsymbol{r}} = \boldsymbol{R}_Z(-\theta)\boldsymbol{R} + \dot{\boldsymbol{R}}_Z(-\theta)\boldsymbol{R} \tag{9.6}$$

$$\boldsymbol{R} = \boldsymbol{R}_Z(\theta)\boldsymbol{r} \tag{9.7}$$

$$\dot{\boldsymbol{R}} = \boldsymbol{R}_{Z}(\theta)\dot{\boldsymbol{r}} + \dot{\boldsymbol{R}}_{Z}(\theta)\boldsymbol{r} \tag{9.8}$$

where $\theta = \omega_e\ (t - \text{toe})$; $R_Z(\theta)$ represents the rotation of rotation matrix around the $Z$-axis by $\theta$ and can be expressed as

$$\boldsymbol{R}_Z(\theta) = \begin{pmatrix} \cos\theta & \sin\theta & 0 \\ -\sin\theta & \cos\theta & 0 \\ 0 & 0 & 1 \end{pmatrix} \tag{9.9}$$

$$\underset{x}{\dot{\boldsymbol{R}}}(\theta) = \begin{pmatrix} -\sin\theta & \cos\theta & 0 \\ -\cos\theta & -\sin\theta & 0 \\ 0 & 0 & 0 \end{pmatrix} \cdot \omega \tag{9.10}$$

In Eqs. (9.4) and (9.5), the nonuniformity of the Earth's rotational angular velocity is ignored.

2) Conversion between ECI coordinates and Type-II non-singular variable orbital element

In the orbital dynamics of a spacecraft, four types of state variables are usually optional, including position and velocity, classic Kepler element, Type-I non-singular variable, and Type-II non-singular variable. The four types have different applicability and performance. Considering the algorithm universality, the algorithm shall be applicable to both the MEO/IGSO satellites with small eccentricity and the GEO satellites with small eccentricity and inclination. The algorithm is based on the Type-II non-singular variable, i.e.,

$$\boldsymbol{X} = \left(a, \xi = e\cos\tilde{\omega}, \eta = e\sin\tilde{\omega}, h = \sin\frac{i}{2}\cos\Omega, k = \sin\frac{i}{2}\sin\Omega, \lambda = M + \tilde{\omega}\right) \tag{9.11}$$

where $a$ is the semi-major axis of orbit; $e$ is orbital eccentricity; i is orbit inclination; $\Omega$ is the right ascension at the ascending node; $\tilde{\omega} = \omega + M$, where $\omega$ is the argument of perigee and M is mean anomaly.

For the interconversion methods of position and velocity, classic Kepler element, and Type-II non-singular variable, see the references.

(3) Onboard calculation method for the analytical solution of the state transition matrix

The state transition matrix here is matched with the abovementioned uplink long-time forecast ephemeris parameters, and can be calculated with a numerical integration method. Its algorithm cooccurs with the ground longtime precise orbit prediction algorithm. Similarly, considering the uplink constraints and onboard storage constraints, the state transition matrix needs to be simplified.

In general, the data size of the state transition matrix is six times as big as orbital state variables. Considering the strong constraints of the satellite-ground data uplink and onboard data storage, the analytical solution of the onboard state transition matrix is adopted in the algorithm here. Next, the calculation formula of the onboard state transition matrix will be addressed, and its accuracy will be evaluated.

The variables of state transition matrix are also Type-II non-singular variables. The state transition matrix $\mathbf{\Phi}$ consists of two parts, namely

$$\mathbf{\Phi} = \mathbf{\Phi}^{(0)} + \mathbf{\Phi}^{(1)} \tag{9.12}$$

where $\mathbf{\Phi}(0)$ is the two-body motion part, and $\mathbf{\Phi}(1)$ is the perturbation part of Earth variability term J2. The calculation formulas of the two parts are as follows:

$$\mathbf{\Phi}^{(0)} = \begin{pmatrix} 1 & 0\,0\,0\,0\,0 \\ 0 & 1\,0\,0\,0\,0 \\ 0 & 0\,1\,0\,0\,0 \\ 0 & 0\,0\,1\,0\,0 \\ 0 & 0\,0\,0\,1\,0 \\ -\frac{3n}{2a}(t-t_0) & 0\,0\,0\,0\,1 \end{pmatrix} \tag{9.13}$$

$$\mathbf{\Phi}^{(1)} = \begin{pmatrix} 0 & 0 & 0 & 0 & 0 & 0 \\ b_{21} & b_{22} & b_{23} & b_{24} & b_{25} & 0 \\ b_{31} & b_{32} & b_{33} & b_{34} & b_{35} & 0 \\ b_{41} & b_{42} & b_{43} & b_{44} & b_{45} & 0 \\ b_{51} & b_{52} & b_{53} & b_{54} & b_{55} & 0 \\ b_{61} & b_{62} & b_{63} & b_{64} & b_{65} & 0 \end{pmatrix} \tag{9.14}$$

where the elements of $\Phi(1)$ are expressed by the following equations:

$$\begin{cases} b_{21} = -\eta_0\left[-\frac{7}{2a}\tilde{\omega}_1(t-t_0)\right] \\ b_{22} = -\eta_0\left[-\frac{4\xi_0}{1-e^2}\tilde{\omega}_1(t-t_0)\right] \\ b_{23} = -\left[1+\frac{4}{1-e^2}\eta_0^2\right]\tilde{\omega}_1(t-t_0) \\ b_{24} = -\eta_0\left[\left(-\frac{5}{2}\right)8h_0\left(1-2\left(h_0^2+k_0^2\right)\right)+4h_0\right]\left(\frac{3J_2}{2p^2}\right)n(t-t_0) \\ b_{25} = -\eta_0\left[\left(-\frac{5}{2}\right)8k_0\left(1-2\left(h_0^2+k_0^2\right)\right)+4k_0\right]\left(\frac{3J_2}{2p^2}\right)n(t-t_0) \end{cases} \tag{9.15}$$

$$\begin{cases} b_{31} = \varepsilon_0\left[-\frac{7}{2a}\tilde{\omega}_1(t-t_0)\right] \\ b_{32} = \left[1+\frac{4}{1-e^2}\xi_0^2\right]\omega_1(t-t_0) \\ b_{33} = \xi_0\left[-\frac{4\eta_0}{1-e^2}\tilde{\omega}_1(t-t_0)\right] \\ b_{34} = \xi_0\left[\left(-\frac{5}{2}\right)8h_0\left(1-2\left(h_0^2+k_0^2\right)\right)+4h_0\right]\left(\frac{3J_2}{2p^2}\right)n(t-t_0) \\ b_{35} = \xi_0\left[\left(-\frac{5}{2}\right)8k_0\left(1-2\left(h_0^2+k_0^2\right)\right)+4k_0\right]\left(\frac{3J_2}{2p^2}\right)n(t-t_0) \end{cases} \tag{9.16}$$

$$\begin{cases} b_{41} = -k_0\left[-\frac{7}{2a}\Omega_1(t-t_0)\right] \\ b_{42} = -k_0\left[\frac{4\xi_0}{1-e^2}\Omega_1(t-t_0)\right] \\ b_{43} = -k_0\left[\frac{4\eta_0}{1-e^2}\Omega_1(t-t_0)\right] \\ b_{44} = -k_0\left[4h_0\left(\frac{3J_2}{2p^2}\right)n(t-t_0)\right] \\ b_{45} = -\left[\Omega_1 + 4k_0^2\left(\frac{3J_2}{2p^2}\right)n\right](t-t_0) \end{cases} \quad (9.17)$$

$$\begin{cases} b_{51} = h_0\left[-\frac{7}{2a}\Omega_1(t-t_0)\right] \\ b_{52} = h_0\left[\frac{4\xi_0}{1-e^2}\Omega_1(t-t_0)\right] \\ b_{53} = h_0\left[\frac{4\eta_3}{1-e^2}\Omega_1(t-t_0)\right] \\ b_{54} = \left[\Omega_1 + 4h_0^2\left(\frac{3J_2}{2p^2}\right)n\right](t-t_0) \\ b_{55} = h_0\left[4k_0\left(\frac{3J_2}{2p^2}\right)n(t-t_0)\right] \end{cases} \quad (9.18)$$

$$\begin{cases} b_{61} = -\frac{7}{2a}\lambda_1(t-t_0) \\ b_{62} = \frac{\xi_0}{1-e^2}(3M_1 + 4\tilde{\omega}_1)(t-t_0) \\ b_{63} = \frac{\eta_0}{1-e^2}(3M_1 + 4\omega_1)(t-t_0) \\ b_{64} = \left[\left(-\frac{3}{2}\sqrt{1-e^2} - \frac{5}{2}\right)8h_0\left(1 - 2\left(h_0^2 + k_0^2\right)\right) + 4h_0\right]\left(\frac{3J_2}{2p^2}\right)n(t-t_0) \\ b_{65} = \left[\left(-\frac{3}{2}\sqrt{1-e^2} - \frac{5}{2}\right)8k_0\left(1 - 2\left(h_0^2 + k_0^2\right)\right) + 4k_0\right]\left(\frac{3J_2}{2p^2}\right)n(t-t_0) \end{cases} \quad (9.19)$$

$$\begin{cases} \Omega_1 = -\frac{3J_2}{2p^2}n\cos i \\ \omega_1 = \frac{3J_2}{2p^2}n\left(2 - \frac{5}{2}\sin^2 i\right) \\ M_1 = \frac{3J_2}{2p^2}n\left(1 - \frac{3}{2}\sin^2 i\right)\sqrt{1-e^2} \end{cases} \quad (9.20)$$

The $a$, $e$, and $i$ in the above equations as well as $p = a(1 - e^2)$ are the transient elements at the epoch time to, namely $a_o$, $e_o$, $i_o$, …; $\lambda_1 = M_1 + \tilde{\omega}_1$, where $\tilde{\omega}_1 = \omega 1 + \Omega_1$. The $e_0$ and $i_o$ in the calculation equations of M1, ω1, and Ω1 are calculated by $e_0^2 = \xi_0^2 + \eta_0^2$ and $\sin^2(i_o/2) = h_0^2 + k_0^2$, respectively.

The above calculation method of the state transition matrix only considers the central gravitation term of satellite motion in space and the perturbation term of Earth flattening, while ignoring the rest of the higher order perturbation terms $\Delta\Phi$, such as Earth's higher order nonspherical gravitational perturbation term and solar and lunar third-body perturbation term. Table 9.1 lists the perturbation magnitudes of three types of satellites MEO/GEO/IGSO. With regard to light-pressure perturbation, the typical effective area-mass ratio of the satellite is 109, that is, 10 $m^2$/1000 kg. The magnitudes in the table are given with the gravity of the Earth's center as 1 standard unit.

It can be seen that, for the three types of satellites, the high-order perturbation term neglected by Eq. (9.12) is mainly solar and lunar third-body gravitational perturbation term, whose maximum order of magnitude is $2 \times 10^{-5}$. Because the magnitude of longtime orbital prediction error [i.e., $x$ in Eq. (9.3)] is less than 10 km ($10^{-3}$ in the Earth's dimensionless unit), the magnitude of one-step orbital prediction error given

**Table 9.1** Perturbation magnitudes of MEO/GEO/IGSO satellites

| Perturbative factor | To MEO satellites | To GEO/IGSO satellites |
|---|---|---|
| Perturbation term of Earth flattening, $J_2$ | $10^{-4}$ | $10^{-5}$ to $10^{-4}$ |
| Earth's gravitational perturbation terms except for $J_2$ | $10^{-8}$ to $10^{-7}$ | $10^{-8}$ to $10^{-7}$ |
| Lunar gravitational perturbation term | $10^{-5}$ | $2 \times 10^{-5}$ |
| Solar gravitational perturbation term | $0.5 \times 10^{-5}$ | $10^{-5}$ |
| Perturbation term of Earth's tidal deformation | $10^{-10}$ | $10^{-10}$ |
| Perturbation term of sunlight pressure | $10^{-7}$ | $2 \times 10^{-7}$ |
| Post-Newtonian correction term | $10^{-9}$ | $10^{-9}$ |

by Eqs. (9.3) and (9.12) is as follows:

$$\Delta x_k = \Delta\Phi \times x_{k-1} \sim 2 \times 10^{-8} \tag{9.21}$$

In the international system of units, the above magnitude is 0.13 m, that is, the error of one-step orbital prediction is 0.13 m.

Considering the specific accuracy of inter-satellite ranging, the calculation accuracy of the state transfer matrix in the analytical method is matched with inter-satellite ranging accuracy. Besides, comparing with the state transfer matrix based on ground uplink data, this state transfer matrix can greatly save the data size for satellite-ground transmission and onboard storage. For the satellite platform, a high cost must be paid to implement the above two functions.

2. **Satellite Autonomous Orbit Filtering Correction**

As a real-time recursive algorithm, Kalman filtering is suitable for onboard autonomous orbit determination in various calculation methods. Here, the UD square root of Gram-Schmidt orthogonalization is chosen. The Extended Kalman Filter (EKF) is used to estimate the orbital state variable of the satellite, so as to give the best estimation of the real orbital state variable. The main process is as follows:

(1) One-step prediction, namely time update
Here, the predicted value of satellite orbit state variable is obtained through onboard autonomous orbital prediction. The estimated value of orbital state variable at tk, namely $\widehat{\boldsymbol{X}}^{-}(t_k)$, is calculated with the abovementioned onboard autonomous orbital prediction method.

(2) Kalman gain calculation

$$K_k = P_k^- H_k^T \left[H_k P_k^- H_k^T + R_k\right]^{-1} \tag{9.22}$$

where the subscript $k$ represents the time $t_k$; $R_k$ is the variance matrix of inter-satellite ranging error at $t_k$; $\boldsymbol{P}_k^-$ is the covariance matrix of state variable at $t_k$, i.e., one-step prediction Mean-Square Error (MSE). Its calculation formula is as follows:

$$P_k^- = \Phi(t_{k-1}, t_k) P_{k-1} \Phi^T(t_{k-1}, t_k) + Q_k \tag{9.23}$$

where $\Phi$ $(t_{k-1}, t_k)$ is the state transition matrix of orbital variable, and $Q_k$ is dynamic noise matrix.

$H_k$ in Eq. (9.22) is a measurement matrix, whose calculation equation is

$$\boldsymbol{H}_k = \left(\frac{\partial_{\text{mn}}^-}{\partial \boldsymbol{r}_{\text{m}}^T}\right)_{1\times 3} \left(\frac{\partial \boldsymbol{r}_{\text{m}}}{\partial \boldsymbol{X}_{\text{m}}^T}\right)_{3\times 6} \tag{9.24}$$

where the subscripts m and n represent the two navigation satellites for inter-satellite ranging; $\rho_{\text{mn}}^- = \left\| \boldsymbol{r}_{\text{m}}\left(\widehat{\boldsymbol{X}}_{\text{m}}^-(t_k)\right) - \boldsymbol{r}_{\text{n}}\left(\widehat{\boldsymbol{X}}_{\text{n}}^-(t_k)\right) \right\|$ is the estimated range between the satellites m and n at tk; rm and rn are the position vector estimates of the satellites m and n at tk; $\widehat{\boldsymbol{X}}_{\text{m}}^-$ $(t_k)$ and $\widehat{\boldsymbol{X}}_{\text{n}}^-$ $(t_k)$ are the orbital state variable estimates of the satellites m and n at tk, as defined and calculated in Eq. (9.3). The two matrices in Eq. (9.24) are observation geometry matrix and orbital geometry matrix, respectively, which are calculated with the following methods:

$$\left(\frac{\partial \rho_{\text{min}}}{\partial \boldsymbol{r}_m^{\text{T}}}\right) = \frac{1}{\rho_{\text{rm}}} (\boldsymbol{r}_{\text{m}}^{\text{T}} - \boldsymbol{r}_{\text{n}}^{\text{T}}) \tag{9.25}$$

$$\left(\frac{\partial \boldsymbol{r}_{\text{m}}}{\partial \boldsymbol{X}_{\text{m}}^{\text{T}}}\right) = \left(\frac{\partial \boldsymbol{r}_{\text{m}}}{\partial a_{\text{m}}}, \frac{\partial \boldsymbol{r}_{\text{m}}}{\partial \xi_{\text{m}}}, \frac{\partial \boldsymbol{r}_{\text{m}}}{\partial \eta_{\text{m}}}, \frac{\partial \boldsymbol{r}_{\text{m}}}{\partial h_{\text{m}}}, \frac{\partial \boldsymbol{r}_{\text{m}}}{\partial k_{\text{m}}}, \frac{\partial \boldsymbol{r}_{\text{m}}}{\partial \lambda_{\text{m}}}\right) \tag{9.26}$$

The 6 elements in Eq. (9.26) are expressed by

$$\begin{cases} \frac{\partial \boldsymbol{r}}{\partial a} = \frac{1}{a}\boldsymbol{r}, & \frac{\partial \boldsymbol{r}}{\partial h} = r\cos u \widehat{\mathbf{P}}_h + r \sin u \widehat{\mathbf{S}}_h \\ \frac{\partial \boldsymbol{r}}{\partial \xi} = A\boldsymbol{r} + B\dot{\boldsymbol{r}} & \frac{\partial \boldsymbol{r}}{\partial k} = r\cos u \widehat{\mathbf{P}}_k + r \sin u \widehat{\mathbf{S}}_k \\ \frac{\partial \boldsymbol{r}}{\partial \eta} = C\boldsymbol{r} + D\dot{\boldsymbol{r}}, & \frac{\partial \boldsymbol{r}}{\partial \lambda} = \sqrt{a^3}\dot{\boldsymbol{r}} \end{cases} \tag{9.27}$$

where $r = (x, y, z)\ T$ is satellite position vector, and $\dot{r} = (\dot{x}, \dot{y}, \dot{z})^{\text{T}}$ is satellite velocity vector.

$$\begin{cases} A = \frac{a}{p}\left[-(\cos u + \xi) - \left(\frac{r}{p}\right)(\sin u + \eta)(\xi \sin u - \eta \cos u)\right] \\ B = \frac{ar}{\sqrt{p}}\left[\sin u + \left(\frac{a}{r}\right)\frac{\sqrt{1-e^2}}{1+\sqrt{1-e^2}}\eta + \left(\frac{r}{p}\right)(\sin u + \eta)\right] \\ C = -\frac{a}{p}\left[(\sin u + \eta) - \left(\frac{r}{p}\right)(\cos u + \xi)(\xi \sin u - \eta \cos u)\right] \\ D = -\frac{ar}{\sqrt{p}}\left[\cos u + \left(\frac{a}{r}\right)\frac{\sqrt{1-e^2}}{1+\sqrt{1-e^2}}\xi + \left(\frac{r}{p}\right)(\cos u + \xi)\right] \end{cases} \tag{9.28}$$

$$\widetilde{\mathbf{P}}_{\mathrm{h}} = \begin{pmatrix} 0 \\ 2k \\ 2hk/\cos\frac{i}{2} \end{pmatrix}, \quad \widehat{\mathbf{S}}_{k} = \begin{pmatrix} 2k \\ -4h \\ 2(1-2h^2-k^2)/\cos\frac{i}{2} \end{pmatrix} \tag{9.29}$$

$$\cos\frac{i}{2} = \left[\left(1-\left(h^2+k^2\right)\right)\right]^{1/2} \tag{9.30}$$

where $(\alpha, \xi, \eta, h, \kappa, \lambda)$ are Type-II non-singular variables, as defined in Eq. (9.11). $(p = \alpha)$ $(1 - e^2)$ is the semilatus rectum of the instantaneous elliptical orbit of the satellite; $r = ||r||$ is the distance from the satellite to the Earth's center; $u = f + \tilde{\omega}$ is the latitude angle of the satellite; $f$ is the true anomaly of the satellite; $\tilde{\omega} = \omega + \Omega$, where $\omega$ is argument of perigee and $\Omega$ is the right ascension at the ascending node of the satellite.

(3) Measurement updating

$$\hat{X}_k^+ = \hat{X}_k^- + K_k\left(\rho_{mn} - \rho_{mn}^-\left(X_k^-\right)\right) \tag{9.31}$$

$$P_k = [I - K_k H_k]P_k^- \tag{9.32}$$

where $\rho_{\mathrm{mn}}$ is the inter-satellite range measurement at $t_{\mathrm{k}}$; and $\rho_{\mathrm{max}^-}\left(X_k^-\right)$ is the inter-satellite range at $t_{\mathrm{k}}$ estimated according to the value of satellite state variable given by onboard orbital prediction.

3. **Inter-satellite Two-Way Pseudorange Reduction**

The original measurement information based on inter-satellite link ranging is the transmitting and receiving time of the ranging signal. Considering the problem of clock non-synchronization between the satellites transmitting and receiving the ranging signal, two-way ranging is needed. To correct the signal delay caused by the ionosphere, dual-frequency measurement is required. Therefore, in a pair of linked satellites, either satellite obtains 4 pseudorange observations, including 2 pseudoranges formed in the transmitting process and 2 pseudoranges formed in the receiving process. After the phase center correction, relativistic effect correction, and ionospheric delay correction, these inter-satellite pseudorange measurements are decoupled into mutually independent clock-corrected pseudoranges and ephemeris-corrected pseudoranges. Next, the decoupling process of the two pseudoranges will be addressed.

First, the satellite 1 sends the ranging signal to the satellite 2, then

$$\mathrm{PR}_{21} = D_{21} + \left(a_{f_0}^2 - a_{f_0}^1\right)c + \Delta D_{\mathrm{ion}}^{21} \tag{9.33}$$

where $\mathrm{PR}_{21}$ is the measured pseudorange from satellite 1 to satellite 2; $D_{21}$ is the geometric distance between the satellite 1 at the time of ranging signal transmission and the satellite 2 at the time of ranging signal reception; $\Delta D_{\mathrm{ion}}^{21}$ is the ionospheric delay correction given by double frequency; $a_{f_i}^1$ : and $a_{f_i}^2$ are the satellite clock errors of satellite 1 and satellite 2, respectively; and c is the speed of light in vacuum.

Second, the satellite 2 sends the ranging signal to the satellite 1, then

$$\mathrm{PR}_{12} = D_{12} + \left(a_{f_0}^{1} - a_{f_0}^{2}\right)c + \Delta D_{\mathrm{ion}}^{12} \tag{9.34}$$

where $\mathrm{PR}_{12}$ is the measured pseudorange from satellite 2 to satellite 1; $D_{12}$ is the geometric distance between the satellite 2 at the time of ranging signal transmission and the satellite 1 at the time of ranging signal reception; $\Delta D_{\mathrm{ion}}^{12}$ is the ionospheric delay correction given by double frequency; $a_{f}^{1}$ and $a_{f_i}^{2}$ are the satellite clock errors of satellite 1 and satellite 2, respectively; and $c$ is the speed of light in vacuum.

The above two pseudorange observations corrected by ionospheric delay can be reduced to the pseudorange observations corrected at the same moment. If $D$ is the geometric distance between the transmitting satellite and the receiving satellite at that moment, then

$$D_{12} = D_{21} = D \tag{9.35}$$

So the pseudorange corrected by satellite clock, namely $Z_{\varphi}$, is

$$\begin{aligned} Z_{\varphi} &= (\mathrm{PR}_{21} - \mathrm{PR}_{12})/2 \\ &= \left(a_{f_0}^{2} - a_{f_0}^{1}\right) + \left(\Delta D_{ion}^{21} - \Delta D_{ion}^{12}\right)/2 \end{aligned} \tag{9.36}$$

The ephemeris-corrected pseudorange $Z_{\mathrm{d}}$, namely ρmn in Eq. (9.30), is

$$\begin{aligned} \rho_{\mathrm{mn}} &= Z_{\mathrm{d}} = (\mathrm{PR}_{21} + \mathrm{PR}_{12})/2 \\ &= D + \left(\Delta D_{\mathrm{ion}}^{21} + \Delta D_{\mathrm{ion}}^{12}\right)/2 \end{aligned} \tag{9.37}$$

In Eqs. (9.36) and (9.37), the ionospheric delay corrections $\Delta D_{\mathrm{ion}}^{21}$ and $\Delta D_{\mathrm{ion}}^{12}$ are approximately equal to each other, so they can be canceled out by each other when the pseudorange is corrected by satellite clock. In addition, it takes a short time to measure the two-way pseudorange between the two satellites, generally no more than 15 min. During this period, the variation of clock difference between the two satellites is very small, so it can be treated as a constant.

### *9.1.4 Autonomous Navigation Accuracy Analysis and Simulation Demonstration*

The errors in GNSS autonomous navigation have two main sources. One is the position error of the constellation satellite caused by means of the inter-satellite ranging information in the Earth Center Inertial (ECI) space. The other is the long-term prediction error of Earth Orientation Parameters (EOPs), which are no longer updated by the ground control segment during autonomous navigation.

In the autonomous mode of navigation constellation, the User Equivalent Range Error (UERE) will be affected by the EOP prediction error. The EOPs determine the conversion relationship between ECI and ECEF coordinate systems, and convert the ECF position of a navigation satellite into the ECEF position. The EOPs include the x and y coordinates of the celestial pole (also known as polar motion coordinates) and the UT1[1]-UTC deviation. The prediction error of the polar motion coordinates x and y will introduce the error of rotation around the y- and x- axes, while the prediction error of UT1-UTC will bring the error of rotation around the celestial pole. In the long-term prediction of EOPs, the ephemeris conversion problem in the ECEF coordinate system is usually solved by the long-period polynomial fitting. For the 60-day prediction, the error of the polar motion parameter is about 0.03 mas, and the error of the UT1 parameter is <0.005 s. For the navigation users near the equator, the ground positioning deviation is affected most seriously; the error of the polar motion parameter is about 0.03 mas, equivalent to 0.93 m; and the error of the UT1 parameter is <0.005 s, equivalent to 2.32 m.

In the autonomous mode of a navigation constellation, the UERE is also affected by the accuracy of the ephemeris autonomously updated by the satellite. If the long-term forecast ephemeris uploaded from the ground is adopted, the three-dimensional position error of its 60-day forecast on orbit will be in the order of 1 km, a far gap from the required autonomous navigation accuracy. Usually, the satellite ephemeris is corrected by using the Kalman filter to process the inter-satellite ranging data. The simulation result of an autonomous navigation algorithm is shown in Fig. 9.6.

By using the Kalman filter to process the inter-satellite ranging data, the satellite can autonomously update its ephemeris, thus achieving User Range Error (URE) usually better than 3 m. With the typical spatial transmission error of the navigation signal and the typical noise of the user receiver, the navigation constellation can autonomously operate for 60 days, and the accuracy of the user position can be better than 15 m.

## 9.2 Autonomous Health Management of Navigation Satellites

### *9.2.1 Concept of Autonomous Health Management*

The autonomous health management of a navigation satellite involves three aspects: health, management, and autonomy.

Health is the state of an object, which indicates whether the object works normally or not, if not, how abnormal it is. The description of health is omnidirectional and multidimensional. On object composition, it includes the states of the whole system and its subsystems and components. On object-using time, it includes the

[1]UT1 is the universal time of type 1.

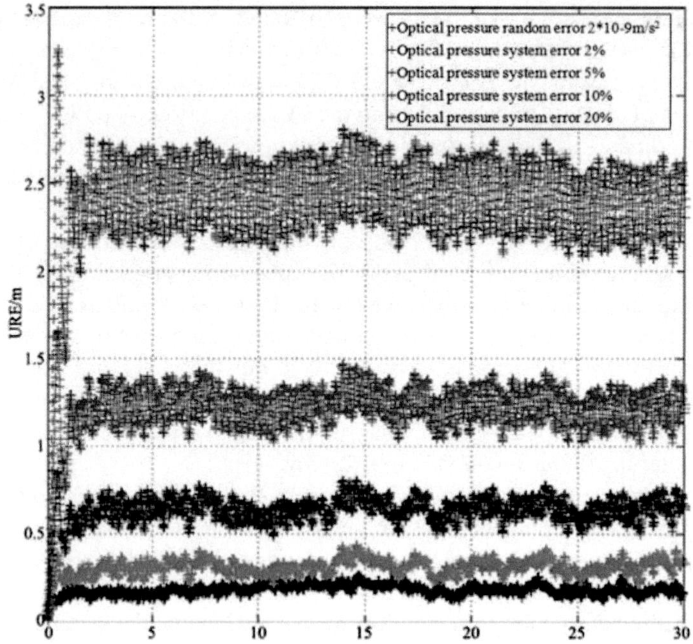


**Fig. 9.6** The simulation result of autonomous navigation algorithm under different light-pressure system errors

past, present, and future states. The description of object health shall clearly indicate the accurate and clear state information of the object system through effective and advanced state monitoring/testing, fault diagnosis and prediction, and other means.

Management is the administrative behavior that organizes, coordinates, and optimizes the activities directly related to the health of an object. It includes the organization and management of state monitoring/testing, fault diagnosis and prediction, and other means to obtain the health information; the organization and management of autonomous maintenance, condition-based assurance, and other activities to derive the assurance decision from the obtained information; and the management of various health data and information as well as maintenance and assurance records. The health management is intended to ensure safe and accurate completion of various tasks, and improve the efficiency of ground maintenance and space-based use and the operational efficiency of the system throughout its life cycle, thereby reducing the cost of operation and maintenance.

Autonomy is the onboard health management ability. A navigation satellite is generally composed of structural subsystem, power supply and distribution subsystem, thermal control subsystem, TT&C subsystem, digital control subsystem (sometimes replaced by integrated electronic subsystem), attitude and orbit control subsystem, propelling subsystem, and various payloads. In order to ensure the normal functioning of the satellite and the delivery of payload services meeting the user requirements, it is necessary to keep the satellite in a specific attitude and on a specific orbit, to ensure the energy supply for the whole satellite and its subsystems, to confine the whole satellite and its onboard equipment to a certain temperature range, and to complete the normal on-orbit operation of spaceborne equipment and components.

For a satellite, autonomous health management is to obtain the data and information on the working status of components, subsystems, and systems by various measuring means, and to judge their working status and performance through the data processing and analysis. The satellite health management is conducted at three levels. The first level is to reduce the ground intervention and the TT&C tasks of the ground system. The satellite autonomously completes various tasks necessary for its normal on-orbit operation, such as thermal control management, satellite management in the Earth shadow period, solar interference protection for infrared sensor, as well as delayed telemetry and state setting during the satellite entry and exit. The second level is to diagnose the occurring failure, determine which parts are working in an abnormal state or with degraded performance, isolate the fault and reconfigure the system according to the fault-handling strategy, bring the system back to normal or put it into degraded use, verify the recovered performance, and ensure no potential negative impacts on later system operations. The third level is the autonomous monitoring of navigation signal, namely the autonomous monitoring of satellite integrity, which will be detailed in Sect. 9.3.

### *9.2.2 Objectives and Workflow of Autonomous Health Management*

For a navigation satellite, the objectives of its autonomous health management are as follows:

(1) In the absence of ground support, the satellite can work normally. In other words, the normal attitude and orbit of the satellite are maintained, the energy supply is guaranteed, the thermal environment for the whole satellite and its onboard equipment is ensured, and the management of onboard equipment and components is accomplished.
(2) In case of any recoverable fault in any redundant equipment, the satellite can still work normally.
(3) In the event of failure, the sudden change of satellite orbit and the interruption of a navigation mission should be avoided as far as possible.
(4) The autonomous health management of a navigation satellite should ensure the safety and life of the satellite. That is to say, in the relevant processing, the optimization of fuel consumption should be considered, and the losses of system configuration and components should be minimized to the largest extent.

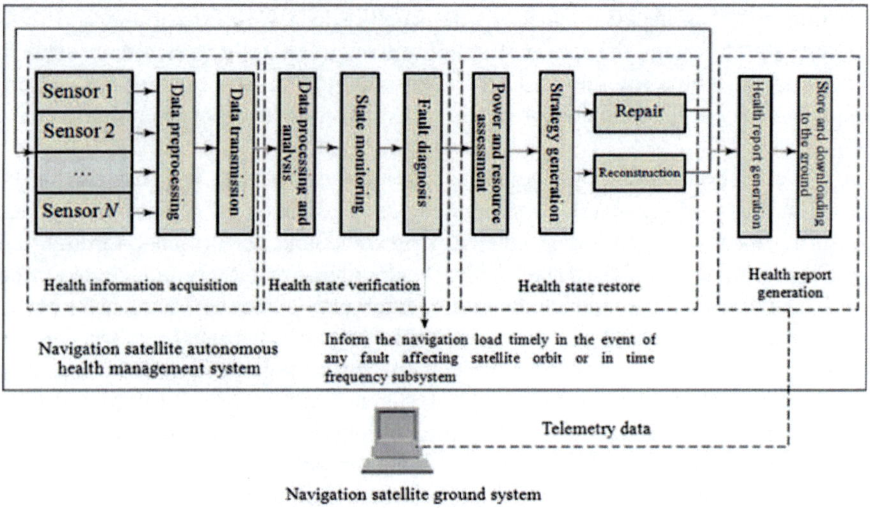


**Fig. 9.7** Autonomous health management workflow of navigation satellite

In principle, the objectives, requirements, and design of autonomous health management of navigation satellites are the same as those of other satellites. Meanwhile, compared with other satellites, there are two points of particular concern. First, in case of a fault that may affect the satellite orbit or time-frequency subsystem and even affect normal navigation service, the navigation payload should be timely notified. Second, spatial integrity exceptions should be monitored (i.e., autonomous integrity monitoring). When the navigation signal cannot be used for positioning, a timely alarm shall be sent to the user.

The process of autonomous health management can be divided into the following four stages: health information acquisition, health status identification, health status recovery, and health report generation, as shown in Fig. 9.7.

Through the above process, the autonomous health management system can learn the current health status of the navigation satellite from the obtained information through processing, analysis, and evaluation, and then develop and implement the corresponding maintenance strategy, in order to keep the satellite in normal working condition, or to recover it from the failure state, or to degrade the satellite use and wait for ground handling while ensuring safety.

## *9.2.3 Framework of Autonomous Health Management*

According to the control mode, the autonomous health management structures of navigation satellites can be classified into the following three types: master-slave control structure, hierarchical control structure, and peer-to-peer control structure.

The principles and characteristics of the three control structures are shown in Table 9.2.

Both the master-slave control structure and the hierarchical control structure are composed of one master node and multiple slave nodes. The difference between them

**Table 9.2** Comparison of three control structures

| Type of control structure | Control principles | Control characteristics |
|---|---|---|
| Master-slave control structure | 1. This structure is composed of one master node and multiple slave nodes<br>2. The master node processes and makes decisions on all the health management information, and autonomously controls all the slave nodes<br>3. Each slave node receives a command from the master node, executes the command, and sends its state information back to the master node for unified processing | 1. Good global performance, and easy global coordination and control<br>2. Poor dynamic response due to the limited communication bandwidth<br>3. Complex master-node logic |
| Hierarchical control structure | 1. This structure is composed of one master node and multiple slave nodes<br>2. The master node releases the plan of autonomous health management. Part of the plan is executed by itself, while the rest by the slave nodes<br>3. Each slave node can accomplish its own health management, and feed the results back to the master node to reduce the burden on the master node | 1. Each slave node is not a simple actuator, but an intelligent actuator and diagnotor<br>2. Good global and local control performance, and open-type system<br>3. Good expansibility, and applyies to a complex dynamic environment |
| Peer-to-peer control structure | 1. All the nodes are in a peer-to-peer relationship. Each node has a complete set of the autonomous control system, and can make and execute the plans autonomously<br>2. A loose cooperative relationship is established between nodes and coordinated by the task manager | 1. Bad global control performance and complex control relationships<br>2. Good local performance |

is that, in the master-slave control structure, all the health management information is processed and decided by the master node, while the slave nodes just passively execute the command rather than undertaking the health management task; in the hierarchical control structure, the slave nodes participate in the processing of health management information, and feed the results directly back to the master node for decision-making and overall control. Different from the nodes in master-slave and hierarchical control structures, the nodes in the peer-to-peer control structure are in equal—rather than master and slave—relationship.

The three control structures are all distinctive. For small satellites under centralized management, the master-slave structure is generally used. For the "centralized + decentralized" two-level distributed satellites under integrated electronic or digital management, the master-slave or team-style control structure is generally used. For the subdivision spacecraft with multiple central processors, the peer-to-peer structure is a better fit.

With the continuous improvement of the onboard data processing capability, integrated electronics has become the development direction of the satellite data management platform. Due to the modular and hierarchical design of integrated electronics, the central management unit serves as the main center to manage the whole satellite, and other modules also have a certain processing capacity. Therefore, better matching and compatibility can be achieved between master-slave (or team-style) control structure and integrated electronic subsystem.

In the design of autonomous health management, autonomous integrity monitoring is relatively independent and implemented by a special module designed in the navigation payload. The satellite completes various tasks on orbit autonomously mainly through the onboard data platform, where the corresponding software module is designed to implement the requirements of autonomous management that have been analyzed by the whole satellite and its subsystems according to their own actual design and tasks. In addition to the above two aspects, all kinds of onboard faults are also handled by autonomous health management. In this process, the methods of fault diagnosis and recovery are involved.

### *9.2.4 Fault Diagnosis Algorithm*

Fault diagnosis algorithm is the core of autonomous health management. The fault diagnosis algorithms in common use fall into three types: the fault diagnosis algorithm based on signal processing, the fault diagnosis algorithm based on the analytical model, and the fault diagnosis algorithm based on the qualitative model.

1. **Fault Diagnosis Algorithm Based on Signal Processing**

The principle of this diagnostic algorithm is to detect the signal with a limited or definite normal working range. When the set threshold is exceeded, the equipment corresponding to the signal is considered as failed. Normally, $y(t)$, an important characteristic parameter reflecting the tested object, is checked by the following rule:

$$m_{\min}(t) < m(t) < m_{\max}(t) \tag{9.38}$$

where $m_{\min}$ (t) and $m_{\max}$ (t) are the upper and lower limits of the characteristic parameter $m(t)$, respectively; and $t$ is time variable.

If the tested object satisfies Eq. (9.38), no fault will occur; otherwise, the measured signal will exceed the threshold so that a fault will occur.

2. **Fault Diagnosis Algorithm Based on Analytical Model**

This fault diagnosis algorithm is to generate a residual error based on a mathematical system model by using the observer (group), Kalman filter, and other methods, and then evaluate or make decisions on the residual error according to a rule. Its core idea is to replace the hardware redundancy with analytic redundancy. The implementation process of this algorithm is divided into the following four steps:

Step 1: establish the fault model of the diagnosed object.
Step 2: use the filtering algorithm to estimate the state of a diagnosed object in the satellite at the next moment. Usually, Kalman filter, observer (group), and other methods are used for state estimation.
Step 3: derive the filtering residual from the state estimation in step 2.
Step 4: set a rational threshold, and judge the residual. If the residual exceeds the threshold, the fault occurrence is considered true; otherwise, the diagnosed object is considered in proper working order.

Taking the control system as an example to explain the above steps. The actuator, sensor, and controlled object of this system are modeled. The resulting fault model and residual generation filter are shown in Fig. 9.8.

Usually, the controlled object can be expressed by Eqs. (9.39) and (9.40):

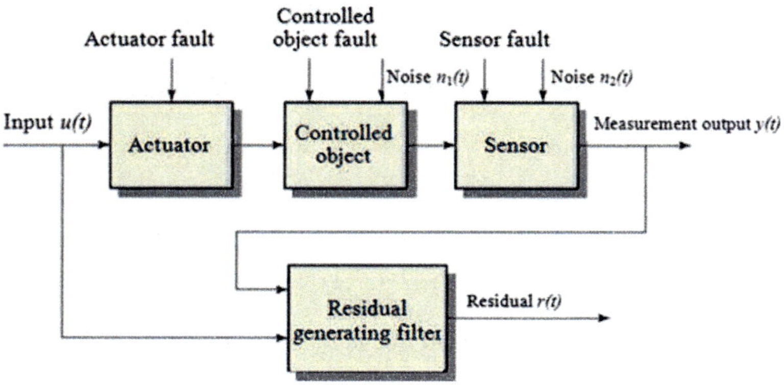


**Fig. 9.8** Fault model and residual generation filter

$$\boldsymbol{x}(t+1) = (\boldsymbol{A} + \Delta\boldsymbol{A})\boldsymbol{x}(t) + (\boldsymbol{B} + \Delta\boldsymbol{B})\boldsymbol{u}(t) + \boldsymbol{E}_1\boldsymbol{n}_1(t) \tag{9.39}$$

$$\boldsymbol{y}(t) = (\boldsymbol{C} + \Delta\boldsymbol{C})\boldsymbol{x}(t) + (\boldsymbol{D} + \Delta\boldsymbol{D})\boldsymbol{u}(t) + \boldsymbol{E}_2\boldsymbol{n}_2(t) \tag{9.40}$$

where $x(t)$ is the state vector; $\mu(t)$ is the input vector of the system, namely the control vector output by the actuator; $y(t)$ is the output vector directly obtained by sensor measurement; $A$, $B$, $C$, and $D$ are the coefficient matrices of the controlled object model; $\Delta A$, $\Delta B$, $\Delta C$, and $\Delta D$ are the uncertainty matrices of the controlled object model; and $E_1$ and $E_2$ are noise matrices.

According to Fig. 9.8 and Eqs. (9.39) and (9.40), the following fault models can be obtained:

$$\boldsymbol{x}(t+1) = (\boldsymbol{A} + \Delta\boldsymbol{A} + \Delta\boldsymbol{A}_{\mathrm{c}})\boldsymbol{x}(t) + (\boldsymbol{B} + \Delta\boldsymbol{B} + \Delta\boldsymbol{B}_{\mathrm{c}})\boldsymbol{u}(t) + \boldsymbol{E}_1\boldsymbol{n}_1(t) + \boldsymbol{B}\boldsymbol{f}_{\mathrm{a}}(t) \tag{9.41}$$

$$\boldsymbol{y}(t) = (\boldsymbol{C} + \Delta\boldsymbol{C} + \Delta\boldsymbol{C}_{\mathrm{c}})\boldsymbol{x}(t) + (\boldsymbol{D} + \Delta\boldsymbol{D} + \Delta\boldsymbol{D}_{\mathrm{c}})\boldsymbol{u}(t) + \boldsymbol{E}_2\boldsymbol{n}_2(t) + \boldsymbol{f}_{\mathrm{s}}(t) \tag{9.42}$$

where $f_{\mathrm{a}}(t)$ is the fault vector of the actuator; $f_{\mathrm{s}}(t)$ is the fault vector of the sensor; and $\Delta A_{\mathrm{c}}$, $\Delta B_{\mathrm{c}}$, $\Delta C_{\mathrm{c}}$, and $\Delta D_{\mathrm{c}}$ are the fault matrices of the controlled object model.

If the uncertainties and faults of the controlled object model are comprehensively considered and expressed by the fault vector $f_{\mathrm{c}}$ $(t)$, then the fault models described by Eqs. (9.40) and (9.41) can be rewritten into Eqs. (9.43) and (9.44).

$$\boldsymbol{x}(t+1) = \boldsymbol{A}\boldsymbol{x}(t) + \boldsymbol{B}\boldsymbol{u}(t) + \boldsymbol{E}_1\boldsymbol{n}_1(t) + \boldsymbol{B}\boldsymbol{f}_{\mathrm{a}}(t) + \boldsymbol{F}_1(t)\boldsymbol{f}_{\mathrm{c}}(t) \tag{9.43}$$

$$\boldsymbol{y}(t) = \boldsymbol{C}\boldsymbol{x}(t) + \boldsymbol{D}\boldsymbol{u}(t) + \boldsymbol{E}_2\boldsymbol{n}_2(t) + \boldsymbol{f}_{\mathrm{s}}(t) + \boldsymbol{F}_2(t)\boldsymbol{f}_{\mathrm{c}}(t) \tag{9.44}$$

where $F_1$(t) and $F_2$ $(t)$ are the coefficient matrices of the fault vector $f_{\mathrm{c}}$ $(t)$.

If the noise vector is $n(t) = [n_1(t)n_2(t)]^{\mathrm{T}}$ and the fault vector is $f(t) = [f_{\mathrm{a}}(t)f_{\mathrm{s}}(t)f_{\mathrm{c}}(t)]^{\mathrm{T}}$, then Eqs. (9.43) and (9.44) can be simplified as

$$\boldsymbol{y}(t) = \boldsymbol{G}(z)\boldsymbol{x}(t) + \boldsymbol{E}(z)\boldsymbol{n}(t) + \boldsymbol{F}(z)\boldsymbol{f}(t) \tag{9.45}$$

where $G(z) = C(zI - \mathrm{A})\text{-}1B + D$; $F(z) = [(zI - \mathrm{A}) - 1E_1E_2$; $E(z) = [(zI - A) - 1BI(zI - A) - 1F1 + F2]$; $z$ is a variable matrix; and $I$ is the unit matrix.

The residual $r(t)$can be written as

$$\boldsymbol{r}(t) = \boldsymbol{y}(t) - \hat{\boldsymbol{y}}(t) \tag{9.46}$$

where $\hat{\boldsymbol{y}}(t)$ is the output vector obtained by the residual filter algorithm.

Set a rational threshold $S$, and diagnose the mean value $E[r(\mathrm{t})]$ of the generated residual $r(t)$. If $E[r(t)] > S$, the diagnosed object is considered failed; if $E[r(\mathrm{t})] \leq \mathrm{S}$, the diagnosed object is considered under normal condition.

By using the above method, fault detection based on the analytic model can be realized.

3. **Fault Diagnosis Algorithm Based on Qualitative Model**

The design idea of this fault diagnosis algorithm is that if the object is frequently diagnosed as being in a certain working state within a period of time, then it can be considered really in this working state. The core of this algorithm is to identify the state of the diagnosed object according to the qualitative observation results. The implementation process of this algorithm is divided into the following four steps:

Step 1: establish the qualitative model of the diagnosed object.
Step 2: qualitatively model the observation results of the diagnosed object.
Step 3: adopt Bayes' statistical mathematics theory, that is, to deduce the probability of the object in each working state from qualitative observation results.
Step 4: integrate the above probability over a period of time, and then use the integration result to identify the working state of the diagnosed object.

4. **Comparative Analysis of Three Diagnostic Algorithms**

In Table 9.3, the diagnostic ideas, advantages, and disadvantages of the three algorithms are compared and analyzed.

### *9.2.5 Methods of Recovery and Reconfiguration*

The spaceborne products and equipment with faults or problems can usually be repaired by backup switching, functional redundancy, product-level and system-level reconfiguration, and other methods.

1. **Backup switching**

Backup switching is to autonomously switch to the backup of a faulty component with redundant configuration and with the onboard autonomous fault diagnosis function.

2. **Functional redundancy**

Functional redundancy includes the functional redundancies of spaceborne sensors, actuators, TT&C channels, and information.

(1) Sensor function redundancy
The redundant sensors are managed by software. The central management unit collects the output information of all the sensors, diagnoses the sensors according to the sensor self-check status and the system information, removes the sensor information suspected to be faulty, and autonomously realizes the redundancy management of sensor information. When one sensor fails, its measurement information can be reconstructed by using the output of other sensors.

**Table 9.3** Comparison of three fault diagnosis algorithms

| Fault diagnosis algorithm | Diagnosis idea | Advantage | Disadvantage |
|---|---|---|---|
| Fault diagnosis algorithm based on signal processing | Diagnose the fault according to the directly measurable input/output and its variation trend, or the relationship between the directly measurable signal and the fault | Simple, no need to model the diagnosed object, and easy to implement | The diagnostic process and result cannot reflect the internal working condition of the system |
| Fault diagnosis algorithm based on the analytical model | At first, establish the system model with fault (including state equation and measurement equation), and process it through some algorithms; after deriving the residual error distribution, judge it by fault judgment rule | It can make full use of the knowledge deeply embedded in the system, and can help accurately detect whether there is a fault | It is overdependent on the mathematical model of the system, and sensitive to the modeling error, parameter perturbation, noise, and interference; usually only applicable to linear objects and single-fault analysis |
| Fault diagnosis algorithm based on a qualitative model | Convert the quantitative model and fault judgment rule into a qualitative description. Establish the probability transfer relationship between different system states, and diagnose the fault by using the probability | No need to establish a mathematical system model, low computational complexity, and good real-time performance | It is difficult to establish a qualitative model, which, even when established, will ignore many internal details of the system, resulting in inaccurate diagnosis results |

(2) Actuator function redundancy
When one actuator fails, its control function can be reconstructed by using the redundant actuator of the same type or the actuator of a different type.
(3) TT&C channel function redundancy
The function redundancy of TT&C channels can be realized by setting up multiple TT&C systems and channels in the navigation satellite.
(4) Information function redundancy
The uplink and downlink information of the navigation satellite realizes the redundancy and backup of functions through multiple channels and paths.

3. **Product-level and system-level reconfiguration**

(1) Product-level reconfiguration

In case of the abnormality and failure of attitude control performance, the control function chain can be reconfigured according to the relevant strategy. In case of energy emergency, the function chain of power supply will be reconstructed and the related payload management will be carried out. In case of the failure of Central Processing Unit (CPU), a dedicated reconfiguration module should be used for the CPU reconfiguration.

(2) System-level reconfiguration
For some system-level faults that can't be located or recovered by the CPU, the satellite will take safety as the priority to enter the corresponding safety mode and wait for ground analysis and processing.

## 9.2.6 Autonomous Health Management of BDS Platform

1. **Autonomous daily management**

   (1) Thermal control management

For the thermal control of a navigation satellite, passive thermal control measures, namely the inherent physical properties of thermal control components, are taken as much as possible to meet the temperature requirement of the whole satellite. Meanwhile, some active thermal control measures are also taken to satisfy special temperature requirements and improve the adaptability of the thermal control subsystem. Active thermal control measures are generally realized by controlling the heater ON-OFF and the related parameters.

Most of the self-regulating heaters in the satellite follow a simple ON/OFF control rule. When the temperature of the controlled object is below the threshold, the thermal control software will give a command to power ON the corresponding heater for heating. When the temperature of the controlled object is above the threshold, the thermal control software will give a command to power OFF the heater. The structure of the self-regulating heating circuit is shown in Fig. 9.9.

The self-regulation of the onboard atomic clock heater is the combination of proportional control and ON-OFF control. When the temperature is within the threshold, the proportional control law will govern. When the temperature is beyond the threshold, the ON-OFF control law will govern (Fig. 9.10).

In addition, the contents of daily thermal control management also include the autonomous switching of temperature control threshold, the adjustment of temperature control mode in the heating circuit, and the enabling/disabling settings of primary and backup heating circuits.

(2) Energy management

Energy management is mainly to achieve the battery capacity calculation, battery charge control, battery discharge control, battery settings based on the operational requirements (under different on-orbit conditions), and various protections.

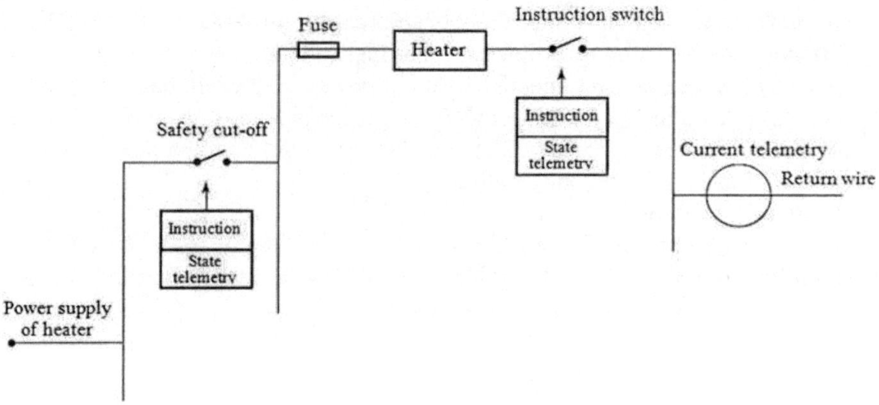


**Fig. 9.9** Structural diagram of self-regulating heating circuit

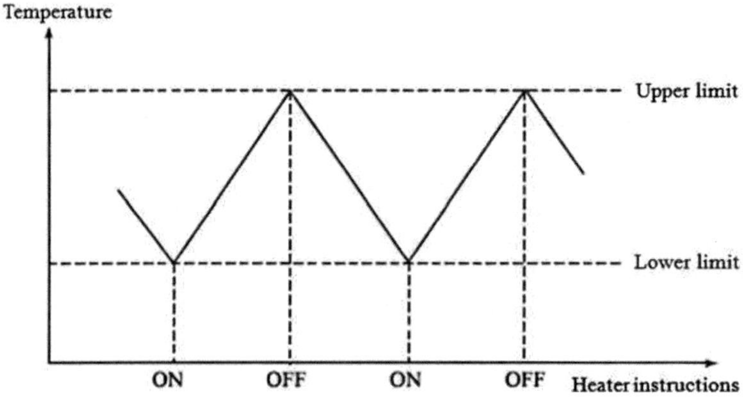


**Fig. 9.10** Heater ON/OFF control law

Taking the management of lithium batteries as an example, the autonomous energy management of a satellite is achieved in two modes: Earth shadow season mode and long light period mode. In the mode of Earth shadow season, the autonomous energy management software is mainly to realize the battery charge management (including overcharge management), discharge management (including overdischarge management), charging-discharging capacity calculation, and other functions. In the mode of the long light period, the software is mainly to realize the battery half-charge layup, SOC detection, and other functions. Meanwhile, in either of the above two modes, the software will manage the battery equalization, while collecting, transmitting, monitoring, and giving an alarm against the power supply data.

The autonomous energy management software works in the Earth shadow season mode by default, which can be switched by the ground system giving a command. The criterion for the conversion between Earth shadow season mode and long light period mode is as follows. When receiving the out-of-Earth-shadow-season symbol from the satellite system, the software will autonomously switch to the long light period mode. When receiving the in-Earth-shadow-season symbol from the satellite system, the software will autonomously switch to the Earth shadow season mode.

(3) Attitude management

The satellite can autonomously control its attitude and orbit. The spaceborne central computer, together with the attitude sensor and the control actuator, constitutes a closed-circuit control system. Under normal circumstances, this system can accomplish the attitude control, the sun-pointing tracking, and control of solar array as well as the management of various sensors and actuators to meet the requirements for thermal control and energy.

Take the interference protection of the Earth sensor as an example. Earth sensor in a navigation satellite (Fig. 9.11) can automatically ban a detector channel, that is, when one probe fails or is interfered with, its use shall be prohibited, and the other three probes shall be used. When the interference is over, the interfered probe needs to be reactivated to work with the other three probes.

The possible interference sources are mainly the sun, the moon, the Earth's north and south poles, and others. The interference is analyzed on the satellite according to the position of its orbit and interference source and the field of view. It shall prohibit the use of the corresponding probe when the interference occurs, and reactivate it after the interference ends.

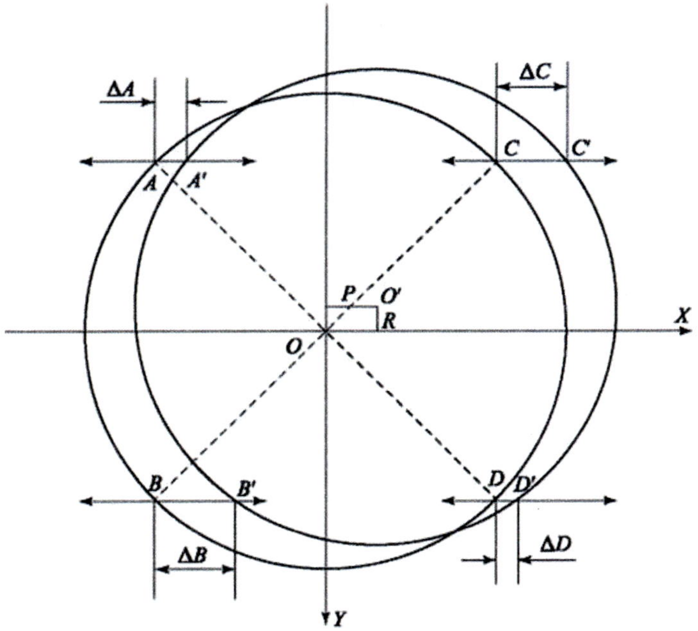


**Fig. 9.11** Operating principle of infrared Earth sensor

(4) House keeping

The house-keeping functions mainly include the dynamic organization and scheduling of navigation satellite telemetry, the storage and downlink of delayed telemetry, the processing and distribution of uplink direct-discrete telecontrol command and injected data, the processing and execution of time-scale command, and the time management.

The satellite orbit has some untraceable arcs, in which the satellite telemetry data cannot be transmitted to the ground in real time. The delayed telemetry data in the invisible arcs is processed in the following two ways.

(1) When the satellite networking is not completed or the local satellite data cannot be transmitted to other satellites through an inter-satellite link, the telemetry data of the satellite in the non-visible arcs will be stored in the local satellite memory. When the satellite passes territory, the stored telemetry data will be transmitted to the ground TT&C station.
(2) When the satellite networking is completed and the inter-satellite link is successfully established, the data of the local satellite in the non-visible arcs will be transmitted, via the inter-satellite link, to a satellite in the observable and controllable arc for downlink.

During the autonomous operation, the key telemetry parameters of the satellite are stored. After the satellite-ground TT&C link is restored, those parameters will be transmitted down to the ground so that the health status of the satellite system can be analyzed.

(5) Management of various equipment and components

In the navigation satellite, various equipment and components shall be managed onboard according to the contents of a flight mission, so as to meet the requirements for equipment/component states in each flight stage. For example, in the launch period, the fixed amplifier for TT&C should be set at low power in order to prevent the low-pressure discharge. When the vacuum degree reaches the required level on the orbit at a certain altitude, the fixed amplifier should be set at high power to lower the requirements for ground receiving equipment. For another example, during the satellite flight, the antenna should be set to either directional working mode or omnidirectional working mode as required.

To achieve the above management, the requirements of the whole satellite as well as the operational requirements of equipment and components should be analyzed, and a reasonable management strategy should be designed and implemented in the software.

2. **Autonomous Fault Management**

The autonomous fault management system on the BeiDou satellite platform is a hierarchical structure, where the faults at different levels are monitored and processed by different modules. According to the possible influence on components, satellite

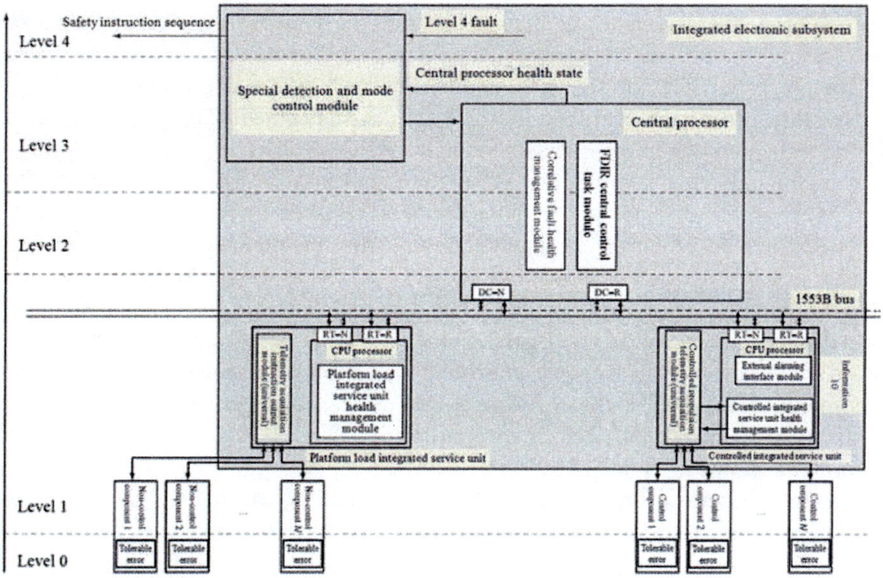


**Fig. 9.12** BDS 5-level autonomous fault management framework

system and functions, and satellite CPU, the faults can be divided into 5 levels (Fig. 9.12).

**Level-0**: tolerable error. They can be handled and recovered autonomously, without the intervention of integrated electronic software/hardware. The occurrence of a level-0 fault has no impact on the functions of stand-alone units, subsystems, and even the whole satellite.

**Level-1**: component-level independent fault. The level-1 fault refers to the independent faults in components (including stand-alone units, single boards, modules, etc.).

The level-1 fault is also divided into two sublevels:

Level 1-a: the faults in non-control units and Integrated Service Unit (ISU).
Level 1-b: the faults in the sensor or actuator of attitude and orbit control subsystem.

**Level-2**: correlative faults. The level-2 fault refers to the correlative fault related to satellite energy or attitude. They are detected and restored by the corresponding health management module, and can lead to the reconfiguration of the relevant satellite function chain.

**Level-3**: the fault in the hardware or software of the central manager unit.

**Level-4**: top-level fault. The level-4 fault refers to the loss of the Earth sensor signal on the satellite. This fault cannot be detected and recovered by low-level (levels 0–3) fault mechanisms, but can force the satellite into the sun-pointing safety mode, and then wait for the ground analysis and handling.

The processes of dealing with the faults at every level are as follows.

Handling process of level-0 fault: The level-0 fault occurs in the components, and can be recovered by the components autonomously.

Handling process of level-1 fault: After the occurrence of a level-1 fault, the faulty equipment will be detected and isolated and then replaced by redundant equipment. This fault has no effect on the navigation function of the satellite. The handling process of a level-1 fault is shown in Fig. 9.13.

Handling process of level-2 fault: When a level-2 fault occurs, the satellite platform cannot keep its indicators normal. In the process of fault isolation and recovery, the related platform performance indicators will decline. However, after this fault is correctly dealt with by the software of the central management unit, the whole satellite can regain normal performance indicators. A level-2 fault can be handled in the process shown in Fig. 9.14.

Handling process of level-3 fault: The level-3 fault is detected and handled by a module dedicated to detection and mode control. After this fault occurs, the backup or emergency central computer will be switched according to the fault handling logic.

Handling process of level-4 fault: The level-4 fault mainly includes the energy fault and the loss of the Earth sensor signal. After such a fault occurs, a special detection and mode control module will generate the corresponding command sequence, which will enable the spacecraft into the sun-pointing directional safety mode to wait for ground handling.

Through the autonomous health management of the satellite, faults can be detected and handled to ensure satellite safety and minimize the impact of an occurring fault on navigation tasks.

## 9.3 Navigation Satellite Autonomous Integrity Monitoring

For navigation satellites, it is essential to ensure the integrity of the user navigation service. Therefore, the service information of the navigation signal must be completely monitored by itself, that is, the navigation satellite autonomous integrity monitoring. Integrity is a confidence evaluation index of the services provided by the system. When the navigation system cannot be used for certain scheduled operation, it shall be able to send timely and effective alerts to users. Integrity is especially important for users in the life safety field, such as civil aviation and automatic driving. The navigation system service with poor integrity will probably cause significant casualties or property damage to users in the life safety field using the system.

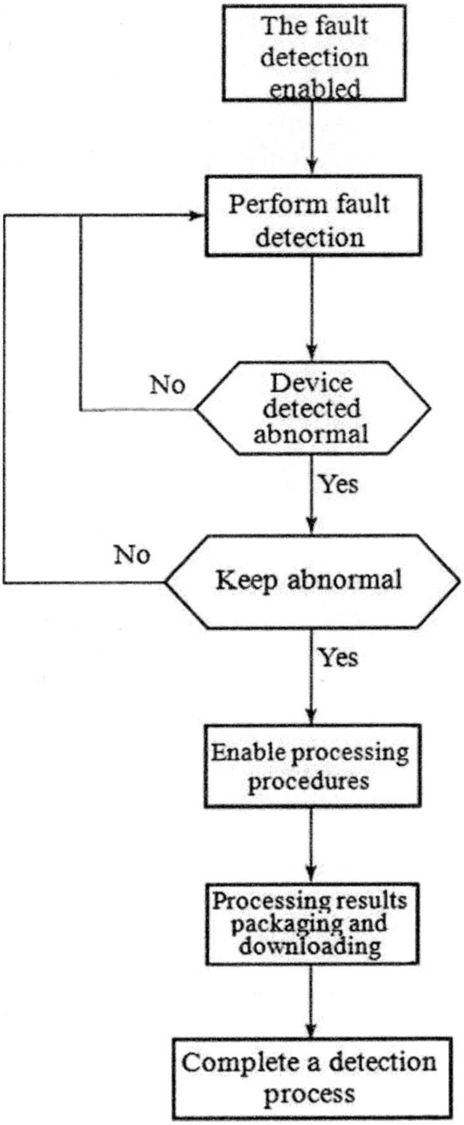


**Fig. 9.13** Handling process of level-1 fault

## *9.3.1 Integrity Monitoring Demands and Methods*

With the development of GNSS, the navigation accuracy demand of users has been met. However, in some special applications, such as the precision approach of aviation, vessel approach, and automatic driving, users have very high requirements for the integrity of the navigation system. According to different service stages, when

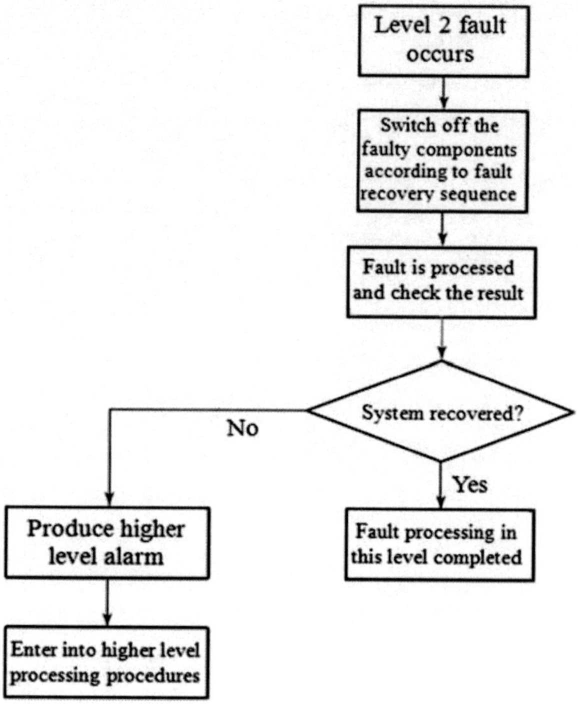


**Fig. 9.14** Handling process of level-2 fault

the navigation service error exceeds the alarm threshold, an alarm message must be sent within several minutes to 1 s.

In most navigation applications, there is not a clear quantitative requirement on integrity. At present, only aviation puts strict requirements on integrity as well as performance on accuracy, continuity, and availability.

Most of the GNSS systems only provide by itself the integrity services from the transoceanic crossing to the route stage. Integrity services for approach in non-precision and precision Class I are provided by the corresponding satellite-based augmentation system (SBAS), which is developed by various countries or regions, such as the Wide Area Augmentation System (WAAS) by the USA and the System Differential Correction and Monitor (SDCM) by Russia, European Geostationary Navigation Overlay Service (EGNOS) by Europe, Multi-Functional Satellite Augmentation System (MSAS) by Japan, and GPS-Aided GEO Augmented Navigation (GAGAN) by India. As BDS-2 and BDS-3 provide GNSS as well as SBAS service, they also provide integrity services at the same time.

SBAS implements the wide-area differential technology with wide-area differential integrity monitoring techniques; meanwhile, it makes use of the observation data of the deployed monitoring stations to analyze and process the integrity of the satellite ephemeris, the clock difference, and the ionospheric grid corrections, and obtain the corresponding integrity of UDRE, GIVE, and compensation parameters. This integrity information will be broadcasted to the user along with the wide-area differential corrections. By the user side, the corresponding integrity is analyzed and processed while processing wide-area differential positioning. The wide-area differential integrity monitoring alarm time is generally 6 s, and the integrity risk probability is $2 \times 10^{-7}$/150 s, which meets Class I precision approach performance requirements.

The Class II/III precision approach with extremely severe alarm time of less than 2 s is generally serviced by ground-based augmentation systems, which implements the local area differential technology by the local area differential integrity monitoring technology; meanwhile, it makes use of the observation data of the deployed monitoring stations to analyze and process the pseudorange observations' differential correction, and obtain integrity information, which will be broadcasted to the user along with the local area differential correction. By the user side, the corresponding integrity is analyzed and processed while processing the local area differential positioning. The local area differential integrity monitoring alarm time is generally 2 s, and the integrity risk probability is $1 \times 10^{-9}$/15 s, which meets the requirements of Class II and III precision approach performance.

In the Receiver Autonomous Integrity Monitoring (RAIM) technology, in the user receiver, positioning is processed based on observations from multiple navigation satellites; meanwhile, the anomaly satellites are detected, analyzed, and excluded based on multi-satellite redundant observations to obtain positioning integrity information. There is almost no alert delay in this technology, but a monitor significant redundant satellite is needed at the same time. With the relatively weak integrity monitoring performance, it is generally only used in the route stage. In the user receiver, the supplemental information, such as inertial navigation or barometric altitude, are combined to implement integrity monitoring. This technology is called Aircraft Autonomous Integrity Monitoring (AAIM), which complements the advantage of the satellite-based augmentation system and is called as the "Relative" Receiver Autonomous Integrity Monitoring (RRAIM). The integrity monitoring and processing technology based on a multi-satellite system combination is called the Advanced Receiver Autonomous Integrity Monitoring (ARAIM), whose integrity performance is significantly improved.

In recent years, the Satellite Autonomous Integrity Monitoring (SAIM) technology has emerged. In this technology, the integrity monitoring receivers are equipped in navigation satellites to directly monitor the navigation signals broadcasted by the same satellite. It generates and broadcasts alert signs immediately to the users once the anomaly is detected. Since the integrity is monitored directly on the satellite, the time delay of the ground monitoring is avoided and the time is shortened to alert. This is the advantage of SAIM and an important supplementation of the integrity monitoring systems. With SAIM payloads, the system integrity service capability of BDS-3 has been enhanced.

### 9.3.2 Analysis of Navigation Satellite Integrity Anomaly Events

The navigation satellite integrity anomaly event refers to the event in which the navigation satellite has a fault or performance degradation for some reason, causing the user receiver to generate an excessive positioning error, and which is not monitored by the system.

SAIM with multiple satellite receivers can simultaneously monitor the downlink navigation signal and generate alert information by redundant voting. Once multiple receivers detect an anomaly, an alert message is immediately generated and broadcasted to the user. However, SAIM can only monitor the satellite itself, but not able to detect the anomalies generated in the space transmission and user segment.

Navigation satellites may have a variety of faults, such as abnormal power supply to the platform, satellite clock in-orbit failure, and navigation signal interruption, and not every fault has a threat to the integrity for the user. The navigation receiver positioning error can be defined by formula (9.47):

$$\sigma_{\mathrm{u}} = \mathrm{GDOP} \times \sigma_{\mathrm{UERE}}, \tag{9.47}$$

where $\sigma_{\mathrm{u}}$ represents the user positioning error, GDOP represents the geometric accuracy factor, and $\sigma_{\mathrm{UERE}}$ is the ranging error.

The source of satellite service information integrity anomaly events in the space segment mainly includes satellite clocks, navigation signals, and a navigation message.

(1) The satellite clock provides frequency signals for navigation signals. Once anomalies occur, all navigation signals will be out of order. The main types of failure include phase hopping, phase slip, frequency hopping, frequency slip, and random phase jitter. Phase hopping will cause a fast change of the pseudorange, and other faults will cause distortion of the pseudorange, which will affect users.
(2) Navigation signal anomalies mainly include signal power, signal delay, code/carrier phase inconsistency, and pseudocode distortion anomalies. The abnormal reduction of signal power will reduce the C/N of the received signal and increase the measurement errors. The signal delay anomaly will cause pseudorange deviation, the code/carrier phase inconsistency will affect the high-precision users, and the pseudocode distortion anomalies will lead to the distortion of correlation peaks. Different front-end filter bandwidth and correlator intervals will lead to the inconsistency of receiver ranging, which cannot be eliminated by the differential method.
(3) Navigation message anomalies mainly include navigation data bit inversion and navigation message error. Abnormal navigation message can lead to incorrect interpretation of navigation information, and affect positioning.

Figure 9.15 shows the anomaly events of the navigation satellite service

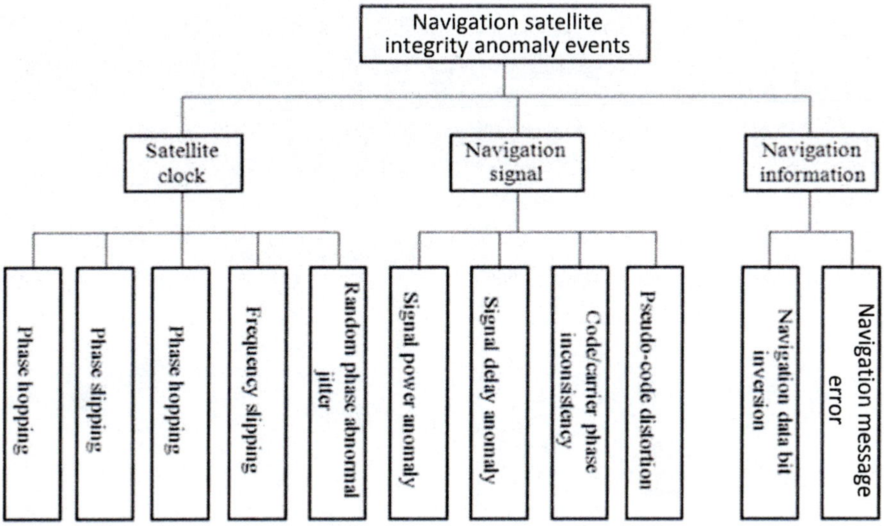


**Fig. 9.15** Space Integrity Anomaly Events

information.

### 9.3.3 *Principle of SAIM*

SAIM monitors satellite integrity by the satellite itself. Therefore, every satellite in the constellation is under monitoring. Once an abnormal situation occurs, the alarm information will be converted into the navigation signal and the user is alerted at the next signal update after confirming that there is no failure of the monitoring receiver. In this way, the integrity risks caused by satellite anomaly are greatly reduced.

The targets of satellite autonomous integrity monitoring are satellite clocks, navigation signals, and navigation message. The satellite SAIM equipment receives navigation signals, obtains the pseudorange, carrier phase measurement values, monitoring correlation values, signal power, and receives time-frequency signals to obtain satellite clock phase hopping measurement value and frequency hopping measurement value. These values are then processed and synthesized to obtain a Signal Integrity Flag (SIF). At the same time, the satellite transmits the observation data, such as pseudorange and carrier phase measurements, back to the ground to evaluate the performance of SAIM. The overall process of BDS satellite SAIM is shown in Fig. 9.16.

Currently, BDS satellite SAIM only generates the alert sign for rapidly varying anomalies, while, for slowly-varying anomalies, the data is analyzed by the ground system to determine whether to make alert signs. If the satellite clock phase or frequency is abnormally hopped and exceeds the preset alarm threshold, all signals

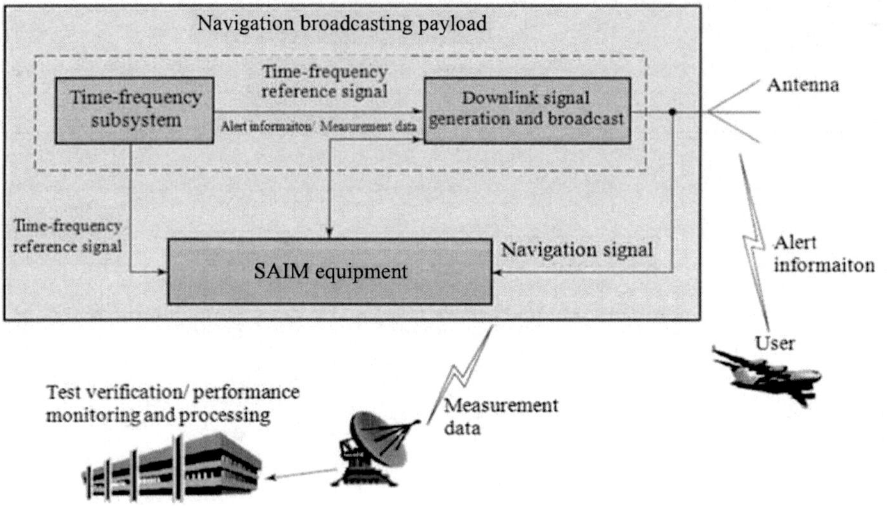


**Fig. 9.16** Overall process of BDS satellite SAIM

will sound an alarm. The signal branch will alert if the delay or power of the signal branch is abnormally hopped. At the same time, the satellite transmits the SAIM measurements to the ground for further analysis and processing, such as long-term monitoring of pseudocode/carrier phase consistency, and evaluation of S-curve zero-crossing deviation caused by correlation peak distortion.

SAIM alarm mode includes information alarm and signal alarm. Information alarm is to embed the alert signs into navigation information and broadcast them to users along with other navigation message. Signal alarm is to set the navigation signal spreading code to 0101 sequence, that is, to use the nonstandard code for the alarm. The alert time of alarm information is mainly limited by the broadcast period of the navigation message, and the time length of the alarm signal is the delay time from the occurrence of a fault to the generation of an alarm signal.

Currently, SAIM mainly monitors rapidly varying faults of satellite clocks and navigation signals. In fact, in addition to the rapidly varying anomalies, there are also slowly varying anomalies, such as slowly varying anomalies of orbits and clock difference. Monitoring orbit and clock difference to further improve the User Range Accuracy (URA) will be an important technological development trend of SAIM in the future.

By analyzing the on-orbit operation of the navigation satellite, it is found that the slowly varying anomaly of the satellite clock is the most common failure in an on-orbit operation of the navigation satellite. The orbit and the clock difference will be monitored on the basis of two-way ranging and communication of the inter-satellite link, where the local satellite clock is connected with other satellite clocks in the constellation for comparison with multiple atomic clocks. In such a way, the slowly varying fault monitoring of the orbit and the clock difference will be supported, forming the monitoring capability of the URA.

### 9.3.4 Methods of SAIM

1. **Navigation Signal Monitoring**

(1) Signal Power Monitoring

The signal power will reduce the $C/N_0$ of the user receiver when DAC or amplifier is abnormal. Ideally, the signal power fluctuates in a certain range, usually less than 0.5 dB. The power measurement can be obtained by the following equation:

$$\mathrm{PW} = 10 \times \lg\left(I^2 + Q^2\right) + \Delta\mathrm{PW}, \tag{9.48}$$

where PW is the signal power, $I$ is the in-phase branch correlation value of the received signal, $Q$ is the orthogonal branch correlation value of the received signal, and $\Delta$PW is a constant for power zero calibration.

Any abnormal power changing will be detected immediately, and an alarm message will be reported if the abnormal value exceeds the preset threshold.

(2) Signal Delay Monitoring

The signal can be delayed by a pseudocode abnormal hopping, instruction abnormality, or other payload faults. Under normal circumstances, the pseudorange of the monitoring receiver shall be a Gaussian distribution that fluctuates within a certain range. However, the frequency of the receiver is different from satellite clocks, and the measured pseudorange has a linear drift. Therefore, the clock difference shall be compensated during the signal delay measurement. The actual signal delay measurement can be obtained by the following equation:

$$T_{\mathrm{Delay}} = \mathrm{PR} + \Delta T, \tag{9.49}$$

where $T_{\mathrm{Delay}}$ is the signal delay measurement, PR is the pseudorange measurement, and $\Delta T$ is clock difference.

If the signal delay is changed due to an abnormality occurring in the payload, $T_{\mathrm{Delay}}$ will be rapidly changed; if the abnormal value exceeds the threshold, an alert message will be generated immediately.

(3) Code/Carrier Phase Inconsistency Monitoring

In general, for high-precision users, the carrier phase measurement value should not exceed the observation range of navigation solution. The inconsistency of the code/carrier phase observed on the ground is mainly caused by the ionosphere, but it may also be possible that the code/carrier phase inconsistency occurs on the satellite, because the frequency signal generating pseudocode phase and carrier phase are from different paths, and the abnormality in either path will lead to the difference of pseudocode phase and carrier phase, even if both the pseudocode and carrier can be traced to the satellite clock.

Code-Carrier Bias (CCB) monitoring method is to calculate whether the difference between code pseudorange increment (in m) and carrier pseudorange increment (in m) exceeds the threshold value. The specific steps are as follows:

a) Calculate the original code pseudorange increment (in m) and carrier pseudorange increment (in m):

$$\Delta \mathrm{PR}(k) = \mathrm{PR}(k) - \mathrm{PR}(k-1) \tag{9.50}$$

$$\Delta \varphi(k) = \varphi(k) - \varphi(k-1) \tag{9.51}$$

(b) Calculate the difference between the code pseudorange increment and the carrier pseudorange increment (in m):

$$\Delta I(k) = \Delta \mathrm{PR}(k) - \Delta \varphi(k) \tag{9.52}$$

(c) Compare the code/carrier increment difference with the threshold value to determine whether they are consistent.

If $\Delta I(k) > \mathrm{TH}_{\Delta I}$, the pseudocode increment is inconsistent with the carrier increment, and if $\Delta I(k) \leq \mathrm{TH}_{\Delta I}$, the pseudocode increment is consistent with the carrier increment.

2. **Satellite Clock Monitoring Technology**

The clock of the navigation satellite is usually a rubidium clock or hydrogen clock. The reference clock used for autonomous integrity monitoring is generally a high-stability crystal oscillator. High-stability crystal oscillator has good short-term stability, but its long-term stability is worse than a satellite atomic clock. Therefore, autonomous integrity monitoring is used to monitor the rapid change of satellite clock phase or frequency, rather than the reduction of long-term stability, which can be detected through the long-term observations of the ground monitoring and evaluation system.

Note that $t_k$, $t_{k-1}$, and $t_{k-2}$ are the clock differences between the satellite clock and the local reference crystal oscillator for three consecutive moments, where $k$ represents the current moment. It is assumed that the satellite clock has a phase change $x$ at the current moment. In order to eliminate the phase shift caused by the frequency offset of the satellite clock from the local reference frequency, subtraction is performed on the adjacent measured values:

$$\Delta T_1 = t_k + x - t_{k-1} = \Delta t_{f(k,k-1)} + x \tag{9.53}$$

While continuing the evolution and progress of satellite navigation systems, various countries develop the space-based and ground-based augmentation systems to greatly improve the navigation accuracy, integrity, continuity, and availability, thus promoting the rapid development of the navigation application industry.

At present, all the countries building the satellite navigation systems are also constructing their Satellite-Based Augmentation Systems (SBASs) and expanding their Ground-Based Augmentation Systems (GBASs). With the advantages of wide coverage and low operating cost, the SBAS can greatly improve the positioning accuracy, integrity, availability, and continuity of satellite navigation systems. All the countries are developing the inter-satellite augmentation system. In recent years, the navigation augmentation system based on the low-orbit mobile communication system has also become a research hotspot.

As an efficient and reliable means to improve the service quality of the satellite navigation system, the GBAS has been recognized by the global GNSS community and has become an important development direction in the GNSS field. The combination of GBAS and "Internet+" will eventually evolve into a cross-industry, cross-region, and cross-country "cloud" service network. This network will bring great changes to surveying and mapping engineering, engineering construction, land management, urban planning, national condition census, and other fields so as to provide strong support for national construction.

(3) Communication and navigation integrated service

While continuously improving the PNT service performance, each GNSS actively develops its special services and gradually achieves the multi-task and multisystem integration. The multitask integration is the integration of such tasks as search and rescue, data processing and transmission, electromagnetic monitoring and nuclear explosion detection, and scientific testing. The multisystem integration is the integration of the navigation system with other PNT systems. In this process, the navigation system gradually becomes the core and basis, and integrates with other space-based systems (such as reconnaissance, communication, and monitoring systems) into a space-based information system.

In addition to providing basic navigation services, the GPS is also equipped with a nuclear explosion detection system, disaster relief system, Blue Force identification system, and other payloads. The Galileo carries the search and rescue system, laser reflector, space particle detector, and other payloads. China's BDS is provided with the payloads such as laser reflector, space environment detector, electromagnetic environment monitoring system, search and rescue system, and data link to continuously develop incremental services.

(4) Global GNSS toward compatibility and interoperability

The current GNSSs mainly include four systems: GPS, GLONASS, Galileo, and BDS. In 2004, Galileo and GPS signed an agreement to achieve mutual compatibility and interoperability. Through adjusting the signal modulation mode, the Galileo system has reduced the interference to GPS military signals. Through the cooperation between Galileo and GPS, the modulation modes such as BOC and MBOC as well as the error correction codes such as Weil code, random spread spectrum code, convolution code, and LDPC code have been proposed to improve the interoperability and robustness of the signals from both sides. Considering the compatibility and interoperability with Galileo and GPS in the construction of Russian GLONASS, the spread spectrum signal system will operate on the subsequent GLONASS-K satellite and in the L1 and L5 frequency bands of GPS.

Under the background of the competition and cooperation among GNSS systems and the construction of China's Navigation Satellite System (NSS), China's NSS has also achieved good compatibility and interoperability with other systems to cope with the situations in the future such as the same frequency band occupied by various GNSS systems, a large number of design constraints, and sophisticated coordination.

(5) Research and construction of national integrated Positioning, Navigation, and Timing (PNT) system

Traditional navigation means, for example, inertial navigation, astronomical navigation, ground radar navigation, and Roland/Omega navigation system, have more or less the problems such as limited application areas and occasions, low precision, and complicated and expensive user equipment, so they cannot fully meet the corresponding requirements. Compared with traditional navigation means, the service area of the satellite navigation systems has covered all the land, sea, and air locations around the world, up to the low-earth satellite orbit at the attitude of 1000 km, and is expanding toward high orbit. The positioning accuracy of about 10 m and the time accuracy of tens of nanoseconds can be achieved at almost any time and anywhere in the world. The Satellite navigation (Satnav) chip is embedded in almost every smartphone, in a way that we don't even feel it. The measuring receiver can achieve the centimeter-level positioning accuracy and meet the requirements of basic surveying and Earth parameter measurement.

Even with so excellent performance, the satellite navigation system still cannot fully meet the increasing needs of various users as the product of social economy development. This is mainly reflected in the following areas. In the areas with physical barriers, such as forests, urban valleys, and indoor, underground and underwater areas, the system still cannot provide satisfied services. In the areas with electromagnetic interference, the service quality of this system will be greatly reduced. The space beyond the geosynchronous orbit still can't be effectively covered by the current navigation signal. Moreover, the system itself faces the risks in availability and stability.

So far, a large number of PNT technologies and systems have been developed and deployed all over the world, but have not become a unified whole to give full play to the system advantages. To this end, it is expected that the existing PNT capabilities can be integrated at the national level and their development direction can be explored in order to meet the future needs of military, civil, and commercial high-precision PNT services. This gives birth to the concept of national PNT system that integrates various PNT means, resources, and methods and can fully meet all the PNT needs.

At the end of the 1990s, the initial design of the EU Galileo program already incorporated the main basic features of the PNT system. But later, due to the limitations of the budget and management system, this program had to shift the focus of work to the direction of satellite navigation. While developing and deploying the new generation of GLONASS satellite, Russia upgrades its ground-based radio navigation system, and explores and coordinates the navigation function based on land-based radar signal to actively develop the PNT system. The UK has put forward the concept of "flexible PNT system", which mainly includes the "Guard" and "Sentinel" initiatives. It is also systematically considering the construction and technical development of the PNT system. The US is the first country to put forward the concept of the national PNT system and define this system as an important national infrastructure. At present, specific development goals and plans have been developed.

With the evolution and innovation of PNT technology and the growth of user demand, the PNT system will be continuously developed and updated. It is a realistic way to meet the needs for future PNT services systematically through the inheritance, integration, and development of the existing PNT system. On this basis, the concept of national integrated PNT system has been proposed as a national information infrastructure that serves national defense, economy, and society, establishes and maintains national spatial and temporal references, broadcasts and acquires spatial and temporal information, applies the PNT services, and performs other tasks, under national unified organization and coordination.

The PNT system has the following main features. It has evolved from the mode where the NSS in current wide use plays a dominant role in the mode where various navigation techniques are integrated and systematically developed on the basis of NSS. Through this process, the service coverage has been significantly expanded. The coverage of satellite navigation, originally including land surface and near space, has been extended to underwater, underground, electromagnetic-interference and physical shielding environments. The navigation indicators such as convenience, continuity, reliability, and accuracy are improved significantly, so is the cost-effectiveness of navigation services. The construction of a national integrated PNT system with satellite navigation as the core is a strategic focus of national security, and has become the consensus of developed countries and military powers.

## 10.2 Development Trend of Satellite Navigation Technologies

The development of satellite navigation technologies is an important support for NSS development, which, in turn, greatly promotes the development of related technologies while improving the NSS performance. Meanwhile, the research and development of emerging technologies will promote the revolution of navigation systems.

According to the development trend and conception of NSS, the technological development in the following aspects will improve the capability and level of NSS in the future to some extent.

### 10.2.1 New Navigation Signal Structure

The design of global satellite navigation signals must satisfy many constraints and have international compatibility. Future navigation signals shall be characterized by multiple service types, high precision, strong signal robustness, strong anti-jamming ability, and good compatibility and expandability. We can learn from the theories of the ground mobile communication network to study the potential of applying new modulation mode to the navigation field, or study new efficient signal modulation mode based on these methods.

In the face of increasingly tight frequency resources, it is particularly urgent to develop the navigation signals of higher frequency, respond to the demand for multi-task integration and task expansion, and comprehensively study the new navigation signal system that can meet the needs of various users.

The new signal system shall focus attention on the research of consumer users' demand for low power consumption and efficient spectrum utilization.

### 10.2.2 Spaceborne Cold Atomic Clock Technology

In the twenty-first century, with the development of laser cooling and confinement technology, the laser-cooled ultracold atomic fountain clock has become the most accurate atomic clock on the ground and the first frequency standard defined as the international second, with an accuracy of $10^{-16}$ orders of magnitude. The most accurate atomic clock in current use is the fountain clock. The fountain clock has such high accuracy mainly due to the application of laser cooling technology. Compared with the previous cesium-beam atomic clock, the cold atomic fountain mode has extended the time of interaction between atoms and microwaves by two orders of magnitude. In the operation of the ground fountain clock, the time interval between two atom-microwave interactions is generally around 1 s, and the width of frequency discrimination spectrum line is limited to about 1 Hz, due to the effect of gravity. So the accuracy and stability can hardly be further improved. However, under the condition of space microgravity, the time of interaction between laser-cooled ultra-cold atoms and microwave cavity can be increased by an order of magnitude, so that the accuracy of the atomic clock can be improved correspondingly. Currently, spaceborne cold atomic clocks are designed with an accuracy of $10^{-17}$ orders of magnitude. The European Space Agency (ESA) and the US National Aeronautics and Space Administration (NASA) have carried out the research programs of spaceborne cold atomic clocks in succession. Currently, the program with certainty is the ACES program of ESA.

In addition to the cold atomic clock working in the microwave band, the cold atomic clock working in the optical band (namely optical clock) is also a research hotspot. Based on theoretical analysis, the short-term stability of the optical clock is several orders of magnitude higher than that of the microwave clock. The representative neutral atomic optical clocks are the cold calcium-atom fountain clocks made by the US National Institute of Standards and Technology (NIST) and the Physikalisch-Technische Bundesanstalt, respectively, with the uncertainties of $4.3 \times 10–14$ and $1.2 \times 10–14$, respectively, and the stability of 10–16 orders of magnitude.

The main focus of the spaceborne cold atomic clock is its adaptability to space environment and its long life and high reliability in the application to navigation satellite space. Moreover, due to the accuracy improvement of this clock, more accurate timing and calibration are required to ensure the time-frequency accuracy of the NSS space segment. Therefore, it is necessary to match the accuracy of inter-satellite and satellite-Earth time-frequency transfer with that of the spaceborne cold atomic clock.

### *10.2.3 MEO/HEO and Deep-Space Satellite Navigation Technology*

In modern society, HEO satellites are playing an increasingly important role in communication, navigation, meteorology, and early warning. Previous space launch missions mainly rely on the support of ground TT&C stations and ocean-going vessels to complete the in-orbit navigation so that the limited ground TT&C resources are under heavy scheduling pressure. The NSS can provide services for the spacecraft in MEO and HEO, thus greatly reducing the workload of ground TT&C stations and ocean-going vessels. Through its application, the service potential of satellite navigation can also be further tapped and utilized. On the other hand, with the continuous expansion of current and future flight missions to outer space, and with the increasing prominence of navigation needs of the probes in MEO and HEO and even in deep space, the research on the application of satellite navigation to those probes in space missions has become extremely valuable.

The topic of GNSS Space Service Volume (SSV) first appeared on the ICG platform at the ICG-5 conference in 2010, where it was put forward by the experts from the United States in the form of a released reference document. In 2011, it was formally reported at the ICG-6B conference. NASA elaborated on the definition of SSV in its report, and introduced the SSV service performance of the GPS system. Besides, from the perspective of space application demand, it suggested all the GNSS systems to release their SSV service performance parameters, improve the GNSS space service standard, and thus enhance the service performance of GNSS systems in

MEO and HEO. Europe and Japan quickly responded to the US initiative at the 2012 ICG-7B conference. The ESA and the Japan Aerospace Exploration Agency (JAXA) introduced the SSV performance of Galileo and QZSS systems, respectively. Both China and Russia started to participate in the discussion of this topic in the B-group from the 2013 ICG-8conference, where they made the relevant reports. Currently, all the GNSS system suppliers are coordinating the release of the SSV Manual. It can be predicted that with the full establishment of the SSV service performance indicator system and the development of MEO/HEO GNSS receivers, the GNSS will provide extensive navigation and timing services for the space vehicles in MEO and HEO and near-Earth space in the near future.

With the deepening of space knowledge, all the countries have quickened their pace of deep-space exploration. The scope of exploration missions is also becoming wider and wider, including the solar system missions (such as Earth missions, moon mission, and Mars mission) and the space missions beyond the solar system. All the countries have carried out systematic research. NASA proposed the space communication and navigation system architecture, a communication and navigation network composed of ground-based units, lunar relay units, and Mars relay units. This architecture can provide communication and navigation services for the spacecraft working in the solar system. At the same time, the X-ray pulsar navigation, Lagrange-point satellite navigation, and other technologies have become the research hotspots. With the advancing of universe exploration, the scope of NSS technologies will be gradually extended.

International interoperability shall be the focus of satellite navigation in MEO and HEO and deep space. Exploring the universe is the common goal of all mankind. It requires the coordination and communication among the GNSS systems of various countries to jointly support the expansion of satellite navigation coverage.

### *10.2.4 Long-Term Operation Technology of Autonomous Constellation*

Autonomous constellation navigation is an implementation process where, without ground system support for a long time, the satellites in the constellation continuously correct the longtime forecasted ephemeris and clock parameters injected by ground stations through the inter-satellite bidirectional measurement and data exchange as well as spaceborne processor filtering, autonomously generate the navigation message, and maintain the basic constellation configuration, in order to meet the user's demand for high-precision navigation and positioning applications.

The utilization of autonomous navigation technology can effectively reduce the number of ground TT&C stations and the number of information injections from ground stations to satellites, bring down the cost of system maintenance, monitor the integrity of satellite navigation information in real time, and enhance the system viability. Inter-satellite measurement and communication link is the core technology of autonomous navigation for satellites. It has become one of the development trends of future NSSs to construct an inter-satellite link through which the navigation constellation can achieve highly autonomous longtime operation.

The development of highly autonomous longtime constellation operation technology shall focus on the high-precision conversion between inertial space coordinates and Earth-fixed coordinates after leaving the ground operation control system, the autonomous correction of spatial ionosphere parameters by navigation constellation, and the control of the deviation between the time reference independently maintained by navigation constellation and the ground UTC time.

### *10.2.5 Advanced Inter-satellite Link Technology*

The inter-satellite link technology makes it possible to achieve autonomous constellation operation, constellation TT&C, satellite-ground combined orbit determination, and other functions. With the increasingly pressing need for multisystem and multitask fusion, the role of the inter-satellite link has become more and more prominent. The development of new technologies such as laser communication, V frequency band, and quantum communication has provided many technical approaches for the implementation of the inter-satellite link.

Owing to significantly reduced system size, mass and power consumption as well as strong electromagnetic interference resistance and confidentiality, the laser technique will become the key choice for the inter-satellite link in the near future. With a precisely pointing high-speed laser inter-satellite link, the satellites will provide a faster inter-satellite link network to achieve a large inter-satellite information transmission capacity and accurate range measurement so as to provide a wide space for the expansion of system functions.

The development of an advanced inter-satellite link shall focus on high-precision measurement and high-capacity transmission, as well as application-oriented high-precision frequency transfer. Especially when establishing an inter-satellite link with the laser technique, a synchronous relationship shall be built between laser frequency and navigation satellite time reference. Femtosecond comb is a technique necessary for future navigation satellites equipped with a laser inter-satellite link.

### *10.2.6 X-Ray Pulsar Navigation Application Technology*

Pulsar is the remnant of massive star evolution and collapse and supernova explosion. It is a celestial body under extreme physical conditions such as ultrahigh temperature, pressure and density and ultra-strong magnetic, electric, and gravitational fields. It has a typical radius of 10 km, a mass equivalent to that of the sun, and a core density up to 1012 kg/cm$^3$. Pulsar is a neutron star in high-speed rotation and with extremely stable periodicity. In particular, the periodic stability of millisecond pulsar is as high as 10–19 to 10–21 s/s, so this pulsar is reputed as the most stable clock in nature. By using the X-ray signals from pulsars as natural beacons, the spacecraft can autonomously determine the navigation parameters such as position, velocity, time, and attitude. Therefore, X-ray pulsar is called the lighthouse in the universe and can provide stable autonomous navigation information services for LEO, deep-space exploration, and interstellar flight. With very important engineering application value and scientific research significance, X-ray pulsar navigation has received high attention from various aerospace powers.

In 2004, the Defense Advanced Research Projects Agency (DARPA) proposed the "X-ray Navigation and Autonomous Positioning Verification" (XNAV) program. In 2011, the NASA Goddard Space Flight Center and the University Space Research Association (USRA) jointly proposed the "Station Explorer for X-ray Timing and Navigation Technology" (SEXTANT) testing, a demonstration and enhancement mission for "Neutron Star Interior Composition Explorer" (NICER) technology.

In addition, the ESA, Russia, Germany, the UK, Japan, India, Iran, and other countries and organizations have also carried out the research and ground demonstration of X-ray pulsar navigation technology.

Since 2005, CAST has conducted continuous research on the theories and methods of the pulsar navigation system, and made breakthroughs in key technologies. It has developed various types of X-ray detector prototypes and products, ground test systems, and special test satellites for pulsar navigation. In November 2016, the first pulsar experiment satellite (XPNAV-1), developed by the Chinese Academy of Space Technology (CAST), was launched and successfully received the pulsar signals. This launch mission is a revolutionary breakthrough in China's autonomous navigation technology, and a solid step toward the establishment of an accurate pulsar navigation database, on the basis of which the engineering application of pulsar navigation can be carried out to realize the autonomous navigation for spacecraft in a real sense.

In 2017, the Hard X-ray Modulation Telescope (HXMT) satellite developed by CAST, whose user system as well as payload developer was the CAS Institute of High Energy Physics, was successfully launched to study the high-precision orientation of high-energy celestial bodies such as black holes, neutron stars, and active galactic nuclei. It can be predicted that with the accumulation of X-ray photon data, the HXMT will certainly promote the rapid development of pulsar navigation system technology in China.

X-ray pulsar is a key strategic development direction to achieve longtime and high-precision autonomous navigation for spacecraft. The accelerated development of the X-ray pulsar navigation technique will certainly bring revolutionary changes and far-reaching impact on the development of space technology.

To achieve the pulsar navigation, three major scientific problems need to be solved, including pulsar detection, interstellar navigation charting, and large-scale navigation. In general, the autonomous navigation based on X-ray pulsars is to solve the following problems: the "visibility" of X-ray pulsar, the "accuracy" of pulsar TOA, and the "usability" of spacecraft.

To solve the scientific problems of pulsar navigation, the following six key technologies shall be mastered.

The first key technology is X-ray detector. The X-ray detector is used for the longtime cumulative measurement of pulsars and the construction of pulsar navigation data and pulsar time reference, in order to solve the problems of "visibility" and "measurement accuracy".

The second key technology is time system maintenance. By extracting the pulsar signal and phase from the X-ray photon counter, the phase measurement and frequency output of a pulsar are acquired for the adjustment of the basic frequency of the spaceborne clock, in order to achieve the clock correction and time system maintenance. In this way, the problem of "usability" can be solved.

The third key technology is the autonomous navigation algorithm and fault-tolerant processing technique. By using the orbit determination and time synchronization algorithm based on X-ray pulsar, the navigation parameters such as spacecraft position, velocity, and time can be determined. By using the attitude measurement algorithm based on X-ray pulsar and taking the pulsar image information as basic observation, the attitude angle parameter of spacecraft can be determined. Moreover, the robustness of navigation information processing can be improved, and the reliability and integrity of navigation information ensured. This technology is mainly used to solve the problem of "usability".

The fourth key technology is the pulsar navigation database. It serves as the basis of pulsar navigation. Its development aims to establish and refine the pulsar navigation database, construct an electronic navigation map for space flight, and further solve the problem of "measurement accuracy".

The fifth key technology is the establishment and maintenance of spatial and temporal references. The X-ray pulsar navigation belongs to the field of large-scale space-time transformation and space navigation applications. In essence, the establishment and maintenance of spatial and temporal references for X-ray pulsar navigation are to establish a set of basic theoretical data processing methods, a set of self-consistent model algorithms and constants, and a realizable reference framework. This technology is aimed at solving the problem of "measurement accuracy".

The sixth key technology is pulsar TOA conversion. An engineering reduced model for photon TOA conversion is established to obtain the time delay between spacecraft and solar system barycenter, which, in turn, is used to determine the orbital and time clock parameters of spacecraft. Thus, the problems of "measurement accuracy" and "usability" can be further solved.

The research on pulsar navigation and X-ray detection technologies will directly promote the cross-over study among multiple disciplines such as physics, astronomy, electronic science and technology, control science and engineering, and aerospace science and technology. X-ray has been used in autonomous constellation navigation, deep-space exploration and interstellar flight, space-ground integrated pulsar time system, and interdisciplinary cross-over study.

The first application of X-ray is longtime and high-precision autonomous constellation navigation. The autonomous satellite navigation based on X-ray pulsar is free from the accumulation of overall constellation rotation error, so the navigation constellation can autonomously operate with high accuracy for a long time.

This navigation mode is in line with the development direction toward the seamless high-precision autonomous navigation for near-Earth orbit (NEO) spacecraft.

The second application of X-ray is the autonomous navigation and precision control of spacecraft in deep-space exploration and interstellar flight. The rapid development of deep-space exploration and interstellar flight has posed challenges to the ground-based deep-space TT&C network. It is hard for traditional navigation modes and their combinations to meet the autonomous navigation and precise control requirements of deep-space and interstellar vehicles. The X-ray pulsars are suitable for navigation applications throughout the solar system and interstellar space. By measuring the TOA of X-ray photons emitted by pulsars, the navigation parameters of deep-space and interstellar vehicles relative to the solar system barycenter can be determined. The numerical experiments show that if the inertial angular position precision of pulsar is 0.1 mas, and the pulsar timing model accuracy, time transformation model accuracy, and pulse TOA measurement accuracy are all 100 ns, then the spacecraft orbit determination accuracy can be 10 m, the time synchronization accuracy can be 1 ns, and the attitude measurement accuracy will be better than 20 as. As a result, the autonomous navigation and precise control requirements of deep-space and interstellar vehicles can be satisfied.

The third application of X-ray is the research on the space-ground integrated pulsar time system. The advantage of pulsar time scale is long-term stability up to 10–19 s/s, four orders of magnitude higher than the most stable laboratory hydrogen clock in current use. By establishing a space observation platform to observe the radio signals from pulsars for a long time, a pulsar time reference system with long-term stability can be established to serve national defense, scientific research, and national economic construction.

The fourth application of X-ray is interdisciplinary and basic scientific research. The research results of X-ray pulsar navigation and detection technology can directly promote the development of such disciplines as physics, astronomy, geophysics, instrument science and technology, electronic science and technology, information and communication engineering, control science and engineering, computer science and technology, mapping science and technology, and aerospace science and technology. They can help accelerate and push forward the multidisciplinary integration and interdisciplinary research, better develop and utilize the pulsar resources, improve the research capacity and level of basic, leading-edge and exploratory disciplines, and promote the development of human civilization and the scientific and technological progress.

With a potential engineering application value, the X-ray pulsar navigation has become a research hotspot all over the world. The development and application of X-ray pulsar navigation technology can guarantee the longtime high-precision autonomous operation of deep-space and interstellar spacecraft, and provide technical support for the national integrated PNT system to cover the whole solar system.

### 10.2.7 LEO Satellite Navigation Augmentation Technology

The LEO constellation system can improve the usability of the navigation system and increase the number of user satellite observations. Especially in forests, mountains, and urban areas with complex terrain, it can effectively supplement the coverage of GNSS satellites and improve the positioning geometry. Owing to a fast speed relative to the ground, the LEO satellites can help users with quick navigation signal locking and high-accuracy rapid positioning, so as to shorten the initialization time to less than 1 min. The spaceborne high-speed data transmission equipment can provide a wide range of high-precision differential services to achieve the positioning accuracy of lane level and decimeter level. In addition, the LEO satellites have the advantages of high signal transmission power and interference immunity. In recent years, the LEO mobile communication technology has been developing in full swing. The integration of communication and navigation functions in the LEO satellites as well as the compatibility with the GNSS system will greatly improve the existing "vulnerability" of the GNSS system.

The establishment and maintenance of high-precision low-orbit spatial and temporal references is the primary problem to be solved for LEO satellites to broadcast the navigation signals. The establishment and maintenance of high-precision autonomous Spatio-temporal references is the key to achieve the navigation augmentation and backup of LEO satellites. The LEO global augmentation and backup system broadcasts both navigation signals and supplemental augmentation information. In terms of mission nature, the integration with LEO communication satellites is an optimization solution. That is to say, while communicating with LEO satellites, the user can accomplish the navigation and positioning functions without inviting additional overhead. Therefore, the signal system design where navigation and communication are integrated is an important technical route.

With the development of satellite technology, hundreds of small, micro- and nanosatellites are expected to be launched in the future. In actual engineering construction, a large number of navigation augmentation loads can be deployed on various LEO satellite platforms provided that their volume, weight, and power consumption do not affect the original satellite mission. Therefore, the miniaturization, lightweight, and low power consumption of navigation augmentation loads are the basis of large-scale deployment of such loads on various LEO satellite platforms.

### 10.2.8 Multisource Information Fusion Navigation Technology

Multisource information fusion is a procedure for processing the information from multiple sensors or information sources. It processes the measurement information from multiple sources by means such as information correlation, information integration, and filtering to improve the estimation accuracy of the target state and other features. Owing to the advantages of high accuracy, good fault tolerance, low information acquisition cost, and information complementation, this method has been applied in public life, industry, transportation, finance, and other fields. Especially under the military background, it plays a very important role. Compared with single-source data, the multisource data fusion can enhance the system survivability, expand the coverage of time and space, improve the service accuracy and reliability, and reduce the information ambiguity. Besides, its user terminal has strong points such as low cost, small size, and low power consumption.

The application of navigation information is evolving from single-system positioning and timing to "multisource navigation". Through the new combinations of navigation sensors and the use of filtering algorithm, it provides users with PNT services in various environments to meet the changing task demands and environmental requirements. The positioning accuracy of satellite navigation ranges usually 5–10 m, or up to centimeter level in precision positioning, which can basically meet the needs of national economic construction and daily life. However, in the fields of military application and surveying and mapping, the measurement means with higher accuracy are still needed, and the navigation means with better performance and usability are preferred. For example, the signals generated by the inertial navigation system have good continuity, low noise, high data-updating rate, and good short-term accuracy and stability. The barometer can provide altitude information more accurately than satellite navigation. The scenario/terrain matching can obtain the information matching the target. Astronomical navigation means such as pulsar and star sensor can provide the spacecraft with more accurate information on direction and time in High Earth Orbits (HEOs) and deep space. Magnetic/gravitational navigation means can provide underwater and underground position information not accessible to satellite navigation. Even indoors, the ground wireless communication network and wireless local area network (WLAN) can also provide the positioning services. The multisource fusion technology of an integrated navigation system can improve the measurement accuracy of satellite navigation in more areas such as indoor, dense forest, underwater, and deep space.

The architecture of the navigation information system has three fusion levels, namely business fusion, technique fusion, and network fusion. The goals of business fusion are to share the business system on the basis of network communication to provide unified business experience for end users so that they can get access to the same business through different terminals and networks, to implement unified database management, and to realize unified business development and control as much as possible. Technology fusion is to technologically make full use of the complementarities of each navigation sensor system in different environments, make up for their differences in the underlying technology, implement and promote their unique technology fusion, and provide users with higher quality services.

The network integration requires the operators to share network resources, such as mobile communication network and WLAN, so that mobile terminals can be switched seamlessly between the two networks to obtain necessary information such as range measurement and data.

Key technologies of multisource information fusion mainly include data conversion, data association, and fusion algorithm. Among them, the fusion algorithm is the core technique of multisource information fusion.

### *10.2.9 Autonomous Health Management Technology of Navigation Satellite*

Navigation satellites are the core of a navigation satellite system (NSS) and the basic guarantee for NPT services. Navigation satellites aim to ensure their safe and stable operation by timely detecting and handling their faults, and improving autonomous capability is an important cornerstone for building up the usability, continuity, and stability of NSSs. Therefore, the longtime autonomous health management of navigation satellites has become an important direction of satellite technology development.

Autonomous satellite health management is based on, but not equivalent to, satellite fault diagnosis. First, the status of the whole satellite system is evaluated through fault diagnosis and early fault prediction. Then, decisions are made through the health management system to autonomously select the required treatment measures, so as to minimize the impact of faults on the whole satellite system, reduce the system maintenance cost and improve the satellite service life. The autonomous health management of a satellite requires the analysis and prediction of fault occurrence and trend, the timely in-orbit diagnosis and isolation of faulty equipment, and the determination of system reconstruction countermeasures to slow down and avoid the occurrence of serious faults as much as possible, effectively reduce the impact of faults on the whole satellite, and ensure the long-life operation of the satellite in orbit. In addition to ensuring the safe and reliable operation of satellites, this technique is also of great significance to reduce the workload of ground TT&C personnel and the cost of satellite launch and operation.

The autonomous health management technology of navigation satellites usually includes the design and architecture of autonomous health management system, as well as the relevant techniques such as autonomous fault diagnosis, early fault prediction, and fault recovery.

With the development of Artificial Neural Network (ANN), the autonomous management of navigation satellites based on cognitive technology has also gradually entered people's vision. With the operating mode and mechanism of the human brain as reference, this technology can properly process and save the information on the outside world and surroundings, and make decisions or reflections, so as to effectively improve the "behavioral effect feedback" for multiple navigation satellites in the system.

As one of the four cutting-edge scientific technologies in the twenty-first century, cognitive technology is a highly interdisciplinary emerging science occurring at the interfaces of psychological science, computer science, neuroscience, scientific philosophy, and other basic sciences.

The autonomous health management system based on cognitive technology is an intelligent information processing system with the functions of fault detection, fault tolerance, and error correction. On the basis of the operating mode of the human brain, it acquires and determines the information on normal and abnormal working conditions of subsystems or products/equipment, including normal working parameters and the type, position, and occurrence time of faults, through the perception, analysis and understanding, and autonomous learning and memorization of historical and current satellite state data. Then according to the fault information, the system implements the fault position analysis and intelligent reasoning, and makes the decisions such as redundancy replacement, equipment reconfiguration, and degraded system operation to ensure normal system operation. Thus, the navigation satellites in the system can efficiently and autonomously adapt their working condition to the failures possibly caused by complex internal and external environments, thereby achieving the "behavior effect feedback". This technology is one of the key directions for future NSS development.

The autonomous satellite health management system concept based on cognitive technology mainly includes three basic elements:

(1) Intelligent information processing

The main task of intelligent information processing is to obtain and improve the cognition of navigation satellites in the system toward the overall service condition and environment through continuous interaction with the internal state information of navigation satellites, the information of inter-constellation satellite measurement and data, and the information on the external space environment.

(2) Fault information and its learning, storage, and memory

In order to ensure the normal operation of various products, equipment, and subsystems, traditional navigation satellites usually downlink the onboard state data via the telemetry link to the ground personnel, who will interpret and process the collected telemetry data in real time according to the telemetry parameter processing methods and interpretation criteria. Through the accumulation of various types of internal information and the fault information acquired in the space environment, the model based on cognitive technology can extract the common features of autonomous health management, establish the priori knowledge model database, and improve the accuracy of fault cognition.

(3) Behavior effect feedback

In nature, behavior effect feedback is to, through the perception of fault information, extract knowledge from a large amount of telemetry data, measuring and monitoring data and raw sensor data, make a decision based on intelligent reasoning, and then feed the decision back to the onboard information processing system and the corresponding products, equipment, and subsystems, where the behavior effect will be adapted to the fault decision to improve the adaptability of the navigation satellite.

To strengthen the research on key technologies of autonomous satellite health management, attention shall be paid to satellite information system architecture, information monitoring and sensing methods, FMEA analysis (for the existing or potential faults), system multi-parameter correlation model, and intelligent fusion decision protocol or dynamic reconstruction algorithm. Based on the analysis of overall NSS requirements and system service capabilities, a simulation and demonstration system for autonomous health management can be developed and established to help the navigation satellites achieve autonomous fault diagnosis and evaluation, system reconstruction, and fault prediction demonstration.

## 10.3 Development Conception of BDS

As China's major space infrastructure, the BDS system will be mainly characterized by systematization, networking, and intellectualization. Lagrange satellite navigation system, LEO satellite navigation system, and other systems for make-up, augmentation, and backup will be comprehensively developed on the basis of key technologies such as the dynamic maintenance and transmission of high-precision time-space references, navigation and communication integration, and X-ray pulsar navigation to substantially improve the service and operation level of the BDS system.

The key capabilities of the BDS system in the future will include the following:

In terms of coverage, the service scope of this system will extend from the Earth's surface to near-Earth space and even deep space and from open areas to partially sheltered areas such as cities and canyons, and support the seamless switching between indoor navigation and underwater navigation.

In terms of service accuracy, the BDS will have the global positioning accuracy of submeter level, the timing accuracy of nanosecond level, and the integrity that meets the Class-I precision approach requirement in civil aviation.

In terms of autonomous operation, the BDS will have the capability of longtime autonomous constellation operation, without the loss of navigation service accuracy.

In terms of extended services, the BDS will integrate multiple tasks including space-based augmentation, global location reporting, and inter-satellite data transmission, and will collaborate with multiple systems such as ground-based PNT system and space-based systems (e.g., communication and reconnaissance).

According to the future requirements for BDS and the development trend of worldwide NSSs, the future BDS will be an integration of the MEO/HEO NSS system, Lagrange satellite navigation make-up system, and LEO navigation augmentation and backup system. It will be developed in parallel with pulsar navigation, ground-based augmentation, land-based navigation, underwater navigation, indoor navigation, MicroPNT, and other systems to constitute an integrated space-time system that is "more extensive, fused, and intelligent".

# References

1. BeiDou Navigation Satellite System Signal in Space Interface Control Document, Open Service Signal B1I (Version 3.0), February 2019.
2. BeiDou Navigation Satellite System Signal in Space Interface Control Document Open Service Signal B3I (Version 1.0), China Satellite Navigation Office February 2018.
3. BeiDou Navigation Satellite System Signal in Space Interface Control Document Open Service Signal B1C (Version 1.0), China Satellite Navigation Office December, 2017.
4. BeiDou Navigation Satellite System Signal in Space Interface Control Document Open Service Signal B2a (Version 1.0), China Satellite Navigation Office December, 2017.
5. CCSDS130.0-G-3. Overview of space communications protocols [R]. Washington DC:CCSDS Secretariat, 2014.
6. ECSS-E-70-41C. Telemetry and telecommand packet utilization [S]. Noordwijk: ESA Publications Division, 2016.
7. Europen GNSS (Galileo) Open Service Signal In Space Interface Control Document, OS SIS ICD, Issue 1, European Union 2010.
8. Europen GNSS (Galileo) Open Service Signal In Space Interface Control Document, OS SIS ICD, Issue 1.3, European Union 2016.
9. GLONASS Interface Control Document Code Division Multiple Access Open Service Navigation Signal in L1 frequency band Edition 1.0. Russian Space Systems, JSC, 2016.
10. GLONASS Interface Control Document Code Division Multiple Access Open Service Navigation Signal in L2 frequency band Edition 1.0. Russian Space Systems, JSC, 2016.
11. GLONASS Interface Control Document Code Division Multiple Access Open Service Navigation Signal in L3 frequency band Edition 1.0. Russian Space Systems, JSC, 2016.
12. International Telecommunication Union(ITU). 2003. Resolution 741(WRC-03). Protection of the radio astronomy service in the band 4990–5000 MHz from unwanted emissions of the radionaviga tion -satellite service(space -to -Earth) operating in the frequency band 5010–5030 MHz(R). Geneva.
13. NAVSTAR GPS Space Segment/Navigation User Segment Interfaces IS-GPS-200K, March 2019
14. NAVSTAR GPS Space Segment/User Segment L5 Interface. 2013. Interface Specification IS-GPS-705, Global Positioning Systems Directorate.
15. NAVSTAR GPS Space Segment/User Segment L1C Interface. 2013. Interface Specification IS-GPS-800, Global Positioning Systems Directorate.
16. Resolution 1, 13th CGPM. Metrologia, 1968, 4: 41–45.
17. Russian Institute of Space Device Engineering. 2008. GLONASS Interface Control Document, Navigational radiosignal In bands L1, L2 (Edition 5.1). MOSCOW.
18. VDOVIN V D. Global geocentric coordinate system of the Russian federation. http://www.oosa.unienna.org/pdfficg/2012/template/pz90-02-v2.pdf.



J. Xie et al., *Satellite Navigation Systems and Technologies*, Space Science and Technologies, https://doi.org/10.1007/978-981-15-4863-5

19. Avila-Rodriguez, J.A., G.W. Hein, S. Wallner, J.L. Issler, L. Ries, L. Lestarquit, A. Latour, J. Godet, F. Bastide, T. Pratt, and J. Owen. 2008. The MBOC modulation: the final touch to the Galileo frequency and signal plan. *NAVIGATION Journal of The Institute of Navigation* 55 (1): 14–28.
20. Betz, J.W. 1999. The offset carrier modulation for GPS modernization. In *Proceedings of the 1999 National Technical Meeting of the Institute of Navigation*, San Diego, CA, 639–648.
21. Dafesh, P.A., T.M. Nguyen, and S. Lazar. 1999. Coherent adaptive subcarrier modulation (CASM) for GPS modernization. In *Proceedings of the 1999 National Technical Meeting of The Institute of Navigation*, San Diego, CA, January 1999, 649–660.
22. Dafesh, P.A., and C.R. Cahn. 2009. Phase-optimized constant-envelope transmission (POCET) modulation method for GNSS signals. In *Proceedings of the 22nd International Technical Meeting of The Satellite Division of the Institute of Navigation*, Savannah, GA, 2860–2866.
23. Soualle, F., M. Soellner, S. Wallner, J.-A. Avila-Rodriguez, et al. 2005. Spreading code selection criteria for the future GNSS Galileo. In *Proceedings of the European Navigation Conference GNSS 2005*, Munich, Germany.
24. Rushanan, J.J. 2007. The spreading and overlay codes for the L1C signal. In *Proceedings of the 2007 National Technical Meeting of The Institute of Navigation*, San Diego, CA, 539–547.
25. Rushanan, J.J. 2006. Weil sequences: a family of binary sequences with good correlation properties. In *ISIT 2006*, Seattle, USA, 1648–1652.
26. Kaplan, E.D., and C.J. Hegarty. 2005. *Understanding GPS: principles and applications*, 2nd ed. Norwood, MA, USA: Artech House Publishers.
27. Li, P., Z. Chen, Y. Gu, and Y. Si. 2013. Spacecrafts navigation signal research based on GNSS constellation. In *China Satellite Navigation Conference 2013 Proceedings*, vol. 2, 35–48. Beijing: Springer.
28. Li, P., Z. Chen, Z. Wu, W. Hao, and L. Wang. 2016. Spacecrafts navigation demand analysis and simulation on MEO constellation. In *China Satellite Navigation Conference 2016 Proceedings*, 241–255. Beijing: Springer.
29. Liu, Y.J., W.H. Chen, J. Zhou, et al. 2013. Behavioral modeling for concurrent dual-band power amplifiers using 2D hammerstein/wiener models. *International Journal of RF and Microwave Computer -Aided Engineering* 23 (6): 646–654.
30. Liu, Y., W. Chen, J. Zhou, B. Zhou, and F.M. Ghannouchi. 2013. Digital predistortion for concurrent dual- band transmitters using 2-D modified memory polynomials. *IEEE Transactions on Microwave Theory and Techniques* 61 (1): 281–290.
31. Liu, T., J. Liu, X. Deng, et al. 2016. Research on fiber-based time and frequency transfer. *Journal of Time and Frequency* 39 (3): 207–215.
32. Miho, F., H. Takiguchi, and J. Achkar. 2016. Carrier-phase two-way satellite frequency transfer between LNE-SYRTE and PTB. In *30th European Frequency and Time Forum (EFTF)*, 255–259.
33. Rebeyrol, E, C. Macabiau, L. Ries, et al. 2006. Interplex modulation for navigation systems at the L1 band. In *Proceedings of the 2006 National Technical Meeting of The Institute of Navigation*, Monterey, CA, January 2006, 100–111.
34. Rapp, C. 1991. Effects of HPA-nonlinearity on 4-DPSK-OFDM signal for digital sound broadcasting systems. In *Procceedings of the Second European Conference on Satellite Communications (ECSC-2)*.
35. Saleh, A.A.M. 1981. Frequency-independent and frequency-dependent nonlinear models of TWT amplifiers. *IEEE Transactions on Communications* 29 (11): 1715–1720.
36. Journo, S., J. Pourmailloux, J.M. Piéplu, et al. 2006. Producing the Galileo Services from the Ground Mission Segment (GMS). In *ION GNSS 19th International Technical Meeting of the Satellite Division*, Fort Worth, 504–519.
37. Wallner, S., J.-A. Avila-Rodriguez, and G.W. Hein. 2007. Galileo E1 OS and GPS L1C Pseudo RandomNoise Codes—requirements, generation, optimization and comparison. In *Proceedings of the 20th International Technical Meeting of the Satellite Division of the Institute of Navigation (ION GNSS 2007)*, 1549–1563.

38. Soualle, F., and T. Burger. 2007. Radio frequency compatibility criterion for code tracking performance. In ION GNSS 20th International Technical Meeting of the Satellite Division, Fort Worth, TX, 1201–1210.
39. Tan, S. 2007. *Satellite navigation and positioning engineering*. Beijing: National Defense Industry Press. (in Chinese).
40. Tang, Z., H. Zhou, X. Hu, Y. Ran, Y. Liu, and Y. Zhou. 2010. Research on performance evaluation of compass signal. *Sciencetia Sinica Physica, Mechanica & Astronomica* 40 (05): 592–602. (in Chinese).
41. Tang, Z, H. Zhou, and J. Wei. 2011. TD-AltBOC: a new COMPASS B2 modulation. Science China-Physics Mechanics & Astronomy 54 (6): 1014–1021.
42. Yan, T., J. Wei, Z. Tang, Q. Bo, and Z. Zhou. 2015. Unambiguous combined correlation functions for sine-BOC signal tracking. *GPS Solutions* 19 (4): 623–638.
43. Meng, W., H. Zhang, P. Huang, et al. 2013. Design and experiment of onboard laser time transfer in Chinese Beidou navigation satellites. *Advances in Space Research* 51: 951–958.
44. Won, J.H., B. Eissfeller, A. Schmitz-Peiffer, J.J. Floch, F. Zanier, and E. Colzi. 2012. Trade-off between data rate and signal power split in GNSS signal design. *IEEE Transactions on Aerospace and Electronic Systems* 48 (3): 2260–2281.
45. Yao, Z., M. Lu, and Z.M. Feng. 2010. Quadrature multiplexed BOC modulation for interoperable GNSS signals. *Electronics Letters* 46 (17): 1234–1236.
46. Yao, Z., J. Zhang, and M. Lu. 2016. ACE-BOC dual-frequency constant envelope multiplexing for satellite navigation. *IEEE Transactions on Aerospace and Electronic Systems* 52 (1): 466–485.
47. Zhang, K. 2013. Generalised constant-envelope DualQPSK and AltBOC modulations for modern GNSS signals. *Electronics Letters* 49 (21): 1335–1337.